Phosphorus Loss from Soil to Water

Edited by

H. TUNNEY AND O.T. CARTON

Teagasc
Johnstown Castle Research Centre
Wexford
Ireland

and

P.C. BROOKES AND A.E. JOHNSTON

Soil Science Department
IACR Rothamsted
Harpenden
UK

CAB INTERNATIONAL

CAB INTERNATIONAL
Wallingford
Oxon OX10 8DE
UK

CAB INTERNATIONAL
198 Madison Avenue
New York, NY 10016–4341
USA

Tel: +44 (0)1491 832111
Fax: +44 (0)1491 833508
E-mail: cabi@cabi.org

Tel: +1 212 726 6490
Fax: +1 212 686 7993
E-mail: cabi-nao@cabi.org

© CAB INTERNATIONAL 1997. All rights reserved. No part of this publication may be reproduced in any form or by any means, electronically, mechanically, by photocopying, recording or otherwise, without the prior permission of the copyright owners.

A catalogue record for this book is available from the British Library, London, UK.
A catalogue record for this book is available from the Library of Congress, Washington DC, USA.

ISBN 0 85199 156 4

Typeset in Baskerville by Solidus (Bristol) Ltd
Printed and bound in the UK by Biddles Ltd, Guildford and King's Lynn

Contents

Preface		xi
Acknowledgements		xiv
1	**Phosphorus in Agriculture and Its Environmental Implications** A.N. Sharpley and S. Rekolainen	1
2	**Estimating the Contribution from Agriculture to the Phosphorus Load in Surface Water** S.D. Lennox, R.H. Foy, R.V. Smith and C. Jordan	55
3	**Phosphorus Losses from Agriculture to Surface Waters in the Nordic Countries** S. Rekolainen, P. Ekholm, B. Ulén and A. Gustafson	77
4	**Reconstructing Historical Phosphorus Concentrations in Rural Lakes Using Diatom Models** N.J. Anderson	95
5	**The Dynamics of Phosphorus in Freshwater and Marine Environments** C.E. Gibson	119
6	**The Behaviour of Soil and Fertilizer Phosphorus** M.A. Morgan	137
7	**Setting and Justifying Upper Critical Limits for Phosphorus in Soils** E. Sibbesen and A.N. Sharpley	151

8	**Phosphorus Fertilizer Strategies: Present and Future** H. Tunney, A. Breeuwsma, P.J.A. Withers and P.A.I. Ehlert	177
9	**Sources and Pathways of Phosphorus Loss from Agriculture** A.L. Heathwaite	205
10	**Hydrological and Chemical Controls on Phosphorus Loss from Catchments** H.B. Pionke, W.J. Gburek, A.N. Sharpley and J.A. Zollweg	225
11	**Movement of Phosphorus from Agricultural Soil to Water** B. Pommel and J.M. Dorioz	243
12	**Losses of Phosphorus in Drainage Water** P.C. Brookes, G. Heckrath, J. De Smet, G. Hofman and J. Vanderdeelen	253
13	**Sustainable Phosphorus Management in Agriculture** G. Bertilsson and C. Forsberg	273
14	**Phosphorus Requirements for Animal Production** P.B. Lynch and P.J. Caffrey	283
15	**Nutrient-Management Planning** T.C. Daniel, O.T. Carton and W.L. Magette	297
16	**A European Fertilizer Industry View on Phosphorus Retention and Loss from Agricultural Soils** I. Steén	311
17	**European Perspective on Phosphorus and Agriculture** F. Mariën	329
18	**Views on Phosphorus and Agriculture – Paris Commission** S. Sadowski	339
19	**Phosphorus Loss in Runoff, Leaching and Erosion**	

SURFACE RUNOFF

19.1	Phosphorus Fractionation in Grassland Hill-Slope Hydrological Pathways R.M. Dils and A.L. Heathwaite	349
19.2	Soil-derived Phosphorus in Surface Runoff from Grazed Grassland P.M. Haygarth and S.C. Jarvis	351
19.3	Agricultural Phosphorus Load and Phosphorus as a Limiting Factor for Algal Growth in Finnish Lakes and Rivers O.P. Pietiläinen	354

19.4   Increase in Soluble Reactive Phosphorus Transport in   356
       Grassland Catchments in Response to Soil Phosphorus
       Accumulation
       R.V. Smith, R.H. Foy and S.D. Lennox
19.5   Phosphorus Loss to Water from a Small Low-Intensity   358
       Grassland Catchment
       H. Tunney, T. O'Donnell and A. Fanning

LEACHING

19.6   The Relation Between Accumulation and Leaching of   361
       Phosphorus: Laboratory, Field and Modelling Results
       O.F. Schoumans and A. Breeuwsma
19.7   Organically Combined Phosphorus in Soil Solutions and   363
       Leachates
       W.J. Chardon
19.8   Phosphorus Losses in Drainage Water from an Arable   367
       Silty Clay Loam Soil
       G. Heckrath, P.C. Brookes, P.R. Poulton and K.W.T. Goulding
19.9   Evidence of Phosphorus Movement from Broadbalk Soils   369
       by Preferential Flow
       D. Thomas, G. Heckrath and P.C. Brookes
19.10 Phosphorus Leaching in the Brimstone Farm   370
       Experiment, Oxfordshire
       *K.R. Howse, J.A. Catt, D. Brockie, R.A.C. Nicol, R. Farina,
       G.L. Harris and T.J. Pepper*
19.11 Phosphorus Input to a Brook through Tile Drains under   372
       Grassland
       *C. Stamm, R. Gächter, H. Flühler, J. Leuenberger and H.
       Wunderli*

EROSION

19.12 Phosphate Losses in the Woburn Erosion Reference   374
       Experiment
       J.A. Catt, A.E. Johnston and J.N. Quinton
19.13 Storm-Event Transport of Phosphorus in the Absence of   377
       Surface Runoff Generation
       O.S. Hodun and T.P. Burt
19.14 Impact of Different Tillage Practices on Phosphorus   379
       Losses from Agricultural Fields
       A. Klik and J. Rosner

20 Catchment Studies, Modelling and Management

CATCHMENT STUDIES

20.15 Catchment Studies of the Loss of Phosphorus from Agriculture to Surface Water
R. Grant, B. Kronvang and A. Laubel 383

20.16 Sources of Soluble and Particulate Phosphorus in Surface Water in Eastern England: Relative Importance of Agricultural Versus Small Point Sources
D. Harper and G. Evans 385

20.17 Biogeochemical Significance of Membrane and Ultrafilter Separation of Low-Molecular-Weight Molybdate-Reactive Phosphorus in Soil and River Waters
P.M. Haygarth, M.S. Warwick and W.A. House 386

20.18 Contribution of Agriculture to the Phosphorus in Surface Waters in Finland and Measures to Reduce it
S. Rekolainen 389

20.19 Identifying Critical Sources of Phosphorus Export from Agricultural Catchments
A. Sharpley and J. Lemunyon 391

20.20 Improved Measurements of Phosphorus Loss to Watercourses from Agricultural Areas
L. Wiggers 394

20.21 Phosphorus Loss to Water from Agriculture in the UK
P.J.A. Withers 396

MODELLING AND MANAGEMENT

20.22 Phosphorus Concentrations in Surface Water and Drainage Water in the Watershed of the Poekebeek, Flanders, Belgium
R. Hartmann, H. Verplancke, P. Verschoore and M.M. Villagra 398

20.23 Evaluating a Phosphate Saturation Inventory of Soils in Northern Belgium
I. Schoeters, R. Lookman, R. Merckx and K. Vlassak 400

20.24 Developing an Expert System for the Evaluation of Nutrient Losses from Agriculture to Water in Belgium
P. Scokart, P. Nyssen and P. De Cooman 403

20.25 Phosphorus Loads from Agricultural Areas in an Austrian Watershed: Measurements and Estimation Using Geographical Information System Technology
P. Struß and W.E.H. Blum 406

20.26 Dutch Policy towards Phosphorus Losses in Agriculture
D.T. van der Molen, A. Breeuwsma, P.C.M. Boers and C.W.J. Roest 407

20.27	Present and Future Dutch Regulations to Reduce Phosphorus Loss to Water from Agriculture *P. Hotsma*	410
20.28	Hydrological and Chemical Controls on Phosphorus Losses from Catchments – Coordination of Field Research, Geographical Information Systems and Modelling *J.A. Zollweg, W.J. Gburek, A.N. Sharpley and H.B. Pionke*	412

21 Phosphorus Status of Soils and Fertilizer Recommendations

PHOSPHORUS STATUS OF SOILS

21.29	Phosphorus Composition of Soil Solution: Effects of Sample Preparation and Soil Storage *P.J. Chapman, C.A. Shand, A.C. Edwards and S. Smith*	415
21.30	Inventory of the Phosphate Saturation Degree of the Light-Textured Soils in West Flanders, Belgium *J. De Smet, K. Scheldeman, G. Hofman, M. Van Meirvenne, J. Vanderdeelen and L. Baert*	417
21.31	Changes in Two Transport- and Retention-Related Soil-Phosphate Parameters Following Phosphate Addition *R. Indiati and A.N. Sharpley*	420
21.32	The Downward Movement and Retention of Phosphorus in Agricultural Soils *A.E. Johnston and P.R. Poulton*	422
21.33	The Concentrations and Forms of Phosphorus in Manures and Soils from the Densely Populated Livestock Area in North-West Germany *P. Leinweber*	425
21.34	Comparison of Chemical Forms and Distribution of Phosphorus within Cultivated and Uncultivated Soils: Some Implications for Losses *R.O. Maguire, A.C. Edwards and M.J. Wilson*	427
21.35	Threat of Phosphorus Leaching from Intensively Farmed Agricultural Soils in the Central Reaches of the River Elbe *R. Meissner, H. Rupp and J. Seeger*	430
21.36	The Availability in Soil of Phosphorus Released from Poultry Litter *J.S. Robinson and A.N. Sharpely*	431
21.37	Reducing Soil-Phosphorus Availability with By-products from Power-Generation Plants *W.L. Stout, A.N. Sharpley and H.B. Pionke*	436
21.38	Occurrence and Effects of Phosphate-Saturated Soils *A. Breeuwsma, J.G.A. Reijerink and O.F. Schoumans*	438

PHOSPHORUS RECOMMENDATIONS

    21.39 Soil Phosphorus Levels in Dairy Farming   440
N. Culleton, J. Murphy and W.E. Murphy

    21.40 Defining Critical Levels of Available Soil Phosphorus for Agricultural Crops   441
A.E. Johnston and P.R. Poulton

    21.41 Factors Affecting Critical Soil Phosphorus Values   445
A.E. Johnston and P.R. Poulton

    21.42 Distribution of Available Phosphorus in Soil Under Long-term Grassland   448
W.E. Murphy and N. Culleton

    21.43 Comparison of Fertilizer Phosphorus Recommendations in Ireland and England and Wales   449
P.R. Poulton, H. Tunney and A.E. Johnston

    21.44 The Impact of Fertilizer Strategies on the Phosphorus Status of Arable Soils in England and Wales   452
J.L. Salter, B. Higgs and C.J. Dawson

22 Phosphorus Loss from Agriculture to Water: Synthesis and Summary   455
A.E. Johnston, H. Tunney and R.H. Foy

Index   463

Preface

The topic of this book, phosphorus loss from agriculture to water, is becoming ever more important in understanding the forces which drive the process of eutrophication in lakes. Eutrophication, caused by phosphorus enrichment, is not a new environmental problem. It first came to the fore in the 1960s, most notably in the Great Lakes region of the USA and Canada. In the following decade, a host of other lakes, suffering from varying degrees of phosphorus enrichment, were identified throughout Western Europe and elsewhere in the world. Within the British Isles, the continuing relevance of phosphorus enrichment was highlighted by the detection of toxin-producing algal blooms in many reservoirs during the hot summer of 1989 and the realization that previously pristine large Irish lakes had experienced significant phosphorus enrichment during the 1980s.

Initially, lake eutrophication was primarily linked to sewage-derived phosphorus inputs. Control measures relied on reducing these point-source inputs through sewage diversion, removing phosphorus from sewage effluent and/or reducing the phosphorus content of detergents. Although the water quality of many lakes improved in response to these control measures, others have stubbornly remained eutrophic. For example, reduced point-source inputs of phosphorus have been ineffective in permanently reducing phosphorus concentrations in Ireland's largest lake, Lough Neagh, because non-point-source inputs are large and increasing each year. In other countries, water-quality monitoring programmes have shown phosphorus enrichment to be occurring in lakes draining rural or agriculturally dominated catchments.

The persistence of eutrophication in an era when many point-source phosphorus inputs have been curtailed has turned the focus of attention to agricultural phosphorus. This shift in emphasis is of enormous significance because it presents farmers, their support industry, researchers and government agencies with the challenge of addressing phosphorus control in systems which are much more complex than the typical phosphorus point-source input. Controlling agricultural phosphorus losses from lake catchments depends on understanding and managing phosphorus use and its fate and transport within catchments.

Many readers may define phosphorus use in agriculture as a problem with a simple solution – avoid applying excess phosphorus. In reality, excess phosphorus is a complex issue which may be defined from an agronomic, economic or environmental viewpoint. In derivation and application, 'excess' will depend on site characteristics and soil properties. Although phosphorus behaves conservatively in the soil and aquatic environment, in the sense that it does not suffer losses to the atmosphere, it exists in forms and fractions that differ enormously in terms of reactivity, solubility, transferability and ultimately bioavailability. These properties may vary across a catchment, for example as affected by phosphorus fertility status and soil type. The transport mechanisms which define the flow pathways connecting field to stream present a further source of complexity, for, depending on their nature and location, they may inhibit or promote phosphorus loss from field to stream.

Securing appropriate measures for controlling agricultural phosphorus losses requires a good understanding of the processes and mechanisms which determine loss rates from source areas at the field scale, combined with an appreciation of the influence of landscape features at the catchment scale. At present, it is often difficult to quantify and define sources and pathways of phosphorus loss and to obtain a consensus as to the importance of specific processes controlling loss rates. Undoubtedly, the relative importance of phosphorus sources and pathways differs among regions, so that the most effective control measures will vary. For example, repairing or replacing a leaking slurry tank can be a simple, if sometimes costly task, but devising means which restrict phosphorus losses at the field, landscape or catchment scale is a much more difficult undertaking. What is certain is that limiting phosphorus losses from agriculture will have a high priority in areas where such losses are having adverse environmental effects. Ideally, there should be a balance between the costs of control measures and environmental protection.

The island of Ireland possesses a rich resource of lakes – indeed, it has more large lakes than large inland towns – and yet many of these lakes are suffering from the effects of phosphorus enrichment. Visible impacts of eutrophication have recently become apparent on Loughs Conn, Mask and Corrib in the West of Ireland, a region previously associated with pristine lakes and excellent angling.

Recognizing an urgent need to synthesize current knowledge and to identify areas for further research, Teagasc, the Irish Agriculture and Food Development Authority, took the initiative to host a workshop on factors controlling phosphorus losses to water from agriculture. Given that it was timely as a means of addressing a specific and common problem and, in the context of the peace process, as a vehicle for encouraging cooperation between the Irish and British Governments, the Irish Department of Agriculture, Forestry and Food and the UK Ministry of Agriculture, Fisheries and Food agreed to sponsor the workshop. Further generous financial support was provided by the Commission of the European Communities, the European Fertilizer Manufacturers Association and the Irish Fertilizer Manufacturers Association.

At the outset, the organizing committee was keen that the workshop should act as forum for researchers, administrators and representatives from the agricultural and fertilizer industries. All these groups were represented among the 135 delegates, drawn from more than 20 countries, who came to Johnstown Castle Research Centre, Wexford, Ireland from 29 to 31 September 1995. At a ministerial level, Mr Ivan Yates, *Teachta Dála* (TD) (Minister for Agriculture) (Ireland) and Mr Tim Boswell MP (UK) represented their respective governments for part of the workshop. This book presents the proceedings of the workshop and consists of 18 chapters by the invited speakers and three chapters with the 45 poster papers displayed at the workshop. As such, it provides a valuable and up-to-date review of the subject of phosphorus losses from agriculture to water, which has significance for Ireland and beyond.

Acknowledgements

This book is based on an international workshop held at Teagasc, Johnstown Castle Research Centre, Wexford, Ireland.

The following people were associate editors: R.H. Foy, M.A. Morgan, H.B. Pionke, A.N. Sharpley and J. Humphreys. Special thanks are due to the workshop secretary Ms Eleanor Spillane, Teagasc, and to Mrs Deirdre Hughes, Ms Helen Weir and Ms Christine Jaggard of Rothamsted for assistance with the poster manuscripts.

The workshop was sponsored by the Ministry of Agriculture, Fisheries and Food, London, UK; the Department of Agriculture Food and Forestry, Dublin, Ireland; the European Fertilizer Manufacturers Association, Brussels; the Irish Fertilizer Manufacturers Association, Dublin; and the Commission of the European Communities.

1 Phosphorus in Agriculture and Its Environmental Implications

A.N. Sharpley[1] and S. Rekolainen[2]

[1] USDA-ARS, Pasture Systems and Watershed Management Research Laboratory, Curtin Road, University Park, PA 16802–3702, USA; [2] Finnish Environment Institute, Impacts Division, Helsinki, Finland

INTRODUCTION

Phosphorus (P) is an essential element for plant growth and its input has long been recognized as necessary to maintain profitable crop production. Phosphorus inputs can also increase the biological productivity of surface waters. Although nitrogen (N) and carbon (C) are essential to the growth of aquatic biota, most attention has focused on P inputs, because of the difficulty in controlling the exchange of N and C between the atmosphere and water and fixation of atmospheric N by some blue-green algae. Thus, P is often the limiting element and its control is of prime importance in reducing the accelerated eutrophication of fresh waters. As we move from fresh to saline estuaries, through brackish waters, N generally becomes the element controlling aquatic productivity (Thomann and Mueller, 1987).

Advanced eutrophication of surface water leads to problems with its use for fisheries, recreation, industry or drinking, due to the increased growth of undesirable algae and aquatic weeds and oxygen shortages caused by their senescence and decomposition. Also, many drinking-water supplies throughout the world experience periodic massive surface blooms of cyanobacteria (Kotak *et al.*, 1993). These blooms contribute to a wide range of water-related problems, including summer fish kills, unpalatability of drinking-water and formation of trihalomethane during water chlorination (Palmstrom *et al.*, 1988; Kotak *et al.*, 1994). Consumption of cyanobacterial blooms or water-soluble neuro- and hepatoxins released when these blooms die can kill livestock and may pose a serious health hazard to humans (Lawton and Codd, 1991; Martin and Cooke, 1994).

From a fisherman's point of view, accelerated eutrophication of lakes can increase the population of rough fish compared with desirable game fish. This has a negative impact on the value of other recreational aspects of lakes. However, fishery management often recommends a higher productivity to maintain an adequate phytoplankton and zooplankton for the fish food-chain for optimum commercial fish production. This food-chain may be manipulated by stocking of water with certain fish species in addition to P load reductions, in efforts to reduce the incidence of algal blooms and improve overall water quality. For example, stocking lakes with predatory game fish at the top of the food-chain (piscivore fish – bass, pike, trout) can reduce the number of planktivore or coarser fish (yellow perch, crappies) on which they feed (Andersson *et al.*, 1978; Hessen and Nilssen, 1986; Horppila and Kairesalo, 1990). Similar results may be obtained by selectively fishing planktivores. Zooplankton should then thrive, which in turn will reduce phytoplankton populations, improving water quality.

Lake use has an impact on desirable water quality goals, which will require differing management. Catchment management often becomes more complex with multiple-use lakes and streams. For example, a reservoir may have been built for water-supply, hydropower and/or flood control, and, although not a primary purpose, recreation is often considered a benefit, with aesthetic enhancement (including property value) an additional fringe benefit.

Since the late 1960s, point sources of water pollution have been reduced, due to their relative ease of identification and control. Even so, water-quality problems remain, and, as further point-source control becomes less cost-effective, attention is now being directed towards the contribution of agricultural non-point sources to P in surface waters. Generally, the loss of agricultural P in runoff is not of economic importance to a farmer. However, it can lead to significant off-site economic impacts, in some cases occurring many kilometres from P sources. By the time these impacts are manifest, remedial strategies are often difficult and expensive for the farmer to implement; they cross political and regional boundaries, and it can be several years before an improvement in water quality occurs. Thus, a greater understanding of where P is coming from, how much P in soil and water is too much and how and where we can reduce these inputs and losses must be gained through research and extension programmes, in order to develop agricultural resource systems that sustain production and environmental quality, as well as farming communities.

This chapter discusses the general role of agricultural P management in accelerated eutrophication and where our lack of information limits improved P management. More specific details on processes, quantities, and case-studies are presented in other chapters in this volume.

REASONS FOR CONCERN ABOUT PHOSPHORUS LOSSES FROM AGRICULTURE

Amounts of P transported in runoff from uncultivated or pristine land are considered background loading which cannot be reduced. As we try to assess the impact of agriculture on P loss to surface waters, it becomes evident that little quantitative information is available on background losses of P from agriculture, particularly prior to cultivation. Consequently, it is still difficult to quantify any increase in P loss following cultivation.

Several surveys of catchments in the USA have shown that P loss in runoff increases as the portion of the catchment under forest decreases and agriculture increases (Fig. 1.1). The loss of P from forested land tends to be similar to that found in subsurface or base flow from agricultural land (Ryden *et al.*, 1973). In general, forested catchments conserve P, with P input in rainfall usually exceeding outputs in stream flow (Taylor *et al.*, 1971). As a result, forested areas are often utilized as buffer or riparian zones around streams or water bodies to reduce P inputs from agricultural land (Lowrance *et al.*, 1984, 1985).

A similar difference in P loss from forest and agricultural catchments in Finland was noted by Rekolainen (1989), who also documented greater P loads for a sampling in the 1980s than for one conducted in the early

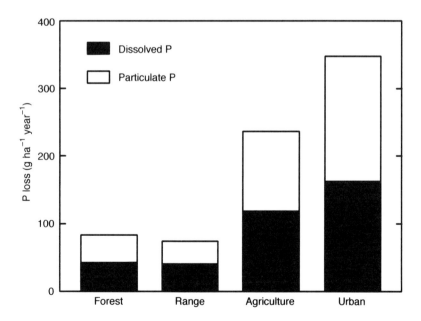

Fig. 1.1. Phosphorus loss in runoff as a function of land use in the USA (adapted from Omernik, 1977).

1970s (Fig. 1.2). The greater P loads for the more recent sampling period may result from an intensification and specialization of agricultural practices in the late 1970s (Rekolainen, 1989). It is possible that an increase in fertilizer P use from 20 kg P ha^{-1} year^{-1} in 1965 to 35 kg P ha^{-1} year^{-1} in 1985 and 50% reduction in land under grass during the same period have been the main factors affecting P loads. In coastal-plain areas of western Finland, several catchments in which 23 to 58% of the area is cultivated are dominated by acid sulphate soils (acid S basins on Fig. 1.2). The small P losses in runoff in these catchments (Fig. 1.2) are probably due to the sorption of P by aluminium (Al) compounds, which dominate the chemical reactivity of these low-pH soils (Rekolainen, 1989).

The potential loss of P from agricultural land is dependent on several factors. These factors include the overall balance of P inputs to and outputs from agricultural systems; amount, form and availability of P in soil; and the relative importance of surface and subsurface runoff in a catchment area.

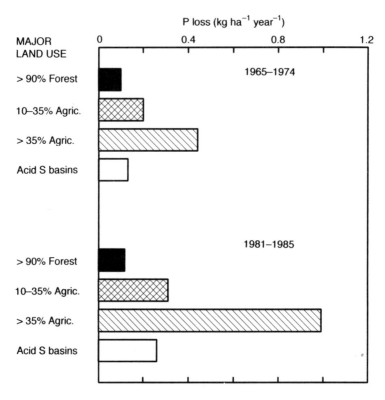

Fig. 1.2. Phosphorus loss in runoff as a function of land use in Finland (data adapted from Rekolainen, 1989).

Agricultural Phosphorus Balance

A generalized P balance and efficiency of plant and animal uptake of P for the USA and several European countries indicates the potential for P accumulation in agricultural systems (Table 1.1). Although the magnitude of P input and output varies among countries, the relative proportions of P uptake in plant and animal products are similar. The efficiency of P uptake by plants depends on a number of edaphic, management and environmental factors. Plant uptake of P increases as soil temperature, moisture, aeration and nutrient status increase. The availability of P to crops is reduced by complexation in soil with calcium (Ca) at high pH, by iron (Fe) and Al at low pH and by high clay content. Liming can increase P availability in soils by stimulating mineralization of organic P or may decrease P availability by the formation of insoluble Ca phosphates at pH > 6.5. In other situations, liming can increase P availability via increased pH (Hartikainen, 1981). A fall in pH or increased biological activity in the rhizosphere, including vesicular–arbuscular mycorrhizal associations with plant roots, can considerably enhance P uptake, especially on low-P soils.

In spite of the relatively efficient recovery of P in crop production of 56–76%, the recovery in animal production is only 10–34%, so that total P recovery by agriculture is only 11–38% (Table 1.1). Thus, the efficiency of P recovery in agriculture is dominated by animal production, as 76–94% of the total crop production is fed to animals (in addition to P additives). Animal-specific studies of P excretion rates substantiate this poor retention of P, with values of 70–80% measured for dairy cows (Aarts *et al.*, 1993), sheep (Haynes and Williams, 1992) and feeder pigs (Archer, 1985) and 87% for poultry (Iserman, 1990). Clearly, agricultural systems which include confined animal operations can determine the overall efficiency of

Table 1.1. Phosphorus balance and efficiency of plant and animal uptake of P for the USA and several European countries. Data for USA adapted from National Research Council (1993) and for European countries from Isermann (1990).

				Efficiency of		
	Input (kg P_2O_5 ha^{-1} year^{-1})	Output (kg P_2O_5 ha^{-1} year^{-1})	Surplus (kg P_2O_5 ha^{-1} year^{-1})	Plant uptake (%)	Animal uptake (%)	Total uptake (%)
USA	39	13	26	56	15	33
Netherlands	143	55	88	69	24	38
East Germany	79	8	71	59	10	11
West Germany	84	29	55	76	34	35

P recycling in agriculture and thereby the magnitude of P surpluses or potential soil accumulations.

Soil Phosphorus

Soil P exists in inorganic and organic forms (Fig. 1.3). In most agricultural soils, 50–75% of the P is inorganic, although this fraction can vary from 10 to 90%. Inorganic P forms are dominated by hydrous sesquioxides and amorphous and crystalline Al and Fe compounds in acidic, non-calcareous soils and by Ca compounds in alkaline, calcareous soils. Organic P forms include relatively labile phospholipids, nucleic acids, inositols and fulvic acids, while more resistant forms are comprised of humic acids. The lability of these forms of P is based on the extent to which extractants of increasing acidity or alkalinity, applied sequentially, can dissolve soil P.

Phosphorus amendments, in either organic or inorganic form, are needed to maintain adequate available soil P for plant uptake in modern agricultural systems. The level of these amendments varies with both soil and plant type (Pierzynski and Logan, 1993). Once applied, P is either taken up by the crop and incorporated into organic P (McLaughlin *et al.*, 1988) or becomes weakly (physisorption) or strongly (chemisorption) adsorbed on to Al, Fe and Ca surfaces (Syers and Curtin, 1988). After the initial adsorption reaction, there is a gradual fixation (absorption) of added P, which renders a proportion of adsorbed P unavailable for plant uptake (Fig. 1.3). Organic P compounds may also become resistant to hydrolysis by phosphatase through complexation with Al and Fe (Tate, 1984).

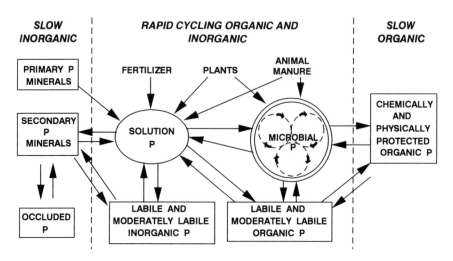

Fig. 1.3. The soil P cycle: its components and measurable fractions (adapted from Stewart and Sharpley, 1987).

With the application of P, available soil P content increases (Table 1.2). The increase in available soil P is a function of certain physical and chemical soil properties, such as clay, organic C, Fe, Al and calcium carbonate ($CaCO_3$) content. The continual application of P can increase soil-test P above levels required for optimum crop yields. In most areas, however, concern focuses on manure application, where amounts of P added often exceed crop removal rates on an annual basis (Table 1.2). As a result, many areas with intensive confined animal operations, such as the Netherlands, Belgium, north-eastern USA and Florida, now have soil P levels that are of environmental rather than agronomic concern (Sharpley *et al.*, 1994b).

Although once considered to be largely irreversible, fixed P can be slowly released back into the soil solution when reserves of less strongly held P are exhausted (Tiessen *et al.*, 1983; Johnston, 1989). As much as 70–85% of P export from sandy soils under pasture in Western Australia originated from residual P from previous fertilizer applications (Ritchie and Weaver, 1993). Thus, the determination of how long high-P soils will remain above crop sufficiency levels is of economic and environmental importance to many farmers integrating manurial P into sustainable management systems. For example, if a field has a high potential to enrich runoff with P due to excessive soil P, how long will it be before manure can be applied without unacceptably increasing the potential for P loss? McCollum (1991) estimated that without further P, 16–18 years of maize (*Zea mays* L.) or soybean (*Glycine max* (L.) Merr.) production would be needed to deplete soil-test P (Mehlich-3 P; Mehlich, 1984) in a Portsmouth fine sandy loam from 100 mg P kg^{-1} to the agronomic threshold level of 20 mg P kg^{-1}.

The rate of decline in available soil P in high-P soils when no further P is added varies with soil type and management (Table 1.3). The rate of decline in available soil P ranged from 0.1 to 30 mg kg^{-1} year^{-1}. With regular P applications, the importance of fixation processes is diminished as soil P sorption capacity becomes slowly saturated and a larger concentration of P is maintained in soil solution (White, 1980). Several authors have found that the rate of decrease in available soil P with depletion by cropping when no P is added is inversely related to the soil's P-buffering capacity (Holford, 1982; Aquino and Hanson, 1984) or P sorption saturation (available soil P/P sorption maximum; Sharpley, 1996).

Even though inorganic P has generally been considered the major source of plant-available P in soils, the incorporation of fertilizer P into soil organic P (McLaughlin *et al.*, 1988) and lack of crop response to fertilizer P due to organic P mineralization (Doerge and Gardner, 1978) emphasize the importance of organic P in soil P cycling. Sharpley (1985a) found that organic P mineralization (15–33 kg P ha^{-1} year^{-1}) in several Oklahoma soils was not completely inhibited by fertilizer P application (20–28 kg P ha^{-1} year^{-1}), with similar amounts of P contributed by both sources. Tate *et al.*

Table 1.2. Available soil P of soil treated with fertilizer or manure for several years and untreated soil.

Soil	Crop	Added P (kg ha^{-1} year^{-1})	Time (years)	Method	Available soil P		Reference and location
					Untreated (mg kg^{-1})	Treated (mg kg^{-1})	
Fertilizer							
Raub, sil	Mixed	22	25	Bray-1	18	24	Barber, 1979; Indiana
		54	25		18	71	
Portsmouth, fsl	Mixed	20	9	Mehlich-1	18	73	Cox et al., 1981; North Carolina and Rothamsted
Batcombe, cl	veg.	27	19	Olsen	16	44	
Richfield, scl	Mixed	20	14	Bray-1	12	54	Hooker et al., 1983; Kansas
	veg.	40	14		12	56	
Pullman, cl	Sorghum	56	8	Bray-1	15	76	Sharpley et al., 1984; Texas
Keith, sil	Wheat	11	6	Bray-1	22	31	McCallister et al., 1987; Nebraska
		33	6		24	47	
Rosebud, sil	Wheat	11	6	Bray-1	10	28	
		33	6		10	48	
Beef manure							
Lethbridge, cl	Barley	160	11	Bray-1	22	424	Chang et al., 1991; Alberta
		320	11		22	736	
		480	11		22	893	

Soil	Crop					Reference
Pullman, cl	Sorghum	90	8	Bray-1	15	63	Sharpley et al., 1984; Texas
		273	8		15	230	
		840	5		15	370	
Poultry litter							
Cahaba, vfsl	Grass	130	12	Bray-1	5	216	Sharpley et al., 1993; Oklahoma
Ruston, fsl		100	12		12	342	
Stigler, sl		35	35		14	239	
Sandstones	Grass		10	Mehlich-1	30	230	Kingery et al., 1994; Alabama
Swine manure							
Norfolk, l	Grass	109	11	Mehlich-1	80	235	King et al., 1990; North Carolina
		218	11		80	310	
		437	11		80	450	
Captina, sl	Grass	101	9	Bray-1	5	121	Sharpley et al., 1991; Oklahoma
Sallisaw, l	Grass	81	15		6	147	
Stigler, sl	Wheat	37	9		15	82	
Cecil, sl	Grass	160	3	Mehlich-1	19	45	Reddy et al., 1980; North Carolina
		320	3		19	100	

cl, Clay loam; fsl, fine sandy loam; l, loam; scl, silty clay loam; sil, silt loam; sl, sandy loam; vfsl, very fine sandy loam.

(1991) also found that labile organic P mineralization was an important source of P to pasture in both low- and high-P-fertility soils in New Zealand. Amounts of P mineralized range from 5 to 20 kg P ha^{-1} year^{-1} in temperate soils and from 67 to 157 kg P ha^{-1} year^{-1} in the tropics, where distinct wet

Table 1.3. The decrease in available soil P following P application.

Soil	Crop	Time (years)	Method	Available soil P Initial (mg kg^{-1})	Available soil P Final (mg kg^{-1})	Decline (mg kg^{-1} year^{-1})	Reference and location
Thurlow, l	Small grains	9 9 9	Olsen	13 20 60	4 4 6	1.0 1.8 6.0	Campbell, 1965; Montana
Georgeville, scl	Small grains	7 7	Mehlich-1	3 7	1 2	0.1 0.6	Cox et al., 1981; North Carolina and Saskatchewan
Haverhill, cl	Wheat/ fallow	14 14 14	Olsen	40 74 134	25 33 69	1.1 2.9 4.6	
Portsmouth, fsl	Small grains	8 9	Mehlich-1	23 54	18 26	0.6 3.1	
Sceptre, c	Wheat/ fallow	8 8 8	Olsen	45 67 147	18 18 40	3.4 6.1 13.4	
Williams, l	Wheat/ barley	16 16	Olsen	26 45	8 14	1.1 1.9	Halverson and Black, 1985; Montana
Richfield, scl	Maize	8 8	Bray-1	12 22	8 14	0.5 1.0	Hooker et al., 1983; Kansas
Carroll, cl	Wheat/ flax	8 8 8	Olsen	71 135 222	10 23 50	7.6 14.0 21.5	Spratt et al., 1980; Manitoba
Waskada, l	Wheat/ flax	8 8 8	Olsen	48 88 200	9 23 49	4.9 8.1 18.9	
Waskada, cl	Wheat/ flax	8 8	Bray	140 320	50 80	11.3 30.0	Wagar et al., 1986; Manitoba

cl, Clay loam; fsl, fine sandy loam; l, loam; scl, silty clay loam; sil, silt loam; sl, sandy loam; vfsl, very fine sandy loam.

and dry seasons and higher soil temperatures enhance microbial activity (Stewart and Sharpley, 1987).

With the development of fumigation–extraction techniques to measure soil microbial biomass P (Fig. 1.3; Brookes *et al.*, 1982; Hedley and Stewart, 1982), its importance in P cycling has been quantified (Stewart and Tiessen, 1987; McLaughlin *et al.*, 1988). In a study of P cycling through soil microbial biomass in England, Brookes *et al.* (1984) measured annual P fluxes of 5 and 23 kg P ha^{-1} year^{-1} in soils under continuous wheat and permanent grass, respectively. Although biomass P flux under continuous wheat was less than P uptake by the crop (20 kg P ha^{-1} year^{-1}), annual P flux in the grassland soils was much greater than P uptake by the grass (12 kg P ha^{-1} year^{-1}). Clearly, microbial P plays an important intermediary role in the short-term dynamics of organic P transformations and suggests that management practices maximizing the build-up of organic matter during autumn and winter may reduce external P requirements for plant growth during the following spring and early summer. This accumulation may also contribute to higher P losses in runoff in early spring and autumn than in summer months (Sharpley, 1980; Yli-Halla *et al.*, 1996).

Phosphorus Transport

Phosphorus is transported in dissolved (DP) and particulate (PP) forms (Fig. 1.4). Particulate P includes P sorbed by soil particles and organic matter eroded during flow events and constitutes the major proportion of P transported from cultivated land (60–90%; Pietilainen and Rekolainen, 1991; Sharpley *et al.*, 1992). Runoff from grass or forest land carries little sediment, and is therefore generally dominated by DP. While DP is, for the most part, immediately available for biological uptake (Walton and Lee, 1972; Nurnberg and Peters, 1984), PP can provide a long-term source of P for aquatic biota (Wildung *et al.*, 1974; Carignan and Kalff, 1980). The bioavailability of PP can vary from 10 to 90%, depending on the nature of the eroding soil (Sharpley, 1993b). Together, DP and bioavailable PP constitute bioavailable P (BAP) or P available for uptake by aquatic biota (Fig. 1.4).

During the transport of P from the edge of the field to the receiving lake or ocean, DP and PP fractions continuously alter as a result of in-stream processes (Fig. 1.4). These processes include uptake of DP by aquatic biota, transformations between PP and DP caused by changes in the equilibrium stream DP concentration, deposition of suspended PP and resuspension of stream-bed or stream-bank PP (Meyer, 1979; House and Casey, 1988). The direction and extent of these P transformations during transport depend on the time of year, the relative amounts of P entering from different sources and, in particular, the rate of flow.

On arrival at the receiving lake, further exchanges of P at the

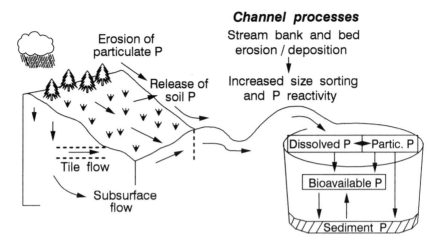

Fig. 1.4. The transport of P from agricultural land to surface waters.

sediment–water interface affect the amount of P available for biological productivity (Fig. 1.4). For example, Theis and McCabe (1978) found that the DP concentration in the lake water of two shallow hypereutrophic lakes in Indiana was reduced by sorption during aerobic periods and increased by release of sediment-bound P during anaerobic periods. In shallow hypereutrophic lakes, an increase in pH of interstitial water can take place due to high algal assimilation close to the lake sediment surface. This can result in release of soluble P from the sediment (Rippey, 1977; Knuuttila *et al.*, 1994). The transformations of P within water bodies must be considered in assessing the impact of P transported from agricultural land on the potential biological productivity and eutrophication risk of the receiving lake.

Amounts of P transported from catchments are a function of catchment hydrology, in terms of when and where surface runoff occurs, soil P content, and amount of P added as fertilizer or manure. This assumes that, in most cases, P export from catchments occurs in surface rather than subsurface runoff, although it is recognized that in some regions, notably Florida, Western Australia and the Netherlands, most P is transported in drainage waters.

Generally, the P concentration in water percolating through the soil profile is small, due to sorption of P by P-deficient subsoils. Exceptions occur in acid organic or peaty soils, where the adsorption affinity and capacity for P are low, due to the predominantly negative charged surfaces and the complexing of Al and Fe by organic matter (Duxbury and Peverly, 1978; Miller, 1979; White and Thomas, 1981). Similarly, P is more susceptible to movement through sandy soils with low P sorption capacities;

in soils which have become waterlogged, leading to conversion of Fe(III) to Fe(II) and the mineralization of organic P; and with preferential flow through macropores and earthworm holes (Ozanne *et al.*, 1961; Gotoh and Patrick, 1974; Sharpley and Syers, 1979b; Bengston *et al.*, 1992). Because of the variable path and time of water flow through a soil with subsurface drainage, factors controlling DP in subsurface waters are more complex than for surface runoff. Subsurface runoff includes tile drainage and natural subsurface flow, where tile drainage is percolating water intercepted by artificial systems, such as mole and tile drains (Fig. 1.4). In general, the greater contact time between subsoil and natural subsurface flow than tile drainage, results in lower losses of DP in natural than in tile flow.

Hydrological controls
Several decades of research have provided an understanding of the mechanisms controlling soil P dynamics and release to runoff at the point or field scales. However, the hydrological controls linking spatially variable P sources, sinks, temporary storages and transport processes within a catchment are less well understood. This information is critical to the development of effective management programmes addressing the reduction of P export from agricultural catchments.

Runoff production in many catchments in humid climates is controlled by the variable-source-area concept of catchment hydrology (Ward, 1984). Here, surface runoff is usually generated only from limited source areas within a catchment (Fig. 1.5). These source areas vary in time, expanding and contracting rapidly during a storm as a function of precipitation, temperature, soils, topography, groundwater and moisture status over the catchment (Gburek and Pionke, 1995). Surface runoff from these areas is limited by soil water storage rather than infiltration capacity. This situation usually results from high water tables or soil moisture contents in near-stream areas (Fig. 1.5).

The boundaries of surface runoff-producing areas will be dynamic both within and between rainfalls (Gburek and Pionke, 1995; Zollweg *et al.*, 1995). During a rainfall event, area boundaries will migrate upslope as rainwater input increases. In dry summer months, the runoff-producing area will be closer to the stream than during wetter winter months, when the boundaries expand away from the stream channel.

The location and movement of variable source areas of surface runoff in a catchment are also influenced by soil structure, geological strata and topography. Fragipans or other layers, such as clay pans of distinct permeability changes, can determine when and where perched water-tables occur. Shale or sandstone strata may also influence soil moisture content and location of saturated zones. For example, water will perch on less permeable layers in the subsurface profile and become evident as surface flow or springs at specific locations in a catchment. Converging topography

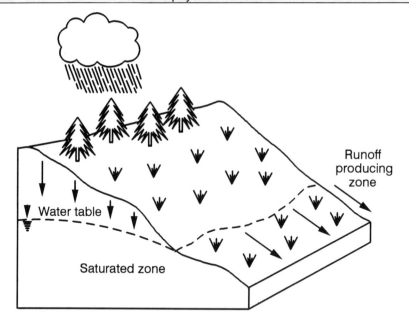

Fig. 1.5. The concept of variable source areas of runoff in hill-land catchments.

in vertical or horizontal planes, slope breaks and hill-slope depressions or spurs also influence variable-source-area hydrology within catchments. Net precipitation (precipitation – evapotranspiration) governs catchment discharge and thus total P (TP) loads to surface waters. This should be taken into account when comparing the load estimates from different regions. It is also one reason why there seems to be more concern with P in humid regions than in more arid regions, although the amount and intensity of agriculture can be higher in arid regions.

In catchments where surface runoff is limited by infiltration rate rather than soil water-storage capacity, areas of the catchment can alternate between sources and sinks of surface flow. This again will be a function of soil properties, rainfall intensity and duration and antecedent moisture condition. As surface runoff is the main mechanism by which P is exported from most catchments (Sharpley and Syers, 1979a), it is clear that, if surface runoff does not occur, P export is negligible (Brookes *et al.*, Chapter 12, this volume). Thus, consideration of hydrological controls and variable source areas is critical to a more detailed understanding of P export from an agricultural catchment.

Physical and chemical controls
As soil P content increases, the potential for PP and DP transport in runoff increases. Sources of PP in streams include eroding surface soil, stream banks and channel beds. Thus, processes determining soil erosion also control PP transport (Fig. 1.4). In general, the P content and adsorption capacity of eroded particulate material is greater than that of source soil, due to preferential transport of clay-sized material. The transport of DP in runoff is initiated by the desorption, dissolution and extraction of P from soil and plant material (Fig. 1.4). These processes occur when rainfall interacts with a thin layer of surface soil (1–5 cm) before leaving the field as runoff (Sharpley, 1985b). Although the proportion of rainfall and depth of soil involved are difficult to quantify in the field, they will be highly dynamic, due to variations in rainfall intensity, soil tilth and vegetative cover.

Erosion is a function of rainfall amounts and intensities and soil texture, topography and management. Management practices should be designed so that erosion reduction can be maximized by considering these factors. Basically, approaches to reducing erosion involve detachment and transport control. Often detachment control (e.g. more crop or residue cover) is reported to be more effective than transport control (e.g. vegetative filter strips).

Several studies have reported that the loss of DP in runoff is dependent on the soil P content of surface soil. For example, a highly significant linear relationship was obtained between the DP concentration of runoff and soil P content (Mehlich-3 P) of surface soil (5 cm) from cropped and grassed catchments in Arkansas, Oklahoma and Texas (Sharpley *et al.*, 1986; Sibbesen and Sharpley, Chapter 7, this volume). A similar dependence of the DP concentration of runoff on Bray-1 P was found by Romkens and Nelson (1974) for a Russell silt loam in Illinois ($r^2 = 0.81$) and on water-extractable soil P ($r^2 = 0.61$) of 17 Mississippi catchments by Schreiber (1988) and 11 Oklahoma catchments by Olness *et al.* (1975) ($r^2 = 0.88$).

These and similar studies related DP in runoff to soil determined by traditional soil-test methods for estimating plant available P. While they showed promise in describing the relationship between the level of soil and runoff P, they are limited for several reasons. First, while DP is an important water-quality parameter, it only represents the dissolved portion of runoff P readily available for aquatic plant growth. It does not represent adsorbed soil P that can become available through desorption. To overcome this limitation, an approach using Fe-oxide-impregnated strips of filter-paper has been developed to estimate the BAP or algal-available P in runoff (Sharpley, 1993a). Acting as a P sink, Fe-oxide strips may have a stronger theoretical basis than chemical extraction in estimating BAP.

Using simulated rainfall (2.54 cm h^{-1} for 30 min), Sharpley (1995) found Fe-oxide soil P was related to the BAP concentration of runoff from ten Oklahoma soils, ranging from sandy loam to clay in texture (Fig. 1.6).

Thus, for a given soil P level, the concentration of P maintained in runoff will be influenced by soil type, because of differences in P-buffering capacity among soils, caused by varying levels of clay, Fe and Al oxides, carbonates and organic matter. For example, a surface soil Fe-oxide P content of 200 mg kg^{-1} would support a runoff BAP concentration of 0.53 mg l^{-1} for the clay but 1.65 mg l^{-1} for the sandy loam (Fig. 1.6).

Another approach, developed in the Netherlands to determine the potential for DP movement in drainage water, estimates soil P sorption saturation as the percentage of P sorption capacity as extractable soil P (Breeuwsma and Silva, 1992; Sibbesen and Sharpley, Chapter 7, this volume). This approach is based on the fact that more P is released from soil to runoff or leaching water as P saturation or amount of P sorbed increases with P additions. Soil P saturation is used in the Netherlands, where farm recommendations for manure management are designed to limit the loss of P in surface water and groundwater (Breeuwsma and Silva, 1992). For Dutch soils, a critical P saturation of 25% has been established as the threshold value above which the potential for P movement in surface water and groundwater becomes unacceptable (Van der Zee *et al.*, 1987, 1990; Breeuwsma and Silva, 1992).

When the P sorption saturation of the Oklahoma soils was calculated using Fe-oxide P as extractable soil P, a single relationship described the dependence of BAP in runoff on P saturation for all soils (Fig. 1.6). Thus, P saturation better describes the effect of soil type in the differential release of soil P to runoff and potential for P loss in runoff than traditional soil-test P measures.

Increases in P loss in runoff have been measured after the application

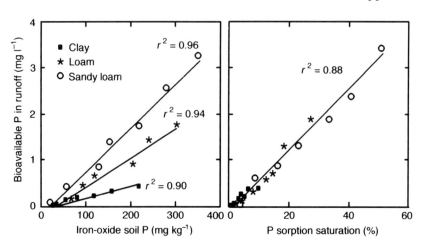

Fig. 1.6. Relationship between iron-oxide strip P and P sorption saturation of several Oklahoma soils and the bioavailable P concentration of runoff (data adapted from Sharpley, 1995).

of fertilizer P and manure (Table 1.4). These losses are influenced by rate, time and method of application; form of fertilizer or manure, amount and time of rainfall after application; and vegetative cover. The portion of applied P transported in runoff was greater from conventional than conservation-tilled catchments (Table 1.4). However, McDowell and McGregor (1984) found fertilizer P application to no-till maize reduced P transport, probably due to an increased vegetative cover afforded by fertilization. As expected, the loss of applied P in subsurface tile drainage is appreciably lower than in surface runoff (Table 1.4). Although it is difficult to distinguish between losses of fertilizer or manure and native soil P without the use of expensive and hazardous radiotracers, losses of applied P in runoff are generally less than 10% of that applied, unless rainfall immediately follows application or runoff has occurred on steeply sloping, poorly drained and/or frozen soils. The high proportion of manurial P in runoff may result from high manure application and often less flexibility in application timing than for fertilizer (Table 1.4). This inflexibility often results from the continuous production of manure throughout the year and a frequent lack of manure storage facilities.

Critical-source-area controls
From the above discussion, it is clear that sources of P export from catchments are determined by hydrological and chemical controls. This chapter and that of Pionke *et al.* (Chapter 10, this volume) have shown that these controls are dynamic and highly variable, both temporally and spatially. This is the basis of the generalization that 90% of annual P export from catchments occurs from only 5% of the land area during only one or two storms. For example, more than 75% of annual runoff from catchments in Ohio (Edwards and Owens, 1991) and Oklahoma (Smith *et al.*, 1991) occurred in one or two severe storm events. These events contributed over 90% of annual P export (0.2 and 5.0 kg ha^{-1} year^{-1}, respectively). However, there has been no attempt to date to couple variable source area hydrological processes with P status over a catchment.

The importance of the linkage between runoff-producing variable source area and areas of high soil P in determining P export from catchments has been clearly demonstrated for a 26-ha catchment in east central Pennsylvania by Zollweg *et al.* (1995). Phosphorus export from the agricultural catchment is dominated by storm flow, which contributed approximately 70% of annual total P export (Pionke and Kunishi, 1992). Zollweg *et al.* (1995) integrated a variable-source-area storm-runoff model (soil-moisture-based runoff model; Zollweg, 1994) within the geographical research analysis support system (GRASS) and the geographical information systems (GIS) framework. Data layers were also developed to represent available soil P (Bray-1 P) over the catchment. Algorithms based on soil P export coefficients developed by Daniel *et al.* (1994) and Sharpley and Smith (1989) were incorporated to predict runoff P from soil P where

Table 1.4. Effect of fertilizer and manure application on P loss in surface runoff and fertilizer application on P loss in tile drainage.

Land use	P added (kg ha^{-1} year^{-1})	P loss Dissolved (kg ha^{-1} year^{-1})	P loss Total (kg ha^{-1} year^{-1})	%*	Reference and location
Surface runoff					
Fertilizer					
Grass	0	0.02	0.22		McColl et al., 1977;
	75	0.04	0.33	0.1	New Zealand
NT maize	0	0.70	2.00		McDowell and
	30	0.80	1.80		McGregor, 1984;
CT maize	0	0.10	13.89		Mississippi
	30	0.20	17.70	12.7	
Wheat	0	0.20	1.60		Nicholaichuk and Read,
	54	1.20	4.10	4.6	1978; Saskatchewan, Canada
Grass	0	0.50	1.17		Sharpley and Syers,
	50	2.80	5.54	8.7	1979a; New Zealand
Dairy manure†					
Alfalfa	0	0.10	0.10		Young and Mutchler,
Spring	21	1.90	3.70	17.1	1976; Minnesota
Autumn	55	4.80	7.40	13.3	
Maize	0	0.10	0.20		
Spring	21	0.20	0.60	1.9	
Autumn	55	1.00	1.60	2.5	
Poultry manure					
Grass	0	0.00	0.10		Edwards and Daniel,
	76	1.10	2.10	2.6	1992; Arkansas
	304	4.30	9.70	3.2	
Grass	0	0.10	0.40		Westerman et al., 1983;
	95	1.40	12.40	12.6	North Carolina
Swine manure					
Fescue	0	0.00	0.00		Edwards and Daniel,
	19	1.50	1.50	7.4	1993a; Arkansas
	38	4.80	4.80	12.6	

Table 1.4. *continued*

Tile drainage					
Maize	0	0.13	0.42		Culley et al., 1983;
	30	0.20	0.62	0.7	Ontario, Canada
Oats	0	0.10	0.29		
	30	0.20	0.50	0.7	
Alfalfa	0	0.12	0.32		
	30	0.20	0.51	0.6	
Grass	0	0.08	0.17		Sharpley and Syers,
	50	0.44	0.81	1.3	1979b; New Zealand

CT, conventional till; NT, no till.
* Percentage P applied lost in runoff.
† Manure applied in either spring or autumn.

runoff occurs, to give P export from the catchment (Fig. 1.7). Reliable estimates of P export were obtained compared with measured values from the catchment.

Runoff of fairly uniform depth was generated in the near-stream areas (Fig. 1.7). Runoff generation further away from the stream channel tended to be at slope breaks, in areas of converging subsurface flow or where slopes were generally shallow. Although runoff generation was high near the stream, soil P was lowest in this area (Fig. 1.7). Thus, P loss in runoff from these areas tended to be low. Loss was greater from the upper regions of the catchment, where storm runoff originated within cropped fields with high soil P contents (Fig. 1.7). Overall, P loss tended to reflect soil P content in areas where runoff occurred (Zollweg *et al.*, 1995). This study demonstrates the type of information needed to delineate critical-source-area controls of P export from agricultural catchments and where remediation efforts may be best directed. Also, better digital geographical data are needed to be able to scale up these results to make management plans for larger regions.

TECHNIQUES FOR REDUCING PHOSPHORUS LOSSES

The overall goal of our efforts to reduce P losses from agriculture to surface waters should aim to balance inputs of P in feed and fertilizer with outputs in produce, together with managing soils to maintain soil P resources at adequate levels. Increasing the use efficiency of P in agricultural systems may be brought about by source and transport control strategies. Although we know how to, and have generally been able to reduce the transport of P from agricultural land in runoff and erosion, less attention has been directed toward source management. This is mainly a result of greater labour and economic constraints on farming with source than with transport management. For example, it is clear from the extent of soils with

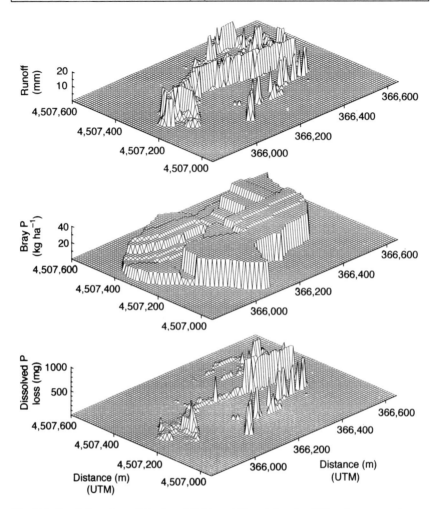

Fig. 1.7. Simulation of runoff depth, soil P (Bray-1 P) and dissolved P loss from an east-central Pennsylvania catchment during a storm in April 1992 (data adapted from Zollweg et al., 1995).

P in excess of levels sufficient for optimum crop yields that more attention should be paid to avoiding soil P build-up via P source management. We will thus give more attention to discussion of source than of transport management.

In the USA, marginal farm profits and cost-share programmes often influence implementation of control measures. Model simulation and field studies indicate that the cost-effectiveness of several best management practices (BMPs) varies (Table 1.5). Although vegetative filter strips and manure management (chemical amendments, storage, waste treatment and

Table 1.5. Cost-effectiveness of BMPs on reducing P losses from continuous maize with a 5% slope and 140 kg P ha^{-1} year^{-1} manure broadcast (data adapted from USDA-ASCS, 1992, and Heathman et al., 1995).

Best management practice	P loss (kg ha^{-1} year^{-1})	Cost-effectiveness ($ kg^{-1} P saved)
None	10.0	–
Contour cropping	6.3	1.7
Terraces	3.2	4.7
Conservation tillage	3.9	0.8
Vegetative buffer areas*	2.5	1.1
Manure management	2.8	3.3
All BMPs	1.8	4.9

* Cost-effectiveness includes the cost of land taken out of production.

barnyard runoff control) can reduce runoff P more than tillage management, conservation tillage may often be a more cost-effective measure. These generalized examples emphasize the need to determine the load reduction required for a given catchment and water body to select appropriate BMPs. Clearly, construction of terraces, which are initially expensive, may in some cases be a viable option. However, careful selection and integration of different practices can improve overall cost-effectiveness.

Strategies to minimize P loss in runoff will be most effective if sensitive or source areas within a catchment are identified, rather than widespread implementation of general strategies over a broad area. Thus, more attention should be paid to avoiding soil P build-up via P source management. However, before cost-effective control measures can be targeted we must identify critical source areas vulnerable to P loss from a catchment, account for spatial and temporal variations in catchment response to implemented measures and ultimately the response of affected waters.

Source Management

Reducing off-farm inputs of phosphorus in feed

In many regions where P has been designated a management priority due to eutrophication concerns, P inputs in feed and fertilizer exceed production outputs. This situation exists in the north-east USA, for example, where dairy farming plays a major role in the agricultural economics of the region. In this region, several farms are reducing the amount of feed imported by intensive grazing of dairy herds (Ford, 1994). Use of intensively grazed pasture has the potential to increase dairy-farm profits, provide labour savings and, as environmental concerns become greater, reduce off-farm inputs of P. For a representative 60-cow Holstein herd farm in Pennsylvania, intensive grazing increased farm income 14–25% compared with no intensive grazing i.e. from $8400 to $12,400 (Elbehri and Ford, 1995). The increased income was mainly due to decreased production costs of feed and fertilizer, as manure was used to supplement fertilizer P. These reduced inputs amounted to an annual 53% reduction in off-farm inputs of P.

Even so, intensively grazed pasture is not 'the answer' for all dairy farms and its success depends to a large extent on the existing crop mix and above-average managerial ability. Thus, there is a need for targeted extension and research programmes to improve optimal feeding under grazing as well as the overall quality of pasture management (Ford, 1994). In the USA, intensive grazing may enable many small dairy farms to reduce production costs to levels that maintain a desired standard of living, while reducing potential P sources for transport in runoff.

Manipulation of dietary P intake by livestock is another aspect of source management which is receiving increasing attention. In the Netherlands, the concentration of P in manure decreased temporarily during the Second World War, when concentrates and fertilizers were less available, and reductions in concentrate P contents are now being similarly implemented to help reduce the amounts of P excreted to land (Wadman *et al.*, 1987). Morse *et al.* (1992) recorded a 23% reduction in excretion of P in faeces and a 17% reduction in total P excretion when dairy cows reduced their daily P intake from 82 to 60 g day^{-1}. Increasing the dietary P intake from 82 to 112 g day^{-1} increased excretion of P in faeces by 49% and total P excretion by 37%. Thus, there is a clear indication that amounts of excreted P can be reduced by carefully matching dietary P inputs to the animal's requirements, especially as P intakes above minimum dietary requirements do not seem to confer any milk-yield advantage (Brodison *et al.*, 1989; Morse *et al.*, 1992).

Enzyme additives for animal feed that will increase the efficiency of P uptake during digestion are also being tested (Lynch and Caffrey, Chapter 14, this volume). Development of such enzymes, which would be cost-effective in terms of animal weight gain, may reduce the P content of manure. One example is the use of phytase, an enzyme that enhances the efficiency of P recovery from phytin in grains fed to poultry. It is now common to supplement poultry feed with mineral forms of P because of the low digestibility of phytin. This contributes to P enrichment of poultry manures and litters. While the phytase enzyme has been shown to decrease the need for mineral P, it is currently too expensive for use as a routine feed additive. Provision of phytase to poultry growers via cost-sharing by federal agencies would represent an innovative approach to controlling non-point-source losses of P at the source, which would be much easier than after-the-fact control programmes for land applied manures.

Soil phosphorus management
Management of P on soil susceptible to P loss involves the use of soil tests based on environmental rather than agronomic considerations to determine P application and suitable application methods. Environmental concern has forced many countries and states in the USA to consider the development of recommendations for P applications and catchment management based on the potential for P loss in runoff, as well as crop P requirements. A major difficulty in the development of these recommendations has been the identification of soil-test P levels that are high enough to raise concerns about the potential for unacceptable levels of P loss in runoff. Examples from several states in the USA are in Table 1.6. Establishing these levels is often a highly controversial process, for two reasons. First, the database relating soil-test P to runoff P is limited to a few soils and crops and there is often a reluctance to rely upon data of this type generated in other regions. Second, the economic implications of establishing soil-test P levels which may limit manure applications are significant. In many areas dominated by animal-based agriculture, there simply is no economically viable alternative to land application. Because of these factors, those most affected by these soil-test P limits are vigorously challenging their scientific basis. Clearly, there is a need to assess the validity of the use of soil-test P values as indicators of P loss in runoff.

Basing manure application on soil P and crop removal of P can reduce the build-up of soil P and at the same time lower the risk of nitrate leaching to groundwater. However, basing manure applications on P rather than N management, could present several technical and economic problems to many farmers. A soil-test-P-based strategy could eliminate much of the land area with a history of continual manure application from further manure additions, as several years are required for significant depletion of high soil P levels. This would force farmers to identify larger areas of land to utilize the generated manure, further exacerbating the problem of local land-area limitations. In addition, farmers relying on manure to supply most of their crop N requirements may be forced to buy fertilizer N to supplement forgone manure N.

There may also be a limited time period over which P applications can be omitted on soils with high-test P without affecting yield. Withers *et al.* (1994) found that, after 3 years of withholding P fertilizer on P-rich calcareous soils in England, crop yields of cereals were significantly reduced. It may therefore be more appropriate to reduce P application rates on certain soils rather than omit P altogether. Small amounts of placed starter fertilizer for vegetable crops have successfully reduced the need for much larger P applications being broadcast (Costigan, 1988) and a similar strategy (for example, foliar P applications) may be appropriate for other crops.

Methods of P application are also important. Rotational applications of P designed to streamline fertilizer operations may leave large amounts of

Table 1.6. Soil P interpretations and management guidelines (adapted from Sharpley et al., 1994a, and Gartley and Sims, 1994).

State	Critical value	Management recommendation	Rationale*
Arkansas	150 mg kg^{-1} Mehlich-3 P	At or above 150 mg kg^{-1} soil P: 1. Apply no P from any source 2. Provide buffers next to streams 3. Overseed pastures with legumes to aid P removal 4. Provide constant soil cover to minimize erosion	CV: Ohio sewage sludge data MR: Reduce soil P and minimize movement of P from field
Delaware	120 mg kg^{-1} Mehlich-1 P	Above 120 mg kg^{-1} soil P: Apply no P from any source until soil P is significantly reduced	CV: Greater P loss potential from high-P soils MR: Protect water quality by minimizing further soil P accumulations
Ohio	150 mg kg^{-1} Bray-1 P	Above 150 mg kg^{-1} soil P: 1. Institute practices to reduce erosion 2. Reduce or eliminate P additions	CV: Greater P loss potential from high-P soils as well as role of high soil P in zinc deficiency MR: Protect water quality by minimizing further soil P accumulations
Oklahoma	130 mg kg^{-1} Mehlich-3 P	30 to 130 mg kg^{-1} soil P: Half P rate on > 8% slopes 130 to 200 mg kg^{-1} soil P: Half P rate on all soils and institute practices to reduce runoff and erosion Above 200 mg kg^{-1} soil P: P rate not to exceed crop removal	CV: Greater P loss potential from high-P soils MR: Protect water quality, minimize further soil P accumulation and maintain economic viability

Michigan	75 mg kg^{-1} Bray-1 P	Above 75 mg kg^{-1} soil P: P application must not exceed crop removal Above 150 mg kg^{-1} soil P: Apply no P from any source	CV: Minimize P loss by erosion or leaching in sandy soils MR: Protect water quality and encourage wider distribution of manures
Texas	200 mg kg^{-1} Bray-1 P or Texas A&M P	Above 200 mg kg^{-1} soil P: P addition not to exceed crop removal	CV: Greater P loss potential from high-P soils MR: Protect water quality by minimizing further soil P accumulations
Wisconsin	75 mg kg^{-1} Bray-1 P	Above 75 mg kg^{-1} soil P: 1. Rotate to P-demanding crops 2. Reduce manure application rates Above 150 mg kg^{-1} soil P: Discontinue manure applications	CV: At that level, soils will remain non-responsive to applied P for 2–3 years MR: Minimize further soil P accumulations

* CV, critical value rationale; MR, management recommendation rationale.

available P on the surface and should be avoided in high-risk areas. Efficient management of P amendments on soils susceptible to P loss involves the subsurface placement of fertilizer and manure away from the zone of removal in surface runoff and the periodic ploughing of no-till soils to redistribute surface P accumulations throughout the root zone. Both practices may indirectly reduce the loss of P by increasing crop uptake of P and yield, which affords a greater vegetative protection of surface soil from erosion.

Subsurface application or knifing of P fertilizer or manure may be recommended to minimize P loss in runoff. However, this may be unacceptable if it reduces residue cover below 30%, a cost-share requirement of BMPs to reduce erosion risk. Clearly, assessments of priorities and greater compatibility between different recommended management plans are needed.

It may be possible that, by utilizing residual soil P, careful crop selection will reduce the amount of nutrients potentially available to be transferred to surface waters (Pierzynski and Logan, 1993). 'Scavenger' crops that have a higher affinity or requirement for P may thereby reduce the amount of P amendment needed and reduce soil-nutrient stratification. Alfalfa, for example, has reduced subsoil nitrate accumulations and may reduce soil P accumulations (Mathers *et al.*, 1975).

Manure management
Manure-management strategies strive to increase the marketability and demand for manure as a replacement for fertilizer P. This may be accomplished by a combination of several factors, which include: (i) development of reliable methods to determine the nutritive value of manure; (ii) basing manure applications on N- or P-dependent on-site characteristics (discussed above); (iii) application timing and method; (iv) use of manure amendments that may reduce environmental hazards; (v) establishment of mechanisms that encourage transporting manure greater distances to areas where P can be more effectively used; and (vi) formation of cooperatives that can more efficiently compost, pelletize or concentrate manure to increase the distance it can be economically transported from its source.

Farm advisers and extension agents are now recommending that the P content of both manure and soil be determined by soil-test laboratories before land application of manure. This is important because there is a tendency among farmers to underestimate the nutritive value of manure. Thus, manure analyses are a constructive educational tool, showing farmers that manure represents a valuable source of P. Manure analyses, in combination with soil-test results, can also demonstrate the positive and negative long-term effects of manure use and the time required to build up or deplete soil nutrients. For instance, with P, they can help a farmer identify the soils in need of fertilization, those containing excess P, which should not be manured, and those where moderate manure applications may be of some value.

Field variability in soil properties, crop-yield potential and topography differences that influence leaching and runoff losses can affect the amount of P that should be applied. In the absence of environmental concerns, producers using manure as a nutrient source naturally prefer to err on the side of using excess. The problem is that an application of manurial P can lead to a range from greatly excessive to highly inadequate nutrient availability, because of inherent differences in soil fertility. For example, mixing a few, very fertile subsamples with any number of subsamples with marginal nutrient availability can lead to analytical results that give producers a false sense of security. A recent example from Nebraska, USA, where scientists extensively sampled a 58-ha maize field, clearly illustrates this type of situation (Hergert *et al.*, 1994; Peterson *et al.*, 1994). Collection of more than 2000 cores on a grid basis from a 58-ha maize field showed that approximately 75% of the field would have been expected to respond to P fertilization. Yet composited annual soil samples did not indicate the need for much, if any, P fertilizer. Further investigation disclosed the existence of an old farmstead that included a pig feeding operation about 20 years earlier and another area where sheep had been fed in confinement over 70 years ago. The integration of the spatial variability of soil properties into manure management decisions will receive increased attention as global positioning systems (GPS) and GIS are developed (Haapala, 1995).

Of necessity, some farmers apply manure to the surface of soils during autumn and winter months, when more time is available for this operation. As plant growth is minimal and, if manure is not incorporated, sorption and plant P uptake are small, the potential for P loss during spring rainfalls is greatly increased. The sooner rain falls after application, the greater the potential for P loss in runoff (Westerman and Overcash, 1980). However, P losses are affected little by an interval longer than about 3 days between application and runoff (Edwards and Daniel, 1993b).

Specialized manure-spreading equipment that allows subsurface injection or careful control of application rates is often too expensive for most farmers. Even so, it may not be feasible to develop equipment that can inject manure in the thin rocky soil underlying many of the poultry and pig farms in south-central USA. The formation of farm cooperatives to share equipment costs and cost-share equipment programmes may alleviate some of the economic constraints. Until this becomes more widespread, many farmers will continue to adapt existing equipment for broadcasting manure.

Commercially available manure amendments, such as slaked lime or alum, can reduce ammonia (NH_3) volatilization and P solubility of poultry litter by several orders of magnitude (Moore and Miller, 1994). Also the DP concentration of runoff from fescue treated with alum-amended litter (11 mg l^{-1}) was much lower than from fescue treated with unamended litter

(83 mg l^{-1}; Shreve et al., 1995). Perhaps the most important benefit of manure amendments (for both air and water quality), however, will be an increase in the N:P ratio of manure, via reduced N loss from manure by NH$_3$ volatilization. An increased N:P ratio of manure would approach crop N and P requirements. Thus, additions of manure based on crop N requirements would reduce the amount of P added, thereby minimizing potential soil P accumulations. Increasing the N content of manures also reduces application rates needed to meet crop N requirements if additional land is available, resulting in economic savings for farmers by decreasing the time associated with manure handling and transport.

The cost of transporting low-density manure more than short distances from the site of its production often exceeds its nutrient value. This has limited the area of land available for application of manure with most manure applied in the immediate vicinity of production. Thus, the dominant geology, soils and topography of the local area often cannot be adequately taken into account prior to application. However, innovative measures are being used by some farmers to transport manure from the area of production. For example, following delivery of grain or feed, lorries and railway trucks are transporting dry manure instead of returning empty (Collins et al., 1988). In Delaware, the local poultry trade organization has established a 'manure bank' network that puts farmers in need of manures in contact with small poultry growers that have inadequate land available to use all the manure generated by their operation. Cost-share monies are also made available to subsidize the use of newer and more efficient manure application equipment. Even so, large-scale transportation of manure from producing to non-manure-producing areas is generally not occurring.

Transport Management

Loss by erosion and runoff may be reduced by increasing vegetative cover through conservation tillage, although this may increase the amount of N leaching that occurs, because of the larger volumes of water draining through the soil profile (Tyler and Thomas, 1977; Sharpley and Smith, 1994). Also, DP and BAP losses can be greater from no-till than from conventional till practices, since accumulation of crop residues and added P at the soil surface provide a source of P to runoff that would be decreased during tillage (Barisas et al., 1978; Uusi-Kamppa and Ylaranta, 1992; Sharpley and Smith, 1994). Such water-quality trade-offs must be weighed against the potential benefits of conservation measures in assessing their effectiveness. In some catchments or fields within catchments, the risk of N contamination of groundwater may override the need for widespread measures for controlling potential P contamination of surface waters.

Additional specific measures to minimize P loss by erosion and runoff include buffer strips, riparian zones, terracing, contour tillage, cover crops and impoundments or small reservoirs. However, these practices are generally more efficient at reducing PP than DP. For example, several studies have indicated little decrease in lake productivity with reduced P inputs following implementation of conservation measures (Young and DePinto, 1982; Gray and Kirkland, 1986). The lack of biological response was attributed to an increased bioavailability of P entering the lakes, as well as internal recycling. Clearly, effective remedial strategies must address the management of P sources and applications, as well as erosion and runoff control.

Implementing Remedial Measures

Since the late 1970s, several studies have investigated the long-term (7–10-year) effectiveness of BMPs to reduce P export from agricultural catchments. These studies quantified P loss prior to and after BMP implementation or attempted to use untreated catchments as controls. Overall, these studies showed that BMPs reduced P export. For example, water-quality improvements have been demonstrated following BMP implementation in several areas of the USA (National Water Quality Evaluation Project, 1988; USDA-ASCS, 1992; Goldstein and Ritter, 1993; Richards and Baker, 1993; Bottcher *et al.*, 1995). With this experience, however, it is evident that several factors are critical to effective BMP implementation. These factors include targeting catchments that will respond most effectively to BMPs and identifying source areas of runoff and erosion, as well as accounting for both catchment and water-body response time and equilibration.

Identifying critical catchments
With limited resources, it will be necessary to select catchments that will provide the greatest reduction in P loss following BMP implementation. Otherwise, overall P inputs to an affected water body may not be decreased sufficiently.

The effect of several manure BMPs on P export from two catchments in the LaPlatte River basins draining into Lake Champlain, Vermont, were evaluated (Meals, 1990). These included barnyard runoff control, milking shed waste treatment, and construction and use of manure storage facilities (Table 1.7). Post-BMP losses of P were lower than before BMPs. For both catchments, barnyard runoff control resulted in the greatest reduction in P export and was the most cost-effective BMP (Table 1.7). The results of this simple cost-effectiveness analysis have important implications for formulating remediation strategies. If a catchment project is being formulated with limited funding, the cost-effectiveness analysis can help target a catchment that will provide the greatest impact for the money invested (Meals, 1990). For instance, if a choice had to be made between the two catchments of

Table 1.7. Cost-effectiveness of animal-waste management BMPs in the LaPlatte River basin project, Vermont, for 1980 to 1989 (data adapted from Meals, 1990).

Management	Catchment 1		Catchment 2	
	P reduction (kg)	Effectiveness ($ kg^{-1} P)	P reduction (kg)	Effectiveness ($ kg^{-1} P)
Barnyard runoff control	311	4	78	14
Milking shed waste treatment	34	12	11	32
Waste storage facility	154	269	14	1963
Total	567	77	103	282

Table 1.7, catchment 1 would have been selected, based on better cost-effectiveness ratios.

In order to make regional assessments to identify critical source areas within large regions, results obtained from experimental plots or fields, as well as model estimates, have to be scaled up. The accuracy of regional estimates depends on how good our experimental results or models are and how reliable available regional data are on the factors governing P transport. The most important data are land use, soil texture, topography and management practices. Once these data are in digital form, GIS techniques can be used to combine them with experimental or model results. In addition to regional assessments, this approach can be used to make comparative studies on the effectiveness of different remedial measures. Using mathematical models to calculate typical P transport values over a wide range of soil textures, slopes and crops can serve as a quick and inexpensive method of making these assessments.

Catchment selection should also consider the affected water bodies and potential for water-quality improvement. While most fresh waters are P-limited, there are notable exceptions where P controls will have marginal to no benefit, e.g. high-elevation lakes in the western USA are N-limited. In some lakes and in many streams, plant productivity is limited by high turbidity, from either anthropogenic or natural sources. In lakes where eutrophication is sustained by internal P recycling mechanisms within the water body, controls on external P sources will have limited effect without some form of in-lake biomanipulation to reduce aquatic bioproductivity, such as the removal of bottom sediments or the introduction of specific fish species. Reducing agricultural P inputs to lakes may not always achieve the desired or even expected water-quality improvements, due to the continued contribution of P from other external sources, e.g. point-source P inputs and rainfall.

To optimize control activities, there is therefore often a need to prioritize management actions to those catchments which provide inputs to P-sensitive lakes. Management agencies are also often required to further target limited financial and human resources to those P-sensitive lakes having the highest public or ecosystem value. Several regions are adopting a catchment approach to target priority lakes and catchments by considering the threat to the water quality and the practicability of alleviating the threat; the practicability of achieving a significant reduction in P inputs; water use; and unique or endangered environmental resources.

Targeting within a catchment
Once an area has been selected for remediation, the next step is selection of appropriate BMPs. Using cost-effectiveness ratios like those outlined in Table 1.7, BMP implementation can be prioritized. For the example of the two catchments in Table 1.7, the most effective BMP installation priority would be barnyard runoff control, milking shed waste treatment and animal-waste storage facilities. Without careful targeting of critical P sources within a catchment, BMPs may not produce expected reductions in P export.

Phosphorus export from the Little Washita River catchment (54,000 ha) and two subcatchments (2 and 5 ha) was measured from 1980 to 1994, while BMPs were installed on about 50% of the main catchment. Practices included construction of flood-control impoundments, eroding-gully treatment and conservation tillage. Following conversion of conventional-till (mould-board and chisel plough) to no-till wheat in 1983, the concentration of TP in runoff from this subcatchment decreased from 3.1 to 0.4 mg l^{-1} and P loss was reduced by 3.3 kg ha^{-1} $year^{-1}$ (Fig. 1.8; Sharpley and Smith, 1994). A year later, eroding gullies were shaped and an impoundment constructed in the other subcatchment. Both P concentration (Fig. 1.8) and loss decreased dramatically (17.1 kg ha^{-1} $year^{-1}$ in 1984 and 3.1 kg ha^{-1} $year^{-1}$ in 1994; Sharpley *et al.*, 1996). There was no effect of BMP implementation, however, on P concentration in flow from the main Little Washita River catchment (Fig. 1.8). A lack of effective targeting of BMPs and control of major sources of P export in the Little Washita River catchment contributed to no consistent reduction in catchment export of P.

Land application of dairy manure in the LaPlatte River basin, Vermont (8832 ha), has been identified as an important source of P to Lake Champlain (2.2 kg P ha^{-1} $year^{-1}$; Meals, 1990). As a result, animal-waste control measures were implemented in the basin during the early 1980s. These BMPs included control of barnyard runoff, milking-shed waste treatment and construction of waste storage facilities (detailed in Table 1.7). There was no apparent reduction in either DP or TP concentration in runoff with increasing percentage of animals in a catchment under a BMP (dashed lines, Fig. 1.9; Meals, 1990). If the runoff P values for catchments where less than 50% of the animals were under BMPs are not considered,

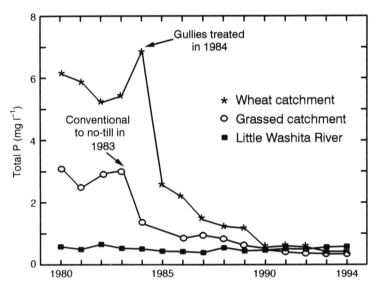

Fig. 1.8. Mean annual total P concentration in flow from the Little Washita River catchment and from two sub-catchments that were in grass with eroding gullies treated in 1984 and wheat converted from conventional to no-till in 1983 (adapted from Sharpley and Smith, 1994; and Sharpley et al., 1996).

both DP and TP in runoff were decreased significantly (r^2 of 0.68 and 0.75, respectively; $P < 0.05$; Fig. 1.9). The low values of implementation (< 42%) represent the initial years of land treatment when BMP implementation was incomplete. Apparently, there is a minimum threshold level of implementation which must be achieved before a significant response to BMPs occurs.

These examples clearly demonstrate that careful targeting of BMPs at an appropriate level of intensity is required to effectively reduce P export from catchments.

Selecting a best management practice
The cost-effectiveness of BMPs for reducing P loss varies both with the type of practice and among catchments (Tables 1.6 and 1.7). Remediation strategies are ongoing processes, in which BMP selection and operation should be continuously re-evaluated to optimize P export reductions.

Meals (1990) investigated the effectiveness of BMP implementation on P loss from Mud Hollow Brook catchment, Vermont, a contributor of P to the P-sensitive Lake Champlain. The catchment is 1682 ha, of which 52% is hay, 21% pasture, 6% maize, 7% idle and 14% rural non-agriculture. Dairy dominates animal activities in the catchment (80%), with 54% of the manure produced in-house being available for controlled field application. In the early 1980s, BMP implementation encompassed over 75% of the

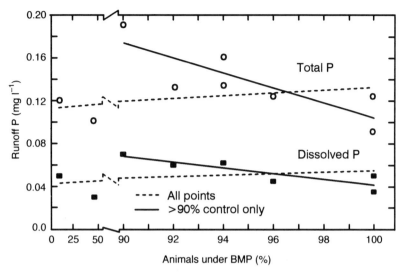

Fig. 1.9. Mean annual P concentration in catchment runoff as a function of the percentage of catchment animals under BMPs in the LaPlatte River basin, Vermont (adapted from Meals, 1990).

catchment and included animal-waste management; conservation contour and strip cropping; pasture management; buffer zones; and stream-bank protection and diversion (Meals, 1990). The effect of BMP implementation on P loss was calculated as the difference between with and without BMPs. Phosphorus losses without BMPs were predicted using pre-BMP concentrations of P and post-BMP flows.

This analysis suggested an increase in P export from Mud Hollow Brook catchment following BMP implementation (Fig. 1.10). Further analysis revealed that annual P export was dominated by one or two extremely high flows (> 95th percentile of all recorded stream flows), which were generally associated with snow-melting or intense rainfall events (Meals, 1990). These few extreme events had a dramatic effect on the overall assessment of BMP impact on P export. When these events were not included in the analysis, BMPs reduced DP and TP export (Fig. 1.10). Meals (1990) concluded that the increased P export following BMP appeared to be associated with very high stream flows and periods of active surface runoff. The capacity of BMPs to reduce P export was probably exceeded during these highly active runoff periods, representing < 5% of the time and high P export. Most of the time, however, BMPs functioned and P export was controlled or reduced. Effective remediation strategies should consider such extreme events in situations where they can dominate P export.

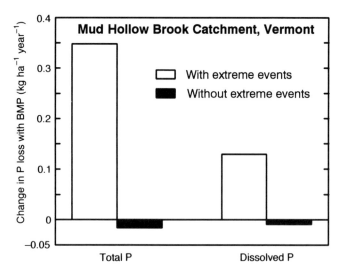

Fig. 1.10. Estimated difference in total and dissolved P loss with and without BMP implementation from Mud Hollow Brook catchment, Vermont, 1985 to 1989, with and without inclusion of extreme storm-flow events (data adapted from Meals, 1990).

Catchment and surface-water response

The response of catchments and lakes to BMPs can often be delayed by catchment re-equilibration, internal recycling of P, incremental land treatment, difficulty in controlling natural sources and short duration of monitoring. For example, a land-treatment programme involving animal-waste management, cropland protection, permanent vegetative cover, nutrient management and stream protection was initiated in 1980 and completed in 1991 on the St Albans Bay catchment (13,000 ha – 85% agriculture and forest) draining into St Albans Bay of Lake Champlain (Meals, 1992). Despite a significant reduction in mean annual P concentration in the main catchment tributary, after urban waste-water treatment was upgraded to include P removal in 1985, little subsequent reduction in P export was seen (Fig. 1.11).

Although BMPs reduced sediment export from St Albans Bay catchment, P concentration increased slightly in most monitored streams and the Bay (Fig. 1.11). The lack of a reduction in P export may have resulted from inadequate timing, type or coverage of BMP (Meals, 1992). Buffer-zone and stream-exclusion practices were not implemented, despite high farmer participation (60% – with 95% treatment of critical source areas).

Also, the lag time between BMP implementation and water-quality improvements may exceed the monitoring period (Clausen *et al.*, 1992). While St Albans Bay sediments appear to act as a reservoir for continued P supply, many catchment soils with high P content will also continue to be a source of runoff P, despite input or management changes. As a result, the trophic state of St Albans Bay appears to be changing from a mesotrophic to a eutrophic condition (Fig. 1.11). Carlson's (1977) trophic-state index considers TP, chlorophyll *a* and Secchi depth values. Internal recycling of P from deposited lake sediments also provided sufficient algal-available P after P inputs had been reduced for lakes in Australasia (Atkins *et al.*, 1993; McComb and Davis, 1993), Canada (Allan and Williams, 1978), Europe (Ryding and Forsberg, 1977; Forsberg, 1985; Bostrom *et al.*, 1988), and the USA (Larsen *et al.*, 1981; Jacoby *et al.*, 1982).

It is difficult for the public to understand or accept this lack of response. When tax funds are invested in remediation programmes, rapid improvements in water quality are usually expected and often required. Thus, effective BMP implementation should consider the re-equilibration of catchment and lake behaviour, where P sinks may become sources of P with only slight changes in catchment management and hydrological response. Education programmes should also be established to highlight the long-term benefits of remedial measures.

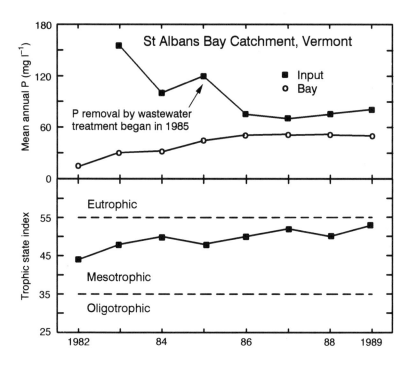

Fig. 1.11. Mean annual P concentration in St Albans Bay, Vermont, and in the main input tributary, with mean annual trophic-state index of the Bay (adapted from Meals, 1992).

FUTURE ACTIONS

It is clear that we have increased our understanding of the chemical and biological processes controlling the fate and dynamics of P cycling in soil–plant–water systems over the last 20 years. We have not been as effective, however, at translating this information to field situations and management of P from environmental rather than production perspectives. As a result, emphasis should be given in the future to development of: (i) realistic water-quality criteria, which involve threshold soil P levels, hydrologically active source areas and site vulnerability to P loss; (ii) education and economic programmes; (iii) nutrient management guidelines, particularly for animal manures; and (iv) how to avoid voluntary programmes becoming regulatory.

Realistic Water-Quality Criteria

Water-quality criteria for P have been established by government agencies in several countries (USEPA, 1976; Anon., 1995). These criteria are based on the early research of Sawyer (1947) and Vollenweider (1968), who proposed critical DP and TP concentrations of 0.01 and 0.02 mg l^{-1}, respectively, which, if exceeded, may accelerate the eutrophication of surface waters. In the UK, Moss *et al.* (1988) considered that most eutrophication problems in lake systems occur when DP concentrations exceed 0.03 mg l^{-1}, while a critical concentration of 0.1 mg l^{-1} has been proposed for river systems (English Nature, 1994). To determine the threshold level of soil P accumulation, Dutch regulators have set a critical limit of 0.1 mg l^{-1} as DP tolerated in groundwater at a given soil depth (mean highest water level) (Breeuwsma and Silva, 1992). In the USA, these concentrations have been modified by state agencies, ranging as high as 0.05–0.1 mg l^{-1}, for DP and TP (USEPA, 1976). For example, Florida recently identified 0.05 mg l^{-1} as the concentration of DP allowable in drainage water entering the Everglades (*USA v. South Florida Water Management District* (1994), US District Court/Southern District, case no. 88–1880–CIV). That state intends to reduce this concentration to 0.01 mg l^{-1} by the year 2000.

These water-quality criteria should not be used as the sole determinant to guide nutrient amendments where P loss in runoff and drainage water is of concern. In some cases, background concentrations of P in runoff from undisturbed areas may exceed the quality thresholds. For example, the mean annual DP concentration of runoff from several wheat catchments in Oklahoma and Texas receiving 20 kg P ha^{-1} $year^{-1}$ of fertilizer P (Smith *et al.*, 1991) and from grassed catchments receiving various types of manure (Heathman *et al.*, 1995; Jones *et al.*, 1995) all exceed a 'generous' critical DP value of 0.1 mg l^{-1} (Fig. 1.12). This was also the case for unfertilized native grass catchments (Fig. 1.12; Sharpley *et al.*, 1986). Thus, it is unlikely that any form of nutrient management would reduce DP in runoff below critical concentrations, particularly where large applications of manure P are regularly made.

A more flexible approach considers the complex relationships between P loadings and physical characteristics of affected catchments (leaching, runoff and erosion potential) and water bodies (mean depth and hydraulic residence time) on a site-specific and recognized water-use basis. Also, water use will influence 'desired or tolerable' nutrient loadings. For example, lakes used principally for water supply, swimming and multipurpose recreation will benefit from low P loadings. However, lakes mainly used for fish production benefit from a moderate degree of biological productivity and thus tolerate higher P inputs.

Clearly, realistic water-quality criteria that guide nutrient management within catchments should encompass more factors than just P concentrations in runoff. Unrealistic or unattainable criteria will not be adopted unless regulated. Thus, it is essential for long-term sustainable management of nutrients that workable water-quality criteria are proposed initially. The phasing in of environmental controls should then receive wider acceptance

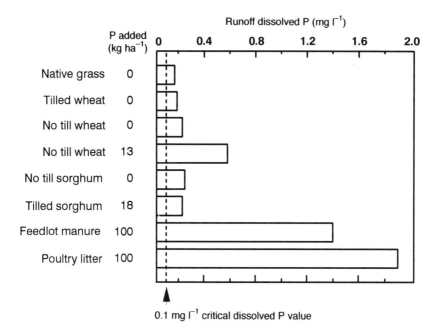

Fig. 1.12. The mean flow-weighted concentration of dissolved P in runoff over 1 year as a function of watershed and manure management in the Southern Plains relative to critical values associated with accelerated eutrophication (data adapted from Heathman *et al.*, 1995, and Jones *et al.*, 1995).

and compliance by farmers without creating severe economic hardships within rural communities.

In areas of P-related water-quality problems, threshold soil P levels are needed to guide fertilizer and manure P applications that account for the build-up of P in the soil and release of soil P to runoff (Table 1.7). However, this is still a technically indefensible and thus controversial process. More field data relating soil P to runoff P should help to alleviate these concerns. Also, soil-testing laboratories can contribute to more environmentally efficient P management by offering special tests for P that would be conducted on samples from areas with high potential for P losses in runoff. Examples of special tests include direct measurements of readily desorbable P, Fe-oxide strip P, biologically available P and P sorption saturation.

Fields for more intensive sampling and testing could be identified based on data available in routine soil tests and supplemental information on site vulnerability to P loss. Lemunyon and Gilbert (1993) recently developed an indexing procedure to rank site vulnerability to P loss based on source (soil P and fertilizer and manure P inputs) and transport factors (runoff and erosion potential). For example, adjacent fields having similar soil-test P level but differing susceptibilities to runoff and erosion, due to contrasting topography and management, should not have similar P recommendations.

Once high-risk areas are known, advisory agencies could conduct more intensive sampling of the upper 0–5 cm, focusing on the most erosion- or runoff-prone areas. Together, these data would not only identify fields where additional P should not be applied, but also specific sites where more intensive soil-conservation practices would be needed because of topographical and hydrological considerations. Despite time, labour and specialized equipment associated with these tests, their incorporation into soil-testing programmes is primarily a matter of realizing the important role these tests can play in improving the environmental efficiency of modern production agriculture.

Education and Economic Programmes

Development of sound extension and education programmes is particularly important for the increasing number of farms integrating confined animal operations. In the USA, farmers with limited land turned to confined animal production as a possible source of steady income to supplement additional off-farm income. In such situations, sound information regarding proper manure-handling techniques needs to get the farmer to minimize any water-quality impacts, because of the vast amounts of manure that must be land-applied. Dissemination of proper manure-handling information is of particular importance to farmers operating on small areas (< 40 ha). Often these farmers turn to confined animal operations to supplement inadequate cash returns on traditional grain and forage production due to inherent low-fertility soils, erratic rainfall and reduced crop prices. Therefore, the local need for P additions for crop production will be lower than in areas of intensive crop or forage production. Many farmers are still not fully aware of the nutritive value of applied manure. This illustrates the need for education and extension programmes as well as an infrastructure that can collect, process and redistribute manures to areas with high local demand for P.

One of the major obstacles facing more farmers is overcoming the economic restrictions of moving manure to a greater acreage, where it could supplement or even replace mineral-fertilizer requirements. The recent trend in the formation of cooperatives that can more cost-effectively compost and compact manure should be encouraged by cost-sharing programmes. Neighbouring farmers and private industry are also developing manure-processing alternatives. Examples of this include centralized storage and distribution networks, regional composting facilities and pelletizing operations, which can produce a value-added processed manure for distribution to other areas. Pelletization is a particularly attractive option because it results in a dried, lightweight material that can be handled, transported and applied in much the same manner as commercial fertilizer.

By composting and compacting, the bulk density of the manure is reduced, as is the cost of transportation. If the consumer is not willing to bear a part of the financial support, it may be necessary to recommend producers and farmers to take part in cooperative manure-treatment programmes. The level of involvement could be linked to the number of animals per farm.

Storage of manure will allow more flexibility in timing applications. A wide range of storage methods and costs are available to farmers (Brodie and Carr, 1988). Inexpensive plastic sheeting can perform well with very low cost for some solid manures. However, all storage methods must be managed carefully to fully realize their potential in an agronomically and environmentally sound BMP.

Nutrient-Management Programmes

Most field evaluations of BMP effectiveness at reducing catchment export of P conclude that nutrient management is the single most effective measure for controlling P losses (USDA-ASCS, 1992; Beegle and Lanyon, 1994; Meals, 1993). This involves the use of regional soil-testing programmes that are flexible enough to accommodate differences among catchments and development of manure-management plans for confined-animal operations.

In many areas, soil P testing is not required with fertilizer or manure applications and recommendations are still largely based on economic, not environmental, options. This has contributed to the localized build-up of soil P (Breeuwsma and Silva, 1992; Behrendt and Boekhold, 1993; Rekolainen *et al.*, 1993; Sharpley *et al.*, 1994b). Even with soil P testing, additional 'insurance' applications of P can be made. For example, 56% of soils surveyed in the Albany Harbour catchment, Australia, had high P status in 1988 and 1989 (Weaver and Prout, 1993). In fact, more than 70% of all fields received P irrespective of soil P status. Further, 90% of fields that would not have responded (high soil P) received P (Weaver and Mlodawski, 1992). If soil-test P recommendations had been followed, P inputs would have been reduced 50–60% in 1990. Clearly, more effective demonstration of the benefits of nutrient-management programmes is needed.

Nutrient-management programmes should be established on a regional rather than local basis to cover areas with similar soil types and growing regions. Several classification systems have defined such ecoregions (Bailey, 1983; Omernik, 1987). Within these ecoregions, attainable water-quality goals vary according to predominant land type and present use (Larsen *et al.*, 1988). As a result, an ecoregional approach to nutrient management may be useful for characterizing attainable water-quality goals. Also, nutrient-management interpretations and guidelines within these regions should be consistent (Gartley and Sims, 1994). Often, inconsistencies in recommendations and interpretation over short distances can lead to farmers questioning the reliability and philosophy of such programmes, as well as a reluctance to use them.

In the Peel–Harvey catchment, Australia, technology transfer failed to consider local community needs, resulting in poor adoption and farmer alienation from remote metropolitan bureaucracy (George and Bradby, 1993). In response, the Department of Agriculture developed cooperative, farmer-based land-care programmes, usually operating at a catchment level. These have been to a large extent successful at integrating nutrient-management programmes into current practices (George and Bradby, 1993; Weaver and Prout, 1993).

In order to develop nutrient-management programmes on a holistic catchment basis, a system of buying and selling pollution credits within a given catchment, similar to that recently adopted for air-quality control, has been suggested. Farmers able to limit P loss below recommended levels could sell credits to a farmer unable to meet these levels. The number of credits a farmer has could be linked to farm area, crop production and, where appropriate, number of animals. As a result, P export from a catchment may be kept within predetermined limits by sharing nutrient-management responsibilities among farmers. It should be noted, however, that 'pollution trading' has been criticized by some environmental groups because it is perceived as allowing wealthier operations to buy the 'right to pollute'. Heated debate is likely to precede the adoption of pollution credits for agriculture, hence careful planning to justify their value and need will be required.

Even so, it is clear that our current technology will not permit an unlimited number of animals in a region, without having an impact on water quality. Thus, it may be necessary to redistribute animals or to limit animal numbers within an area. Several regions now require that new animal facilities which exceed a certain size have an appropriate nutrient-management plan. Thus, it is essential that we develop and transfer technology to implement environmentally sound recommendations for the management of nutrients, particularly in manure.

Voluntary/Regulatory Programmes

It is a well-known fact that controlling non-point source losses is much more complicated than with point sources. Within a large region (state or country) there may be only a few polluters. However, agricultural guidelines have to be developed for several hundred thousand farmers in a region. In addition, processes governing P losses from agricultural soils are numerous, including many natural phenomena, and the relative importance of these processes varies greatly in space and time.

For these reasons, it is not easy to develop any thorough legislative or other regulatory measures to control pollution caused by agriculture. They are simply too formal and inflexible to bring the best set of BMPs to every farm or every parcel within a farm. On the other hand, voluntary measures are more flexible and can be targeted to single farms. They are also psychologically easier for farmers to accept. Thus, they might bring about a sustainable long-term change by affecting attitudes to be more concerned about the environment. However, changes brought about by voluntary programmes can be slow, in many cases too slow to stop accelerating eutrophication of certain water bodies. Very often these voluntary programmes are based on the work done by farm extension and advisory boards, financed partly by governmental or local administration.

Agriculture is more or less subsidized in many developed countries. This has partly resulted in higher and higher yields, using more inputs and thus increased nutrient losses. However, to overcome the problem of overproduction, there has been a recent tendency of converting price support to areal support. When a farmer does not get any greater support with higher yields, a certain shift towards more extensive agriculture is supposed to take place.

A shift from price support to areal support also allows another opportunity. When paid according to the area in cultivation, farmers implementing BMPs can be rewarded with higher premiums. Higher premiums may include not only the possible extra costs and/or losses of income but also a clear incentive. Certain control measures can also be preconditions for receiving a premium. There are several examples of this kind of development, one of the largest being the European Union's

(EU's) new environmental support system which was launched as part of the reform of its common agricultural policy (EEC, 1992). Within this framework, member countries of the EU have prepared monetary programmes targeted to specific environmental problems in each country (Mariën, Chapter 17, this volume). Measures targeted to controlling P losses have been included in the projects of Ireland, Denmark, Sweden and Finland.

Other existing monetary programmes include taxation systems which take environmentally beneficial investments into account, direct support for or lower interest rates on these investments and setting the support based on farm-level nutrient balances. Several countries (and states in the USA) have, however, introduced regulations to control pollution caused by agriculture. In certain cases, such as setting a minimum volume for stored manure to avoid spreading outside the growing season, legislative measures are possible and relatively suitable for every farm. However, many of the most effective measures, such as environmentally sustainable fertilization levels, conservation tillage and more effective manure management, are often too complicated to be suitable for regulations. For example, the EU's nitrate directive (EEC, 1991), which sets an upper limit for N applied with manure (170 kg N ha^{-1}), might be good in certain regions in central Europe. However, in higher latitudes, where yields are only half of those in central Europe (and thus the N uptake much lower), this limit is far too high to be environmentally sound.

CONCLUSIONS

There has been a general belief among many in agricultural research and extension that soils and lake sediments can indefinitely and tightly fix large amounts of P. As a result, many nutrient-management plans for P assume that, if erosion is controlled, P losses will also be controlled (Johnson, 1995). These measures address P control primarily through soil-conservation planning by reducing erosion. This belief has reinforced and maintained the use of N to drive manure management, in particular, and allowed soil P to accumulate. More frequently, increased losses of P in surface runoff and drainage water have been measured. About half of the sandy soils in the Netherlands (about 300,000 ha) are currently considered to be saturated with P, posing an unacceptable level of P leaching to groundwaters. As soil P decline via crop removal is slow, levels are going to be elevated for several years. Also, chemical amendments reduce P solubility, not total amounts in soil, and are thus only temporary measures.

Although the relationship between soil and runoff P has not been quantified over wide areas, it is clear that the potential for P loss in runoff, and thereby accelerated eutrophication, increases as soil P accumulates. To a certain extent, these concerns have not been addressed because managing agricultural inputs and outputs is often much more costly and restrictive to a farmer than general N management. Thus, political and economic issues have tended to fuel these misconceptions. To overcome this, we must convince the agricultural community that management practices can be developed that sustain productivity and profitability as well as water quality. As a reduction in off-farm inputs of P will probably be needed, agribusinesses are adapting to provide nutrient management services for farmers as well as selling fertilizers.

Future programmes should reinforce the fact that all fields do not contribute equally to P export from catchments. In general, about 90% of P export comes from only 5% of the catchment during only one or two storms. Although soil P content is important in determining the concentration of P in runoff, it is likely that runoff and erosion potential will override soil P in determining P export. Clearly, BMPs will be most effective if targeted to hydrologically active source areas of runoff in a catchment.

Consideration of all these factors will be needed in developing extension and demonstration projects that educate farmers and the general public as to what is involved in ensuring clean water. Hopefully, this will help overcome the common perception among end-users of water that it is often much cheaper to treat the symptoms of eutrophication rather than control the diffuse or non-point sources.

REFERENCES

Aarts, H.F.J., Biewinga, E.E. and Van Keulen, H. (1993) Dairy farming systems based on efficient nutrient management. *Netherlands Journal of Agricultural Science* 40, 285–299.

Allan, R.J. and Williams, D.D. (1978) Trophic status related to sediment chemistry of Canadian Prairie lakes. *Journal of Environmental Quality* 7, 99–106.

Andersson, G., Berggren, H., Cronberg, G. and Gelin, C. (1978) Effect of planktivorous and benthivorous fish on organisms and water chemistry in eutrophic lakes. *Hydrobiologia* 59, 9–15.

Anon. (1995) *Nutrient Management Strategy for Victorian Inland Waters*. State Government of Victoria, Melbourne, Australia, 73 pp.

Aquino, B.F. and Hanson, R.G. (1984) Soil phosphorus supplying capacity evaluated by plant removal and available phosphorus extraction. *Soil Science Society of America Journal* 48, 1091–1096.

Archer, J. (1985) *Crop Nutrition and Fertiliser Use*. Farming Press, Ipswich, England.

Atkins, R.P., Deeley, D.M. and McAlpine, K.W. (1993) Managing the aquatic environment. *Fertilizer Research* 36, 171–175.

Bailey, R.G. (1983) Delineation of ecosystem regions. *Environmental Management* 7, 365–373.

Barber, S.A. (1979) Soil phosphorus after 25 years of cropping with five rates of phosphorus application. *Communications in Soil Science and Plant Analysis* 10, 1459–1468.

Barisas, S.G., Baker, J.L., Johnson, H.P. and Laflen, J.M. (1978) Effect of tillage systems on runoff losses of nutrients, a rainfall simulator study. *Transactions of the American Society of Agricultural Engineers* 21, 893–897.

Beegle, D.B. and Lanyon, L.E. (1994) Understanding the nutrient management process. *Journal of Soil and Water Conservation* 49, 23–30.

Behrendt, H. and Boekhold, A. (1993) Phosphorus saturation in soils and groundwaters. *Land Degradation and Rehabilitation* 4, 233–243.

Bengston, L., Seuna, P., Lepisto, A. and Saxena, R.K. (1992) Particle movement of meltwater in a subdrained agricultural basin. *Journal of Hydrology* 135, 383–398.

Bostrom, B., Persson, G. and Broberg, B. (1988) Bioavailability of different phosphorus forms in freshwater systems. *Hydrobiologia* 170, 133–155.

Bottcher, A.B., Tremwell, T. and Campbell, K.L. (1995) Best management practices for water quality improvement in the Lake Okeectiobee Watershed. *Ecological Engineering* 5, 341–356.

Breeuwsma, A. and Silva, S. (1992) *Phosphorus Fertilization and Environmental Effects in the Netherlands and the Po Region (Italy)*. Report 57, Agricultural Research Department, Winand Staring Centre for Integrated Land, Soil and Water Research, Wageningen, the Netherlands.

Brodie, H.L. and Carr, L.E. (1988) Storage of poultry manure in solid form. In: Naber, E.C. (ed.) *Proceedings National Poultry Waste Management Symposium, Columbus, Ohio*. Ohio State University Press, Columbus, Ohio, pp. 115–119.

Brodison, J.A., Goodall, E.A., Armstrong, J.D., Givens, D.I., Gordon, F.J., McCaughey, W.J. and Todd, J.R. (1989) Influence of dietary phosphorus on the performance of lactating dairy cattle. *Journal of Agricultural Science Cambridge* 112, 303–306.

Brookes, P.C., Powlson, D.S. and Jenkinson, D.S. (1982) Measurement of microbial biomass phosphorus in soil. *Soil Biology and Biochemistry* 14, 319–329.

Brookes, P.C., Powlson, D.S. and Jenkinson, D.S. (1984) Phosphorus in the soil microbial biomass. *Soil Biology and Biochemistry* 16, 169–175.

Campbell, R.E. (1965) Phosphorus fertilizer residual effects on irrigated crops in rotation. *Soil Science Society of America Proceedings* 29, 67–70.

Carignan, R. and Kalff, J. (1980) Phosphorus sources for aquatic weeds: water or sediments? *Science* 207, 987–989.

Carlson, R.E. (1977) A trophic state index for lakes. *Limnology and Oceanography* 22, 361–369.

Chang, C., Sommerfeldt, T.G. and Entz, T. (1991) Soil chemistry after eleven annual applications of cattle feedlot manure. *Journal of Environmental Quality* 20, 475–480.

Clausen, J.C., Meals, D.W. and Cassell, E.A. (1992) Estimation of lag time for water quality response to BMPs. In: *Proceedings National RWCP Symposium*, Orlando, Florida.

Collins, Jr, E.R., Halstead, J.M., Roller, H.V., Weaver, Jr, W.D. and Givens, F.B. (1988) Application of poultry manure – logistics and economics. In: Naber, E.C. (ed.) *Proceedings National Poultry Waste Management Symposium, Columbus Ohio*. Ohio State University Press, Columbus, Ohio, pp. 125–132.

Costigan, P. (1988) The placement of starter fertilizers to improve early growth of drilled and transplanted vegetables. In: *Proceedings Fertilizer Society No. 274*. The Fertilizer Society, Peterborough, England.

Cox, F.R., Kamprath, E.J. and McCollum, R.E. (1981) A descriptive model of soil test nutrient levels following fertilization. *Soil Science Society of America Journal* 45, 529–532.

Culley, J.L.B., Bolton, E.F. and Bernyk, V. (1983) Suspended solids and phosphorus loads from a clay soil: I. Plot studies. *Journal of Environmental Quality* 12, 493–498.

Daniel, T.C., Sharpley, A.N., Edwards, D.R., Wedepohl, R. and Lemunyon, J.L. (1994) Minimizing surface water eutrophication from agriculture by phosphorus management. In: *Nutrient Management. Journal of Soil and Water Conservation* 49, 30–38.

Doerge, T. and Gardner, E.H. (1978) Soil testing for available P in southwest Oregon. In: *Proceedings 29th Annual Northwest Fertilizer Conference*, Beaverton, Oregon, pp. 143–152.

Duxbury, J.M. and Peverly, J.H. (1978) Nitrogen and phosphorus losses from organic soils. *Journal of Environmental Quality* 7, 566–570.

Edwards, D.R. and Daniel, T.C. (1992) Potential runoff quality effects of poultry manure slurry applied to fescue plots. *Transactions of the American Society of Agricultural Engineers* 35, 1827–1832.

Edwards, D.R. and Daniel, T.C. (1993a) Runoff quality impacts of pigs manure applied to fescue plots. *Transactions of the American Society of Agricultural Engineers* 36, 81–86.

Edwards, D.R. and Daniel, T.C. (1993b) Drying interval effects on runoff from fescue plots receiving pigs manure. *Transactions of the American Society of Agricultural Engineers* 36, 1673–1678.

Edwards, W.M. and Owens, L.B. (1991) Large storm effects on total soil erosion. *Journal of Soil and Water Conservation* 46, 75–77.

EEC (1991) Council Directive (91/676/EEC) of 12 December 1991 concerning the protection of waters against pollution caused by nitrates from agricultural sources. *Official Journal of the European Communities* No. L375:1–8.

EEC (1992) Council Regulation (EEC) No. 2078/92 of 30 June 1992 on agricultural reduction methods compatible with the requirements of the protection of the environment and the maintenance of the countryside. *Official Journal of the European Communities* No. L215:85–90.

Elbehri, A. and Ford, S.A. (1995) Economic analysis of major dairy forage systems in Pennsylvania: the role of intensive grazing. *Journal of Production Agriculture* 8, 501–507.

English Nature (1994) *Water Quality for Wildlife in Rivers*. English Nature, Peterborough, England.

Ford, S.A. (1994) *Economics of Pasture Systems*. Dairy Economics Fact Sheet No. 1, Pennsylvania State University College of Agricultural Science, Cooperative Extension Service, University Park, Pennsylvania, 9 pp.

Forsberg, C. (1985) Lake recovery in Sweden. In: *EWPCA International Congress: Lake Pollution and Recovery*, April, 1985, Rome, Italy, pp. 352–361.

Gartley, K.L. and Sims, J.T. (1994) Phosphorus soil testing: environmental uses and implications. *Communications in Soil Science and Plant Analysis* 25, 1565–1582.

Gburek, W.J. and Pionke, H.B. (1995) Management strategies for land-based disposal of animal wastes: hydrologic implications. In: Steele, K. (ed.) *Impact of Animal Manure and the Land–Water Interface*. Lewis Publishers, CRC Press, Boca Raton, Florida, pp. 313–323.

George, P.R. and Bradby, K. (1993) The Peel–Harvey catchment management programme. *Fertilizer Research* 36, 185–192.

Goldstein, A.L. and Ritter, G.J. (1993) A performance-based regulatory program for phosphorus control to prevent the accelerated eutrophication of Lake Okeechobee, Florida. *Water Science and Technology* 28, 13–26.

Gotoh, S. and Patrick, Jr, W.H. (1974) Transformations of iron in a waterlogged soil as influenced by redox potential and pH. *Soil Science Society of America Proceedings* 38, 66–71.

Gray, C.B.J. and Kirkland, R.A. (1986) Suspended sediment phosphorus composition in tributaries of the Okanagan Lakes, British Columbia. *Water Research* 20, 1193–1196.

Haapala, H.E.S. (1995) Position dependent control (PDC) of plant production. *Agricultural Science of Finland* 4, 239–250.

Halvorson, A.D. and Black, A.L. (1985) Fertilizer phosphorus recovery after seventeen years of dryland cropping. *Soil Science Society of America Journal* 49, 933–937.

Hartikainen, H. (1981) Effect of decreasing acidity on the extractability of inorganic soil phosphorus. *Journal of the Scientific Agricultural Society of Finland* 5, 16–26.

Haynes, R.J. and Williams, P.H. (1992) Long term effect of superphosphate on accumulation of soil phosphorus and exchangeable actions on a grazed, irrigated pasture site. *Plant and Soil* 42, 123–133.

Heathman, G.C., Sharpley, A.N., Smith, S.J. and Robinson, J.S. (1995) Poultry litter application and water quality in Oklahoma. *Fertilizer Research* 40, 165–173.

Hedley, M.J. and Stewart, J.W.B. (1982) Method to measure microbial phosphate in soils. *Soil Biology and Biochemistry* 14, 377–385.

Hergert, G.W., Ferguson, R.B., Cotway, C.A. and Peterson, T.A. (1994) Developing accurate nitrogen rate maps for variable rate application. *Agronomy Abstracts* 1994, 399. American Society of Agronomy, Madison, Wisconsin.

Hessen, D.O. and Nilssen, J.P. (1986) From phytoplankton to detritus and bacteria: effects of short time nutrient and fish perturbations in a eutrophic lake. *Archiv für Hydrobiologie* 105, 273–284.

Holford, I.C.R. (1982) Effects of phosphate sorptivity on long-term plant recovery and effectiveness of fertilizer phosphate in soils. *Plant and Soil* 64, 225–236.

Hooker, M.L., Gwin, R.E., Herron, G.M. and Gallagher, P. (1983) Effects of long-term annual applications of N and P on corn grain yields and soil chemical properties. *Agronomy Journal* 75, 94–99.

Horppila, J. and Kairesalo, T. (1990) A fading recovery: the role of roach (*Rutilus rutilus* L.) in maintaining high algal productivity and biomass in Lake Vesijarvi, southern Finland. *Hydrobiologia* 200/201, 153–165.

House, W.A. and Casey, H. (1988) Transport of phosphorus in rivers. In: Tiessen, H. (ed.) *Phosphorus Cycles in Terrestrial and Aquatic Ecosystems*, Vol. 1. Europe. SCOPE, UNEP, University of Saskatchewan, Saskatoon, Canada, pp. 253–282.

Isermann, K. (1990) Share of agriculture in nitrogen and phosphorus emissions into the surface waters of Western Europe against the background of their eutrophication. *Fertilizer Research* 26, 253–269.

Jacoby, J.M., Lynch, D.D., Welch, E.B. and Perkins, M.S. (1982) Internal phosphorus loading in a shallow eutrophic lake. *Water Research* 16, 911–919.

Johnson, J. (1995) Why Virginia's poultry industry is committed to nutrient management. *Bay Journal* 6, 14.

Johnston, A.E. (1988) Phosphorus cycling in intensive arable agriculture. In:

Tiessen, H. (ed.) *Phosphorus Cycles in Terrestrial and Aquatic Systems. Proceedings of a SCOPE and UNEP Workshop*, 1 May to 6 May, Czerniejewo, Poland. SCOPE, UNEP, University of Saskatchewan, Saskatoon, Canada, pp. 123–136.

Jones, O.R., Willis, W.M., Smith, S.J. and Stewart, B.A. (1995) Nutrient cycling of cattle feedlot manure and composted manure applied to Southern High Plains drylands. In: Steele, K. (ed.) *Impact of Animal Manure and the Land–Water Interface*. Lewis Publishers, CRC Press, Boca Raton, Florida, pp. 265–272.

King, L.D., Burns, J.C. and Westerman, P.W. (1990) Long-term pigs lagoon effluent applications on 'Coastal' Bermudagrass: II. Effects on nutrient accumulations in soil. *Journal of Environmental Quality* 19, 756–760.

Kingery, W.L., Wood, C.W., Delaney, D.P., Williams, J.C. and Mullins, G.L. (1994) Impact of long-term land application of broiler litter on environmentally related soil properties. *Journal of Environmental Quality* 23, 139–147.

Knuuttila, S., Pietilainen, O.-P. and Kauppi, L. (1994) Nutrient balances and phytoplankton dynamics in two agriculturally loaded shallow lakes. *Hydrobiologia* 275/276, 359–369.

Kotak, B.G., Kenefick, S.L., Fritz, D.L., Rousseaux, C.G., Prepas, E.E. and Hrudey, S.E. (1993) Occurrence and toxicological evaluation of cyanobacterial toxins in Alberta lakes and farm dugouts. *Water Research* 27, 495–506.

Kotak, B.G., Prepas, E.E. and Hrudey, S.E. (1994) Blue green algal toxins in drinking water supplies: research in Alberta. *Lake Line* 14, 37–40.

Larsen, D.P., Schults, D.W. and Malueg, K.W. (1981) Summer internal phosphorus supplies in Shagawa Lake, Minnesota. *Limnology and Oceanography* 26, 740–753.

Larsen, D.P., Dudley, D.R. and Hughes, R.M. (1988) A regional approach for assessing surface water quality: an Ohio case study. *Journal of Soil and Water Conservation* 43, 171–176.

Lawton, L.A. and Codd, G.A. (1991) Cyanobacterial (blue-green algae) toxins and their significance in UK and European waters. *Journal of the Institute of Water Environmental Management* 5, 460–465.

Lemunyon, J.L. and Gilbert. R.G. (1993) Concept and need for a phosphorus assessment tool. *Journal of Production Agriculture* 6, 483–486.

Lowrance, R.R., Todd, R.L., Fail, Jr, J., Hendrickson, Jr, O., Leonard, R. and Asmussen, L. (1984) Riparian forests as nutrient filters in agricultural catchments. *BioScience* 34, 374–377.

Lowrance, R.R., Leonard, R.A. and Sheridan, J.M. (1985) Managing riparian ecosystems to control non-point pollution. *Journal of Soil and Water Conservation* 340, 87–91.

McCallister, D.L., Shapiro, C.A., Raun, W.R., Anderson, F.N., Rhem, G.W., Engelstadt, O.P., Russelle, M.O. and Olson, R.A. (1987) Rate of phosphorus and potassium buildup/decline with fertilization for corn and wheat on Nebraska Mollisols. *Soil Science Society of America Journal* 51, 1646–1652.

McColl, R.H.S., White, E. and Gibson, A.R. (1977) Phosphorus and nitrate runoff in hill pasture and forest catchments, Taita, New Zealand. *New Zealand Journal of Marine and Freshwater Research* 11, 729–744.

McCollum, R.E. (1991) Buildup and decline in soil phosphorus: 30-year trends on a Typic Umprabuult. *Agronomy Journal* 83, 77–85.

McComb, A.J. and Davis, J.A. (1993) Eutrophic waters of southwestern Australia. *Fertilizer Research* 36, 104–114.

McDowell, L.L. and McGregor, K.C. (1984) Plant nutrient losses in runoff from conservation tillage corn. *Soil and Tillage Research* 4, 79–91.

McLaughlin, M.J., Alston, A.M. and Martin, J.K. (1988) Phosphorus cycling in wheat–pasture rotations. III. Organic phosphorus turnover and phosphorus cycling. *Australian Journal of Soil Research* 26, 343–353.

Martin, A. and Cooke, G.D. (1994) Health risks in eutrophic water supplies. *Lake Line* 14, 24–26.

Mathers, A.C., Stewart, B.A. and Blair, B. (1975) Nitrate removal from soil profiles by alfalfa. *Journal of Environmental Quality* 4, 403–405.

Meals, D.W. (1990) *LaPlatte River Watershed Water Quality Monitoring and Analysis Program: Comprehensive Final Report.* Program Report No. 12, Vermont Water Resources Research Center, University of Vermont, Burlington, Vermont.

Meals, D.W. (1992) Water quality trends in the St. Albans Bay, Vermont, watershed following RCWP land treatment. In: *The National Rural Clean Water Program Symposium.* US EPA-625-R-92-006, US EPA, Office of Water, Washington, DC, pp. 47–58.

Meals, D.W. (1993) Assessing nonpoint phosphorus control in the LaPlatte River watershed. *Lake and Reservoir Management* 7, 197–207.

Mehlich, A. (1984) Mehlich 3 soil test extractant: a modification of Mehlich 2 extractant. *Communications in Soil Science and Plant Analysis* 15, 1409–1416.

Meyer, J.L. (1979) The role of sediments and bryophites in phosphorus dynamics in a headwater stream ecosystem. *Limnology and Oceanography* 24, 365–375.

Miller, M.H. (1979) Contribution of nitrogen and phosphorus to subsurface drainage water from intensively cropped mineral and organic soils in Ontario. *Journal of Environmental Quality* 8, 42–48.

Moore, Jr, P.A., and Miller, D.M. (1994) Decreasing phosphorus solubility in poultry litter with aluminum, calcium and iron amendments. *Journal of Environmental Quality* 23, 325–330.

Morse, D., Head, H.H., Wilcox, C.J., van Hern, H.H., Hissem, C.D. and Harris, Jr, B. (1992) Effects of concentration of dietary phosphorus on amount and route of excretion. *Journal of Dairy Science* 75, 3039–3045.

Moss, B., Balls, H., Booker, I., Manson, K. and Timms, M. (1988) Problems in the construction of a nutrient budget for the R. Bore and its Broads (Norfolk) prior to its restoration from agriculture. In: Round, F.E. (ed.) *Algae and the Aquatic Environment.* Biopress, Bristol, England.

National Research Council (1993) *Soil and Water Quality: An Agenda for Agriculture.* National Academy Press, Washington, DC.

National Water Quality Evaluation Project (1988) *Status of Agricultural Nonpoint Source Projects.* North Carolina State University, Raleigh, North Carolina.

Nicholaichuk, W. and Read, D.W.L. (1978) Nutrient runoff from fertilized and unfertilized fields in western Canada. *Journal of Environmental Quality* 7, 542–544.

Nurnberg, G.K. and Peters, R.H. (1984) Biological availability of soluble reactive phosphorus in anoxic and oxic freshwaters. *Canadian Journal of Fisheries and Aquatic Science* 41, 757–765.

Olness, A.E., Smith, S.J., Rhoades, E.D. and Menzel, R.G. (1975) Nutrient and sediment discharge from agricultural watersheds in Oklahoma. *Journal of Environmental Quality* 4, 331–336.

Omernik, J.M. (1977) *Nonpoint Source–Stream Nutrient Level Relationships: A Nationwide Study.* EPA-600/3-77-105, Corvallis, Oregon.

Omernik, J.M. (1987) Ecoregions of the counterminous United States. *Annals Association of American Geologists* 77, 118–125.

Ozanne, P.G., Kirton, D.J. and Shaw, T.C. (1961) The loss of phosphorus from sandy soils. *Australian Journal of Agricultural Research* 12, 409–423.

Palmstrom, N.S., Carlson, R.E. and Cooke, G.D. (1988) Potential links between eutrophication and formation of carcinogens in drinking water. *Lake and Reservoir Management* 4, 1–15.

Peterson, T.A., Schepers, J.S., Chen, C., Cotway, C.A., Ferguson, R.B. and Hergert, G.W. (1994) Interpreting yield and soil parameter maps in the evaluation of variable rate nitrogen applications. *Agronomy Abstracts* 1994, 397. American Society of Agronomy, Madison, Wisconsin.

Pierzynski, G.M. and Logan, T.J. (1993) Crop, soil, and management effects on phosphorus soil test levels. *Journal of Production Agriculture* 6, 513–520.

Pietilainen, O.-P. and Rekolainen, S. (1991) Dissolved reactive and total phosphorus load from agricultural and forested basins to surface waters in Finland. *Aqua Fennica* 21, 127–136.

Pionke, H.B. and Kunishi, H.H. (1992) Phosphorus status and content on suspended sediments and streamflow. *Soil Science* 153, 452–462.

Reddy, K.R., Overcash, M.R., Kahled, R. and Westerman, P.W. (1980) Phosphorus absorption–desorption characteristics of two soils utilized for disposal of animal manures. *Journal of Environmental Quality* 9, 86–92.

Rekolainen, S. (1989) Phosphorus and nitrogen load from forest and agricultural areas in Finland. *Aqua Fennica* 19, 95–107.

Rekolainen, S., Posch, M. and Turtola, E. (1993) Mitigation of agricultural water pollution in Finland: an evaluation of management practices. *Water Science and Technology* 28, 529–538.

Richards, R.P. and Baker, D.B. (1993) Trends in nutrient and suspended sediment concentrations in Lake Erie Tributaries, 1975–1990. *Journal of Great Lakes Research* 19, 200–211.

Rippey, P. (1977) The behavior of phosphorus and silicon in undisturbed cores of Lough Neagh sediments. In: Golterman, H.L. (ed.) *Interactions between Sediment and Fresh Water. Proceedings of an International Symposium held at Amsterdam, the Netherlands, 6–10 September 1976.* Dr W. Junk Publishers, The Hague, the Netherlands, pp. 348–353.

Ritchie, G.S.P. and Weaver, D.M. (1993) Phosphorus retention and release from sandy soils of the Peel–Harvey catchment. *Fertilizer Research* 36, 115–122.

Romkens, M.J.M. and Nelson, D.W. (1974) Phosphorus relationships in runoff from fertilized soil. *Journal of Environmental Quality* 3, 10–13.

Ryden, J.C., Syers, J.K. and Harris, R.F. (1973) Phosphorus in runoff and streams. *Advances in Agronomy* 25, 1–45.

Ryding, S.O. and Forsberg, C. (1977) Sediments as a nutrient source in shallow polluted lakes. In: Golterman, H.L. (ed.) *Interactions between Sediment and Water. Proceedings of an International Symposium held at Amsterdam, the Netherlands, 6–10 September 1976.* Dr W. Junk, The Hague, the Netherlands, pp. 227–234.

Sawyer, C.N. (1947) Fertilization of lakes by agricultural and urban drainage. *Journal of New England Water Works Association* 61, 109–127.

Schreiber, J.D. (1988) Estimating soluble phosphorus (PO_4-P) in agricultural runoff. *Journal of Mississippi Academy of Sciences* 33, 1–15.

Sharpley, A.N. (1980) The effect of storm interval on the transport of soluble phosphorus in runoff. *Journal of Environmental Quality* 9, 575–578.

Sharpley, A.N. (1985a) Phosphorus cycling in unfertilized and fertilized agricultural soils. *Soil Science Society of America Journal* 49, 905–911.

Sharpley, A.N. (1985b) Depth of surface soil–runoff interaction as affected by rainfall, soil slope and management. *Soil Science Society of America Journal* 49, 1010–1015.

Sharpley, A.N. (1993a) An innovative approach to estimate bioavailable phosphorus in agricultural runoff using iron oxide-impregnated paper. *Journal of Environmental Quality* 22, 597–601.

Sharpley, A.N. (1993b) Assessing phosphorus bioavailability in agricultural soils and runoff. *Fertilizer Research* 36, 259–272.

Sharpley, A.N. (1995) Dependence of runoff phosphorus on soil phosphorus. *Journal of Environmental Quality* 24, 920–926.

Sharpley, A.N. (1996) Availability of residual phosphorus in manured soils. *Soil Science Society of America Journal* 60, 1459–1466.

Sharpley, A.N. and Smith, S.J. (1989) Prediction of soluble phosphorus transport in agricultural runoff. *Journal of Environmental Quality* 18, 313–316.

Sharpley, A.N. and Smith, S.J. (1994) Wheat tillage and water quality in the Southern Plains. *Soil and Tillage Research* 30, 33–38.

Sharpley, A.N. and Syers, J.K. (1979a) Phosphorus inputs into a stream draining an agricultural watershed: II. Amounts and relative significance of runoff types. *Water, Air and Soil Pollution* 11, 417–428.

Sharpley, A.N. and Syers, J.K. (1979b) Loss of nitrogen and phosphorus in tile drainage as influenced by urea application and grazing animals. *New Zealand Journal of Agricultural Research* 22, 127–131.

Sharpley, A.N., Smith, S.J., Stewart, B.A. and Mathers, A.C. (1984) Forms of phosphorus in soil receiving cattle feedlot waste. *Journal of Environmental Quality* 13, 211–215.

Sharpley, A.N., Smith, S.J. and Menzel, R.G. (1986) Phosphorus criteria and water quality management for agricultural watersheds. *Lake and Reservoir Management* 2, 177–182.

Sharpley, A.N., Carter, B.J., Wagner, B.J., Smith, S.J., Cole, E.L. and Sample, G.A. (1991) *Impact of Long-term pigs and Poultry Manure Applications on Soil and Water Resources in Eastern Oklahoma.* Technical Bulletin T169, Oklahoma State University, Oklahoma Agriculture Experiment Station, Stillwater, 51 pp.

Sharpley, A.N., Smith, S.J., Jones, O.R., Berg, W.A. and Coleman, G.A. (1992) The transport of bioavailable phosphorus in agricultural runoff. *Journal of Environmental Quality* 21, 30–35.

Sharpley, A.N., Smith, S.J. and Bain, W.R. (1993) Nitrogen and phosphorus fate from long-term poultry litter applications to Oklahoma soils. *Soil Science Society of America Journal* 57, 1131–1137.

Sharpley, A.N., Sims, J.T. and Pierzynski, G.M. (1994a) Innovative soil phosphorus indices: assessing inorganic phosphorus. In: Havlin, J., Jacobsen, J., Fixen, P. and Hergert, G. (eds) *New Directions in Soil Testing for Nitrogen, Phosphorus and Potassium.* American Society of Agronomy Monograph, Madison, Wisconsin, pp. 115–142.

Sharpley, A.N., Chapra, S.C., Wedepohl, R., Sims, J.T., Daniel, T.C. and Reddy, K.R. (1994b) Managing agricultural phosphorus for protection of surface waters: issues and options. *Journal of Environmental Quality* 23, 437–451.

Sharpley, A.N., Smith, S.J., Zollweg, J.A. and Coleman, G.A. (1996) Gully treatment and water quality in the Southern Plains. *Journal of Soil and Water Conservation* 51, 498–503.

Shreve, B.R., Moore, Jr, P.A., Daniel, T.C., Edwards, D.R. and Miller, D.M. (1995) Reduction of phosphorus in runoff from field-applied poultry litter using chemical amendments. *Journal of Environmental Quality* 24, 106–111.

Smith, S.J., Sharpley, A.N., Williams, J.R., Berg, W.A. and Coleman, G.A. (1991) Sediment-nutrient transport during severe storms. In: Fan, S.S. and Kuo, Y.H. (eds) *Fifth Interagency Sedimentation Conference*, March 1991, Las Vegas, Nevada. Federal Energy Regulatory Commission, Washington, DC, pp. 48–55.

Spratt, E.D., Warder, F.G., Bailey, L.D. and Read, D.W.L. (1980) Measurement of fertilizer phosphorus residues and its utilization. *Soil Science Society of America Journal* 44, 1200–1204.

Stewart, J.W.B. and Sharpley, A.N. (1987) Controls on dynamics of soil and fertilizer phosphorus and sulfur. In: Follett, R.F., Stewart, J.W.B. and Cole, C.V. (eds) *Soil Fertility and Organic Matter as Critical Components of Production*. SSSA Special Publication 19, American Society of Agronomy, Madison, Wisconsin, pp. 101–121.

Stewart, J.W.B. and Tiessen, H. (1987) Dynamics of soil organic phosphorus. *Biogeochemistry* 4, 41–60.

Syers, J.K. and Curtin, D. (1988) Inorganic reactions controlling phosphorus cycling. In: Tiessen, H. (ed.) *Phosphorus Cycles in Terrestrial and Aquatic Ecosystems*. UNDP, Saskatchewan Institute of Pedology, Saskatoon, Canada, pp. 17–29.

Tate, K.R. (1984) The biological transformation of P in soil. *Plant and Soil* 76, 245–256.

Tate, K.R., Spier, T.W., Ross, D.J., Parfitt, R.L., Whale, K.N. and Cowling, J.C. (1991) Temporal variations in some plant and soil P pools in two pasture soils of different P fertility status. *Plant and Soil* 132, 219–232.

Taylor, A.W., Edwards, W.M. and Simpson, E.C. (1971) Nutrients in streams draining woodland and farmland near Coschocton, Ohio. *Water Resources Research* 7, 81–90.

Theis, T.L. and McCabe, P.J. (1978) Phosphorus dynamics in hypereutrophic lake sediments. *Water Research* 12, 677–685.

Thomann, R.V. and Mueller, J.A. (1987) *Principles of Surface Water Quality Modeling and Control*. Harper Collins, New York, 644 pp.

Tiessen, H., Stewart, J.W.B. and Muir, J.O. (1983) Changes in organic and inorganic P composition of two grassland soils and their particle size fractions during 60–90 years of cultivation. *Journal of Soil Science* 34, 815–823.

Tyler, D.D. and Thomas, G.W. (1977) Lysimeter measurements of nitrate and chloride losses from conventional and no-tillage corn. *Journal of Environmental Quality* 6, 63–66.

USDA-Agricultural Stabilization and Conservation Service (ASCS) (1992) *Conestoga Headwaters Project, Pennsylvania – Rural Clean Water Program, 10-Year Report 1981–1991*. USDA-ASCS, Washington, DC.

US Environmental Protection Agency (USEPA) (1976) *Quality Criteria for Water*. US Government Printing Office, Washington, DC.

Uusi-Kamppa, J. and Ylaranta, T. (1992) Reduction of sediment, phosphorus, and nitrogen transport on vegetated buffer strips. *Agricultural Science of Finland* 1, 569–575.

Van der Zee, S.E.A.T.M., Fokkink, L.G.J. and van Riemsdijk, W.H. (1987) A new technique for assessment of reversibly adsorbed phosphate. *Soil Science Society of America Journal* 51, 599–604.

Van der Zee, S.E.A.T.M., van Riemsdijk, W.H. and de Haan, F.A.M. (1990) *Het protocol fosfaatverzadigde gronden.* Vakgroep Bademkunde en Plantevoeding, Wageningen Agricultural University, Wageningen.

Vollenweider, R.A. (1968) *Scientific Fundamentals of the Eutrophication of Lakes and Flowing Waters with Particular Reference to Nitrogen and Phosphorus.* OECD Report DAS/CSI/68.27, Paris, France, 182 pp.

Wadman, W.P., Sluijsmans, C.M.J. and De La Lande Cremer, L.C.N. (1987) Value of animal manures: changes in perception. In: Van der Meer, H.G. (ed.) *Animal Manure on Grassland and Fodder Crops.* Martinus Nijhoff, Dordrecht, the Netherlands.

Wagar, B.I., Stewart, J.W.B. and Henry, J.L. (1986) Comparison of single large broadcast and small annual seed-placed phosphorus treatments on yield and phosphorus and zinc content of wheat on chernozemic soils. *Canadian Journal of Soil Science* 66, 237–248.

Walton, C.P. and Lee, G.F. (1972) A biological evaluation of the molybdenum blue method for orthophosphate analysis. *Verhandlungen Interationale Vereinigung Limnologie* 18, 676–684.

Ward, R.C. (1984) On the response to precipitation of headwater streams in human areas. *Journal of Hydrology* 74, 171–189.

Weaver, D.W. and Mlodawski, R.G. (1992) Phosphorus surveying and mapping in the Albany Harbours catchment. In: *Proceedings of Land and Geographic Information Systems Featuring Applications in and around Albany, Western Australia,* November 1992. Australian Urban and Regional Information Systems Association, South Perth, Western Australia.

Weaver, D.M. and Prout, A.L. (1993) Changing farm practice to meet environmental objectives of nutrient loss to Oyster Harbour. *Fertilizer Research* 36, 177–184.

Westerman, P.W. and Overcash, M.R. (1980) Short-term attenuation of runoff pollution potential for land-applied pigs and poultry manure. In: *Livestock Waste – A Renewable Resource. Proceedings 4th International Symposium on Livestock Wastes.* American Society of Agricultural Engineers, St Joseph, Michigan, pp. 289–292.

Westerman, P.W., Donnely, T.L. and Overcash, M.R. (1983) Erosion of soil and poultry manure – a laboratory study. *Transactions of the American Society of Agricultural Engineers* 26, 1070–1078, 1084.

White, R.E. (1980) Retention and release of phosphate by soil and soil constituents. In: Tinker, P.B. (ed.) *Soils and Agriculture, Critical Reports on Applied Chemistry.* Vol. 2, Blackwell Scientific Publications, Oxford, England, pp. 71–114.

White, R.E. and Thomas, G.W. (1981) Hydrolysis of aluminum on weakly acidic organic exchangers: implications for phosphorus adsorption. *Fertilizer Research* 2, 159–167.

Wildung, R.E., Schmidt, R.L. and Gahler, A.R. (1974) The phosphorus status of eutrophic lake sediments as related to changes in limnological conditions – total, inorganic, and organic phosphorus. *Journal of Environmental Quality* 3, 133–138.

Withers, P.J.A., Unnein, R.J. Grylls, J.P. and Kane, R. (1994) Effects of withholding phosphate and potash fertilizer on grain yield of cereals and on plant-available phosphorus and potassium in calcareous soils. *European Journal of Agronomy* 3, 1–8.

Yli-Halla, M., Hartikainen, H., Ekholm, P., Turtola, E., Puustinen, M. and Kallio, K. (1996) Assessment of soluble phosphorus load in surface runoff by soil analyses. *Agriculture, Ecosystems and Environment* 56, 53–62.

Young, R.A. and Mutchler, C.K. (1976) Pollution potential of manure spread on frozen ground. *Journal of Environmental Quality* 5, 174–179.

Young, T.C. and DePinto, J.V. (1982) Algal-availability of particulate phosphorus from diffuse and point sources in the lower Great Lakes basin. *Hydrobiologia* 91, 111–119.

Zollweg, J.A. (1994) Effective use of geographic information systems for rainfall-runoff modeling. PhD Dissertation, Cornell University, Ithaca, New York State.

Zollweg, J.A., Gburek, W.J., Sharpley, A.N. and Pionke, H.B. (1995) GIS-based delineation of source areas of phosphorus within northeastern agricultural watersheds. In: *Proceedings IAHS Symposium on Modeling and Management of Sustainable Basin-Scale Water Resource Systems*, pp. 251–258.

2 Estimating the Contribution from Agriculture to the Phosphorus Load in Surface Water

S.D. Lennox[1], R.H. Foy[2], R.V. Smith[2] and C. Jordan[2]

[1]Biometrics Division, [2]Agricultural and Environmental Science Division, Department of Agriculture for Northern Ireland, Newforge Lane, Belfast BT9 5PX, UK

INTRODUCTION

Agriculture as a source of phosphorus (P) inputs to surface waters is receiving increased attention from water-resource managers. Although point sources of P from urban areas and industry have been reduced, the remaining P inputs from agriculture are often sufficient to maintain eutrophic conditions in the receiving waterways (Kronvang et al., 1993; Sharpley et al., 1994; Foy et al., 1995). There is also evidence suggesting that agricultural P inputs have increased with time, either as a consequence of increased use of P by agriculture, associated with agricultural intensification, or, where agricultural P inputs are stable, from an increase in soil P (Reddy et al., 1980; Tunney, 1990; Rekolainen, 1991; Uunk, 1991; Sharpley and Withers, 1994; Heckrath et al., 1995; Smith et al., 1995).

Phosphorus from agriculture is a highly variable input, from the point of view of spatial and temporal variability, reflecting differences in land use, soil type, management and short- and long-term climate variation. In reviewing measured P loss rates within Great Lakes watersheds of North America, Sonzogoni et al. (1980) highlighted the variability of P loss rates not only between differing land-use categories, but also within a particular land use. Loss rates tended to be greatest from arable cropping and least for low-intensity grassland but, for a given land use, loss rates decreased from coarser (sandy) to finer (clay) textured soils. Stream flow is required to transport P from a catchment, and hydrological processes strongly influence P loss dynamics, while the most important chemical processes having an impact on the transfer of P from land to water are the absorption and desorption of P from the soil (Reddy et al., 1980).

Ryden et al. (1973) proposed that P losses be considered for three types of runoff: surface, storm and base-flow runoff. Although rainfall has a low P content, the highest P concentrations are observed in surface runoff, reflecting both the high P content of surface soils and the occurrence of soil erosion. The latter can be especially important in determining P losses under arable conditions, particularly when there is little crop cover and low soil infiltration rates. In comparison with base flow, storm runoff represents precipitation that moves quite rapidly across or through the upper soil horizons, which have the highest soil P contents, before reaching a drainage channel. Although storm runoff has a limited contact time with the soil, it is often rich in P, so that drainage-water P increases during flood events (Stevens and Smith, 1978; Johnson, 1979). Base flow, originating from groundwater reserves, usually has low drainage-water P concentrations, often less than 20 µg P l^{-1} (Ryden et al., 1973).

It is therefore axiomatic that quantification of agricultural P in drainage water requires that P losses during flood events should be accurately assessed. However, in small catchments, flood events can be short-lived and unpredictable in their occurrence. To cope with this short-term variability in flow rates and P concentrations, the ideal solution is to continuously monitor drainage water for P loss. Remotely controlled samplers allow extended sampling at short time intervals and sampling in a flow-proportional mode. However, an extended sampling programme is expensive, in terms of both service and laboratory personnel costs.

Typically, P loading rates in streams and rivers are estimated by extrapolating the results of a time-averaged sampling programme. Because flood events are short lived, a regular sampling programme will result in a high proportion of base flows being sampled. The simplest but least reliable means of estimating P loading rates is to average P concentrations and multiply that result with the average flow for the period in question. The disadvantage of this method is that the time-based average P concentration is significantly influenced by base-flow concentrations. An alternative to treating P concentration and flow rate as unrelated variables is to use the results of a discrete sampling programme to derive a statistical relationship between P loss rate and flow rate, which can be employed to estimate the P loss rate on days when a water sample was not taken. The most common ratings are based on log loss rate vs. log flow rate regressions, which are flexible in terms of coping with non-linearity of the data (Smith and Stewart, 1977). However, P loss rates predicted from such regression equations have an inherent bias towards underestimating loss rates caused by the transformation from logs to antilogs (Ferguson, 1986, 1987). Although this bias has been known for some time in other scientific fields, its impact on estimating P losses in rivers was not highlighted until the 1980s. It can be overcome by including a simple correction factor in the loss rate vs. flow rate equation (Ferguson, 1986, 1987).

Annual export rates of P from agricultural land are commonly presented as P mass normalized per unit area, e.g. kg P km^{-2}. Treating agricultural P in this manner has the advantages of being directly comparable with agricultural P inputs and easily incorporated into a Vollenweider (1975) type areal lake P loading model, which predicts lake P concentrations. However, the model predicts that the precise impact on lake P of changing P loads will depend on whether the change in loading reflects a change in input P concentrations or a change in flow. There is a case, therefore, for providing information on flow rates and P concentration in addition to export rates, if only to provide a means of accounting for variation in P export among differing areas. When analysing catchment P export rates from an Australian river, Crosser (1989) noted that P export variability between flood events was significantly reduced when water runoff rate (mm) was included in the calculated P export rates (kg P km^{-2}) to give export rates as kg P mm^{-1} km^{-2}. A similar term, g P mm^{-1} ha^{-1}, was employed by Svendsen et al. (1995) to illustrate a similar phenomenon, in which the variability of P loss rates from Danish catchments was reduced when loss rates were normalized for flow and catchment area. An annual P loss rate normalized for flow and area (kg P mm^{-1} km^{-2}) is equivalent to the annual flow-weighted P concentration (mg P l^{-1}) calculated as catchment P mass export divided by catchment flow volume.

In the present chapter, some issues involved in the estimation of agricultural P exports to water are examined for two Irish catchments of contrasting size located in the Lough Neagh drainage basin, each of which is completely or largely devoted to grassland-based agriculture. The analysis has been principally confined to the 0.45 μm filterable molybdate-reactive fraction, which we term dissolved reactive P (DRP) and is sometimes referred to as soluble reactive phosphorus. This fraction is emphasized as it is readily available for algal growth in Lough Neagh (Jordan and Dinsmore, 1985). In contrast, particulate P (PP) in river water entering Lough Neagh is not considered to contribute significantly to algal growth, as it is comparatively resistant to direct assimilation by the phytoplankton and also sediments out of the water column when it reaches the Lough (Stevens and Stewart, 1982b; Gibson et al., 1988). The remaining P fraction in river water, dissolved organic P (DOP), consists mostly of monosubstituted phosphate esters of inositol, which are refractile compounds with limited potential to support algal growth (Stevens and Stewart, 1982a; Stewart, 1991).

STUDY AREAS

The Ballinderry River is one of six major rivers that flow into Lough Neagh, a large eutrophic lake in north east Ireland. It flows in an easterly direction from its headwaters, a region of moorland and mountain (maximum altitude 528 m), but most of the catchment lies below 150 m and is devoted to grassland agriculture (Table 2.1). Catchment characteristics, including land use and animal-stocking rates, obtained from the Northern Ireland Farm Census, are given in Table 2.1. The urban population is concentrated in one town, Cookstown (population 10,000 in 1991). In addition to P present in domestic sewage, other P point sources are considered to originate from a creamery, a meat plant and a rainbow trout fish farm. The rural population exceeds the urban population but it is difficult to ascertain its precise contribution to river P (Table 2.1). The septic tanks that serve the rural population are required to discharge to the soil via soakaways and, in theory, should not be discharged to waterways. As such, their contribution should be proportionally less than from the sewage treatment works (STWs) that serve the urban population. Excluding variation in point sources and the impact of interannual variability in flows, the DRP loss rate from the Ballinderry catchment was shown to be increasing at a rate of 1.45 kg P km^{-2} year^{-1} from 1974 to 1991, an increase which greatly exceeded the potential for increased P output from septic tanks due to rural population growth (Foy et al., 1995). An analysis of the seasonality of

Table 2.1. Catchment characteristics of Ballinderry River and Greenmount farm drain.

	Ballinderry River*	Greenmount farm drain
Area (km^2)	430	0.06
Land use		
Arable (%)	3.6	0
Grassland (%)	79.0	100
Rough grazing (%)	12.3	0
Other land (%)	1.2	0
Forestry (%)	3.8	0
Animal numbers		
Cattle (ha^{-1})	2.02	n/a†
Cattle (DCE ha^{-1})	1.36	n/a
Sheep (ha^{-1})	1.82	n/a
Pigs (ha^{-1})	1.36	n/a
Poultry (ha^{-1})	22.67	n/a
Human population‡		
Urban (km^{-2})	25.9	0
Rural (km^{-2})	30.5	0
Average rainfall (m yr^{-1})	1.096	0.950
Average flow (m yr^{-1})	0.700	0.511

* Ballinderry land use and annual numbers based on results of June 1992 census.
† See text for catchment outputs.
‡ Human population based on 1991 population census.
n/a, not available; DCE, dairy cow equivalent.

DRP losses found that the long-term increase in DRP losses was primarily due to increased winter losses and was greatest under high-flow conditions (Smith *et al.*, 1994).

The other catchment is a lowland (altitude 35 m) farm drain (Irish grid reference: J154836) lying within Greenmount Agricultural College farm estate to the north-east of Lough Neagh. It drains gleyed soils of a medium to heavy loam texture, derived from basaltic till, via a pipe drainage system, which employs a herring-bone layout, using 0.1 m diameter main clay pipes at a depth of 0.8 m fed by 0.075 m diameter clay pipes at approximately 0.7 m depth and spaced at intervals of 15 m. Drains are back-filled with washed stones and sand to about 0.45 m from the surface. The catchment drains 6 ha and is gently sloping. A previous study concluded that surface runoff was not an important drainage pathway (Jordan and Smith, 1985). The drainage area contains three fields which have been under grass since 1965 and are utilized for cattle grazing (2000–3000 cow grazing days per year), some winter grazing by sheep and for two to three silage cuts per year. The annual rate of nitrogen (N) application is approximately 300 kg N ha^{-1}. Phosphorus application consists of slurry and mineral fertilizer. In general, P inputs exceed exports by approximately 25 kg P ha^{-1} year^{-1} (Jordan and Smith, 1985; Smith *et al.*, 1995). Catchment P budgets show that the 1976–1977 period was unusual, with low catchment P inputs in 1977 preceded by high inputs in the previous year due to poultry slurry applications (Table 2.2). An analysis of DRP concentrations in the drainage water has demonstrated a 10 µg P l^{-1} increase in the median DRP concentration between 1981–1982 and 1990–1991 (Smith *et al.*, 1995). Continuously monitored P losses from the catchment were used to assess the errors associated with losses estimated from simulated discrete sample programmes by Stevens and Stewart (1981). They showed that losses based on log–log rating curves consistently underestimated the true losses by around 50%. The bias associated with loss estimates calculated from the product of mean monthly flow and P concentration and linear regression equations was erratic and not predictable. It was concluded that reliable loadings from this catchment required continuous monitoring of P concentration.

METHODS

Water samples were analysed for three P fractions. Total P (TP) and total dissolved P (TDP) were analysed according to the method of Eisenreich *et al.* (1975), while DRP was measured by the molybdate ascorbic acid method of Murphy and Riley (1962). DRP and TDP were measured on filtered samples (0.45 µm membrane filters). Two P fractions were calculated: PP as TP − TDP and DOP as TDP − DRP. Samples were filtered and prepared for analysis on the day of collection.

Table 2.2. Phosphorus balance for Greenmount farm drain (kg P ha^{-1} year^{-1}).

Components	Year					
	1976	1977	1981	1982	1990	1992
Inputs	103.7	7.8	41.0	43.5	51.6	50.9
Precipitation	0.2	0.2	0.2	0.2	0.1	0.3
Slurry	98.9	7.6	30.5	23.3	13.9	6.9
Fertilizer	4.6	0	10.3	20.0	37.6	43.7
Outputs	18.4	21.2	19.5	22.3	23.7	26.5
Drainage	0.0	0.4	1.0	1.8	0.9	0.4
Harvested herbage	14.9	17.9	15.1	15.9	19.6	22.2
Animal products	3.5	2.9	3.4	4.6	3.2	3.9
Inputs − outputs	85.3	−13.4	21.5	21.2	27.9	24.4

Discrete grab samples were collected from the Ballinderry River at a sampling site close to where the river enters Lough Neagh (Irish grid reference H928798). The sampling frequency was approximately weekly throughout the period 1989 to 1992. Ballinderry River flows were based on a rated automatic level recorder at the same site and represent the flow rate at the time of water sampling. Daily P losses were estimated from a P loss rate vs. flow rate relationship (equation (1)), which was derived from the regression of log-transformed data (equation (2)), as recommended for the Lough Neagh rivers by Smith and Stewart (1977). Daily P losses were also calculated from equation (3), which included the correction factor proposed by Ferguson (1987).

$$L_P = aQ^b \qquad \text{equation (1)}$$
$$\log L_P = \log a + b \log Q \qquad \text{equation (2)}$$
$$L_P = \exp(2.65 s^2)\, aQ^b \qquad \text{equation (3)}$$

where

L_P = P loss rate (kg P ha^{-1} day^{-1}),
Q = daily catchment flow normalized for catchment area (mm day^{-1}),
a and b = catchment-specific constants,
s^2 = estimated variance of residuals of regression equation (2).

The Greenmount farm-drain site was continuously monitored from October 1976 to March 1978, January 1980 to December 1981 and January 1990 to December 1991. Water samples were collected by a Manning 54040 flow-proportional sampler. The normal period for which a sample was composited was 24 h, but at weekends and on public holidays the samples were composited over 2- or 3-day periods. Flows were monitored continuously by an Isco 1870 flowmeter during the first two sampling periods and by a Warren Jones FJ 460 FM flowmeter during 1990 and 1991. Losses were calculated as the product of sample P concentration and flow for the composite period.

RESULTS

Ballinderry River

Between 1989 and 1992 the TP loss from the Ballinderry River was estimated to be 50 tonnes P year^{-1}, of which 57% was DRP, 16% DOP and 27% PP. Of the annual DRP loss, only 40% could be attributed to point sources, principally P from the major sewage-treatment works in the catchment at Cookstown (Table 2.3). The remaining 60% DRP was deemed to originate from agriculture, which as a land use accounted for over 95% of the catchment area (Table 2.1). Based on this assumption, the agricultural DRP export in flow was estimated to be 0.36 kg P ha^{-1} year^{-1}. Although point sources were not dominant, they had a major impact on river DRP concentrations. The plot of DRP concentrations vs. sampling date shows a distinct annual cycle, with concentrations in the summer approaching 400 µg P l^{-1} (Fig. 2.1a). However, these high concentrations occurred during low flows when dilution of point sources was least. The accompanying plot of daily DRP loss rate vs. time indicates that amounts lost were strongly flow-dependent and typically lowest in the summer (Fig. 2.1b).

The influence of point sources on river DRP concentrations is demonstrated by the plots of DRP concentration vs. flow (Fig 2.2a,b).

Table 2.3. Sources of DRP in Ballinderry catchment, 1989–1992.

	Tonnes P year^{-1}	%
Point sources		
Urban	7.29	
Creamery	1.41	
Meat plant	1.83	
Fish farm	0.86	
Total point inputs	11.39	40.1
Diffuse inputs (total − point input)	17.01	59.9
Total	28.40	100

Within a limited range of low flows, expressed as daily runoff values from 0.2 to 1.5 mm day^{-1}, DRP concentrations declined with increased flows. At higher flows, this decline was arrested and concentrations were mostly within the range 70–100 µg P l^{-1}. If DRP point-source inputs to the river are assumed to be constant over time, their contribution to river DRP concentration can be calculated from equation (4), in which the constant

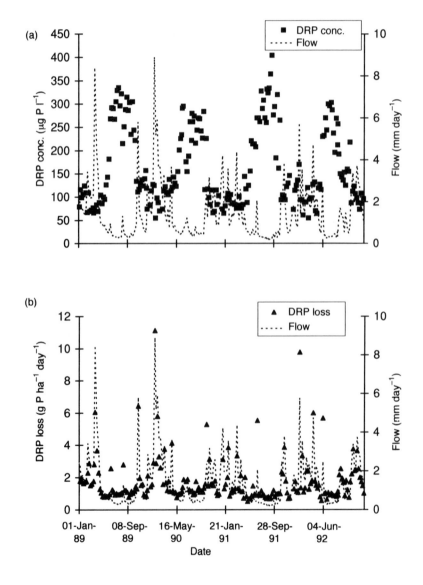

Fig. 2.1. Ballinderry flow rates and (a) DRP concentrations and (b) DRP losses on sample days.

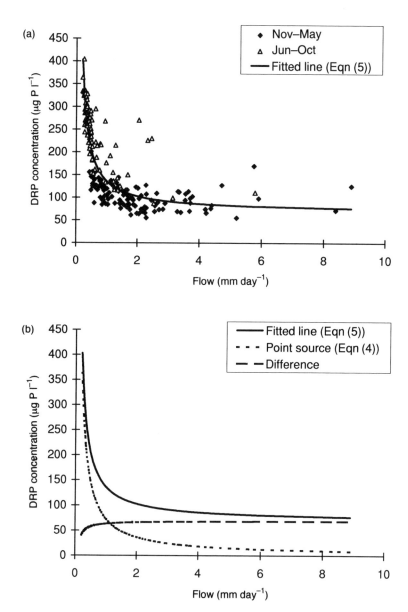

Fig. 2.2. Ballinderry River DRP concentrations vs. flow. (a) Sample data and regression line of equation (5). (b) Impact of runoff on river DRP (equation (5)), point sources (equation (4)) and DRP from diffuse sources.

is calculated from the point-source contribution of 31.2 kg P day^{-1}. The regression of river DRP vs. the reciprocal of flow accounted for 77% of the variation in DRP concentration and the resultant regression equation (5) is plotted on Fig. 2.2a. When equations (4) and (5) are plotted together, the high DRP concentration observed at low flows can largely be accounted for by the influence of point sources, represented by equation (4) (Fig. 2.2b).

$$DRP_{point} = 0.0726/Q \quad \text{equation (4)}$$
$$DRP = 0.0693 + (0.0668/Q) \quad n = 208;\ r^2 = 0.77 \quad \text{equation (5)}$$
Student's t statistic   15.0   26.4

where

DRP_{point} = DRP concentration attributable to point sources (mg P l^{-1}),
DRP = DRP river concentration (mg P l^{-1}),
Student's t statistics obtained from regression of DRP vs. $1/Q$.

The impact of the remaining inputs to the catchment, assumed to originate from agriculture, are calculated as the difference between equations (5) and (4). The resulting curve shows drainage DRP increasing with flow, but within a comparatively modest concentration range of 40 µg P l^{-1} at 0.2 mm flow day^{-1} and 68 µg P l^{-1} at a runoff rate of 10 mm flow day^{-1} (Fig. 2.2b). Over the period of observation, the time-averaged DRP sample concentration of 166 µg P l^{-1} in the Ballinderry was greater than the mean flow-weighted DRP concentration of 96 µg P l^{-1}. Based on equation (4), the time-averaged concentration attributable to point sources was 105 µg P l^{-1} or 63.2% of the mean river DRP concentration, compared with the 40.2% point-source contribution to the flow-weighted DRP concentration.

Despite the strong dependency of daily P losses on flow rate (Fig. 2.1b), there was evidence to suggest that, for a given flow rate, summer river losses were higher than losses at other times of the year. At flow rates between 2 and 3 mm day^{-1}, there were a number of positive outliers from the main DRP vs. flow-rate distribution, which occurred during the June–October period (Fig. 2.2a). When daily DRP losses are log-transformed and plotted against log flow rate, the tendency for two distinct relationships to occur is apparent (Fig. 2.3). The summer and winter log loss rate vs. log flow rate equations each give improved correlation coefficients in comparison with the regression employing the complete data set (Table 2.4). DRP loss rates based on the sum of the two relationships were 4.3% higher than the loss derived from a single equation (Table 2.4). Loss rates estimated from the flow vs. loss relationships in the form given by equation (3) were between 1.9% and 4.6% higher than loss rates estimated using equation (2) (Table 2.4).

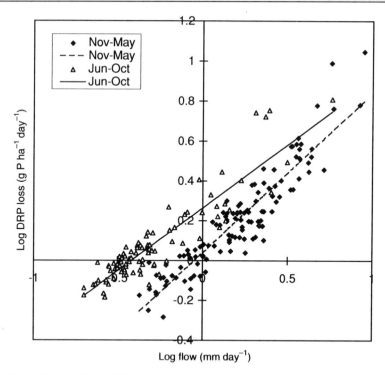

Fig. 2.3. Ballinderry River DRP loss vs. flow rate and regression lines for winter and summer periods.

Table 2.4. Ballinderry DRP loss estimates based on equations (1) and (3).

	r^2	DRP loss (equation (2)) (tonnes P year^{-1})	DRP loss (equation (3)) (tonnes P year^{-1})	Increase (%)
Winter equation	0.85	18.56	19.02	2.5
Summer equation	0.83	9.52	9.70	1.9
Total		28.08	28.72	2.3
Annual equation	0.70	26.33	27.54	4.6

Greenmount Farm Drain

Annual TP exports from the farm-drain catchment varied by a factor of 5.8 compared with a 1.8-fold variation in annual flow rates (Table 2.5). DRP was the largest P fraction and, with the exception of 1982, accounted for more than 40% of TP. There was a high positive correlation between annual flow

Table 2.5. Greenmount farm-drain P losses. Values in parentheses % of TP losses.

Losses (kg P ha^{-1} year^{-1})	Years					Mean
	1977	1981	1982	1990	1991	
TP	0.325	0.980	1.785	0.938	0.377	0.881
DRP	0.140	0.447	0.678	0.591	0.177	0.407
	(43.1)	(45.6)	(38.0)	(63.0)	(46.9)	(46.2)
DOP	0.086	0.149	0.358	0.143	0.090	0.165
	(26.5)	(15.2)	(20.0)	(15.2)	(23.9)	(18.7)
PP	0.099	0.384	0.749	0.204	0.110	0.309
	(30.5)	(39.2)	(42.0)	(21.8)	(29.2)	(35.1)
Flow (mm year^{-1})	358	550	637	526	433	501

and P loss rates but the high-P-loss years of 1981 and 1982 were also when the slurry P inputs were highest (Table 2.2). When slurry application coincided with periods of drain flow within 1 h of application, increased P concentrations were observed in drainage water and could exceed 1 mg P l^{-1}. The time period over which an individual application had an obvious effect on P concentrations was limited, usually no more than 4 weeks, and depended on flow rate. Thus, the impact of slurry inputs on annual P loss rates depended more on flow rates within this period than on the magnitude of annual flow.

The high P loss rate of 1982 was not associated with slurry application but extraordinarily high loss rates following heavy winter grazing by sheep (Jordan and Smith, 1985). This impact is evident from the peak in DRP concentrations during February 1982 (Fig. 2.4). On six occasions in 1981–1982, the measured DRP concentration exceeded 600 µg P l^{-1}, but, for clarity of presentation, these observations have been plotted as concentrations of 600 µg P l^{-1}. During a 4-week period after the removal of sheep in February 1982, daily average TP concentrations of up to 20 mg P l^{-1} occurred and the catchment DRP export rate was 0.46 kg P ha^{-1}, with 0.27 kg P ha^{-1} being lost in 1 week. Rainfall and flow increased over these 4 weeks compared with the period when the catchment was grazed, but were not exceptionally high when compared with peak flows recorded in the autumn of 1982 (Fig. 2.4). In total, over the 4 weeks in February 1982, flow represented 11% of 1982 flow but accounted for 68% of the annual DRP export. This event was responsible for the elevated P loss rates observed in 1982 and, when excluded from the annual total, the 1982 P loss rate for the remainder of the year was less than observed in 1981, despite similar flow rates.

Even allowing for the impact of slurry applications, the relationships

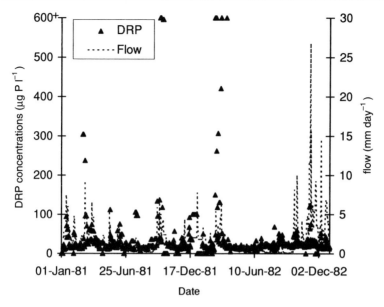

Fig. 2.4. Daily DRP concentrations and flow from Greenmount farm drain for 1981–1982.

between DRP concentration and flow were not precisely defined (Fig. 2.5). For ease of presentation, observations have been excluded when the daily DRP concentration was greater than 600 µg P l^{-1} which followed either slurry applications or the sheep-grazing event. There were six such observations in 1981–1982, two in 1990–1991 but none in 1976–1978. Each relationship shows a high degree of scatter, with a preponderance of low concentrations. In each distribution, the probability of encountering low concentrations decreases as flow rate increases. For the 1990–1991 period, there was also a tendency for more observations to exceed 100 µg P l^{-1} and during this period there was a marked tendency for summer (June–October) DRP concentrations to be higher than those in the remainder of the year (Fig. 2.5c).

DISCUSSION

The justification for using log P loss rate vs. log flow rate relationships to estimate watershed P loss depends on how well the relationships fit the observed data. The high correlations obtained for Ballinderry suggest that the method would accurately estimate loss. This conclusion is supported by similar investigations for the River Main, which also drains into Lough Neagh (Smith and Stewart, 1977; Stevens and Smith, 1978; Stevens and Stewart, 1981). The extended analysis by Stevens and Stewart (1981) of the errors associated with various computational methodologies for estimating

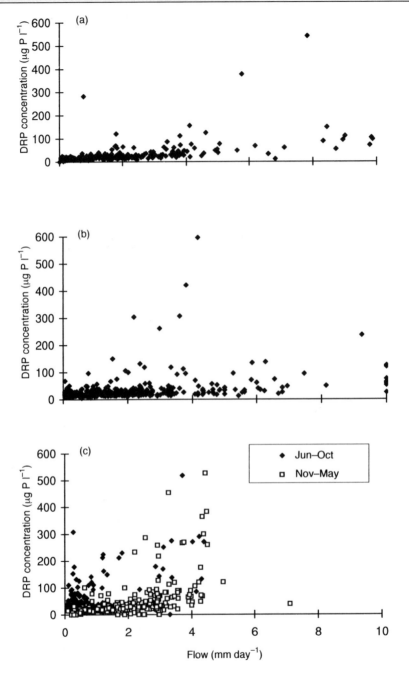

Fig. 2.5. Daily Greenmount mean DRP concentration vs. flow for (a) 1976–1978; (b) 1981–1982; and (c) 1990–1991. Concentrations in excess of 600 µg P l^{-1} excluded.

P loss showed that log vs. log regressions produced loss estimates for P from the River Main which had low variation in respect to sampling day but underestimated true DRP losses by approximately 10%. However, this bias would have been reduced if the correction factor of Ferguson (1986, 1987) had been employed. In the present study, the correction factor and the calculation of separate regression equations increased loss estimates by 5% in comparison with the estimate based on a single log vs. log equation. However, as the agricultural component of Ballinderry P loss was computed as the annual loss less the fixed P inputs from point sources, this small increase equates to a larger increase in diffuse P losses. Stevens and Stewart (1981) found that the most consistently accurate methodology for estimating annual P losses from the River Main was based on the product of annual flow rate times the corresponding flow-weighted P concentration.

Other authors have highlighted complex P loss rate vs. flow rate relationships which are not accurately represented by a single log–log equation. Kronvang (1992) found that DRP concentrations vs. flow rate plots for a Danish river varied depending on the degree to which flow rate was changing at the time of sampling. The highest concentrations occurred when flow rates were increasing rapidly. A similar phenomenon for the Ballinderry could not be detected in this study, although samples taken in summer close to flood peaks had unusually high DRP concentrations. Bailey-Watts and Kirika (1987) showed that, for a tributary of Loch Leven in Scotland, a point-source input reduced the correlation between log P loss rate and log flow rate. A simple log–log equation for the South African Vaal River could not accommodate the observed accelerated increase in P losses at high flow rates (Meyer and Harris, 1991). Although a modified log-transformed quadratic equation was used to accurately estimate P losses, this equation required a more complex correction factor than that employed in equation (3) to remove the bias introduced when converting logs to antilogs (Meyer and Harris, 1991).

In the Ballinderry, improved correlations and predictive capabilities were obtained by using separate summer and winter regression equations. In the Great Lakes watershed, improved accuracy of P loss estimates has also been obtained by employing seasonal log P loss–log flow regression equations (Dolan et al., 1981). Why DRP losses relative to flow should be higher in the summer in both the Ballinderry and the Greenmount farm drain catchments is not known but may reflect the low frequency of summer flood events. This would increase the opportunity for greater P mineralization between events, and this P could be lost. In the winter, due to lower temperatures and the greater frequency of flood events, mineralization is limited. However, some of the Ballinderry increase in summer DRP concentrations must reflect silage effluent entering the river during June and early July, when silage pollution peaks (Foy et al., 1994). This form of pollution occurred frequently in 1989 and 1990 and has been shown to increase flow-weighted P concentrations in farm streams elsewhere in

Northern Ireland (Foy and Withers, 1995). The seasonal variation in flood DRP concentrations has been noted for Canadian streams by Ng *et al.* (1993), who argued that DRP concentrations were higher when there was an absence of plant cover/growth.

When considering the TP losses from the two grassland-based agricultural catchments, the loss rates are high in comparison with the rates quoted for the Great Lakes catchments, which placed an upper limit on TP loss from pasture of 0.5 kg P ha^{-1} (Sonzongoni *et al.*, 1980). They are, however, comparable to a TP loss of 1.67 kg P ha^{-1} for New Zealand pasture, where the annual flow rate was comparable to rates measured in Ireland (Cooper and Thompsen, 1988). The New Zealand site, however, had approximately 50% higher rainfall and DRP made up a smaller component of TP (22%) than observed in the present study. TP export rates of approaching 1 kg P ha^{-1} have been reported from Danish catchments which had a high arable component, lower flow rates and a greater PP component in the P lost from the catchment (Kronvang, 1992; Svendsen *et al.*, 1995).

Comparing the two Irish catchments, there were similarities and contrasts between loss rates. DRP was the largest fraction of TP and present in sufficiently high concentrations to cause eutrophic conditions in receiving waters where intensively managed grassland was the dominant source of runoff (OECD, 1982; McGarrigle, 1993). Each catchment produced evidence to suggest that the summer DRP concentrations were higher than those observed in winter, although the high runoff rates of the winter ensured that winter loadings dominated the annual P loadings, as has been noted previously for the Ballinderry and farm drain (Jordan and Smith, 1985; Smith *et al.*, 1994).

The greater influence of point sources on the time-weighted DRP concentration compared with the flow-weighted concentration from the Ballinderry can be considered in the context of reducing river DRP to prevent excessive algal and plant growth in rivers (Lund and Moss, 1989). In the river situation, annual flow-weighted mean P or annual P loss rates may not estimate the quantity of P available for algal/plant growth, as most P is exported over the winter, when plant growth is minimal. Compared with lakes with a substantial water residence time, in the river situation greater emphasis would be placed on controlling daily spring and summer DRP concentrations and losses when agricultural contribution is lowest. In addition, water-quality standards for European Community (EC) and UK rivers set an upper limit on DRP concentration rather than on an annual average DRP concentration, so that the Ballinderry peak DRP concentrations could only be reduced by reducing point-source P discharges, although diffuse P sources dominate annual losses.

As 77% of the variation in DRP concentrations in the Ballinderry could be attributed to the interaction of flow on point sources of P (e.g. equations (4) and (5)), there is comparatively little variation remaining which can be attributed to non-point sources. After allowing for the effects of annual variation in river flow, Foy *et al.* (1995) and Smith *et al.* (1994) have shown that the increase in annual P loss from agricultural sources over 15 years was gradual rather than erratic. This contrasts with the variability in both daily DRP and annual P losses from the farm drain catchment. Part of the annual variability in P from the drain site reflects the impact of heavy grazing in the winter of 1982. Dramatically higher surface P export from grassland has previously been noted in Devon by Heathwaite (1993) as a response to heavy grazing pressure. Surface runoff is not thought to contribute to flow in the Greenmount drain, which flows underground, but the rapid response of drain-flow P to slurry applications indicates the presence of hydrological pathways which permit the rapid transfer of rainfall water to drain flow.

The absence of marked variability at high flow rates in the Ballinderry could indicate an increased stability in DRP concentrations, either within the river or at the catchment level. An effect of catchment size is that a specific management event, such as slurry spreading, would have an impact on a large percentage of the small drain catchment but probably on only a small fraction of the larger river catchment at any one time. Phosphorus redistribution mechanisms have been proposed by Svendsen *et al.* (1995) for a Danish catchment to reconcile the greater variability of small subcatchment P losses with the lesser variability from the entire catchment. The significance of these processes in a river such as the Ballinderry is unknown. The P absorption kinetics of sediments from the River Main, which also forms part of the Lough Neagh catchment, gave a DRP equilibrium concentration of 40 µg P l^{-1} (Smith and Jordan, 1985). This concentration was lower than the river concentration of DRP and suggests that, in that river, the sediments acted as a sink for soluble P throughout the year. In the Ballinderry, the DRP concentration range was similar to that of the Main and concentrations did not decline to less than 40 µg P l^{-1}. Overall, however, much of the bed of substrate of Ballinderry consists of either cobbles, gravel or sand, each of which offers little potential for P absorption.

CONCLUSIONS

Diffuse P collectively refers to P inputs to surface waters from rural areas but represents a variety of sources, including P present in land-drainage water, surface runoff and what are, in effect, minipoint sources arising from farmyard pollution, rural septic tanks and sewage systems. Quantifying and partitioning these inputs present practical difficulties arising from their variability with time and place. Temporal variability of diffuse P losses was examined using data from two Irish catchments: the Ballinderry River, draining 430 km^2 and dominated by grassland-based agriculture, and a farm-drain catchment of 0.06 km^2, also devoted to grass. Dissolved P fractions accounted for upwards of 50% of diffuse total P loss rate from each catchment, a higher percentage than is commonly recorded for dissolved P from diffuse P sources. Point-source inputs from towns and industry accounted for 40% of the annual DRP loss rate from the Ballinderry catchment and had a major impact on the DRP concentration vs. flow rate relationship, with DRP concentrations decreasing with flow. When the impact of point sources was removed from the DRP concentration vs. flow rate relationship, the DRP concentration attributable to diffuse sources was between 40 and 68 μg P l^{-1}, with concentration increasing with flow rate. Distinct summer (June–October) and winter log P loss rate vs. log flow regression equations, based on spot samples, were obtained for the purpose of estimating annual P losses from the Ballinderry catchment. The transformation of the loss-rate logarithm to antilogs produces an inherent bias towards underestimating loss rate, which was removed by the incorporation of a correction factor based on residual variability of the regression equation. In contrast, the farm drain catchment P loss rate vs. flow rate relationships were poorly defined and the loss rates were best derived from continuous monitoring for drainage P. Annual DRP and TP loss rates varied by factors of 3.8 and 4.8, respectively, between years (DRP, 0.18–0.68 kg P ha^{-1} $year^{-1}$; TP, 0.38–1.8 kg P ha^{-1} $year^{-1}$). Although annual loss rates were positively correlated with annual flow rate, which varied by a factor of only 1.5 between years (runoff range 0.42–0.63 m $year^{-1}$), much of the variation in annual loss rates was related to the interaction of short-term runoff events with slurry spreading or poaching of land by animals in the winter. This short-term variability, dependent on the interaction of catchment farm activities and runoff events specific to the catchment, creates problems in extrapolating the annual P loss rates from a small to a larger catchment.

REFERENCES

Bailey-Watts, A.E. and Kirika, A. (1987) A re-assessment of phosphorus inputs to Loch Leven (Kinross, Scotland): rationale and an overview of results on instantaneous loadings with special reference to runoff. *Transactions of the Royal Society, Edinburgh: Earth Sciences* 78, 351–367.

Cooper, A.B. and Thomsen, C.E. (1988) Nitrogen and phosphorus in stream waters from adjacent pasture, pine, and native forest catchments. *New Zealand Journal of Marine and Freshwater Research* 22, 279–291.

Crosser, P.R. (1989) Nutrient concentration–flow relationships and loads in the South Pine River, South-eastern Queensland. I. Phosphorus loads. *Australian Journal of Marine and Freshwater Research* 40, 613–630.

Dolan, D.M., Kui, A.K. and Geist, R.D. (1981) Evaluation of river load estimation methods for total phosphorus. *Journal of Great Lakes Research* 7, 207–214.

Eisenreich, S.J., Bannerann, R.T. and Armstrong, D.E. (1975) A simplified phosphorus analysis technique. *Environmental Letters* 9, 43–53.

Ferguson, R.I. (1986) River loads underestimated by rating curves. *Water Resources Research* 22, 74–76.

Ferguson, R.I. (1987) Accuracy and precision of methods for estimating river loads. *Earth Surface Processes and Landforms* 12, 95–104.

Foy, R.H. and Withers, P.J.A. (1995) The contribution of agricultural phosphorus to eutrophication. *Proceedings of the Fertilizer Society* 365, 32 pp.

Foy, R H., Smith, R.V., Jordan, C. and Lennox S.D. (1994) The impact of climatic and agricultural variable on the frequency of silage pollution incidents. *Journal of Environmental Management* 41, 105–121.

Foy, R.H., Smith, R.V., Jordan, C. and Lennox S.D. (1995) Upward trend in soluble phosphorus loadings to Lough Neagh despite phosphorus reduction at sewage treatment works. *Water Research* 29, 1051–1063.

Gibson, C.E., Smith, R.V. and Stewart, D.A. (1988) A long term study of the phosphorus cycle of Lough Neagh, Northern Ireland. *Internationale Revue der gesamten Hydrobiologie* 73, 249–257.

Heathwaite, A.L. (1993) The impact of agriculture on dissolved nitrogen and phosphorus cycling in temperate ecosystems. *Chemistry and Ecology* 8, 217–231.

Heckrath, G., Brookes, P.C., Poulton, P.R. and Goulding, K.W.T. (1995) Phosphorus leaching from soils containing different P concentrations in the Broadwalk experiment. *Journal of Environmental Quality* 94, 904–910.

Johnson, A.H. (1979) Estimating solute transport in streams from grab samples. *Water Resources Research* 15(5), 1224–1228.

Jordan, C. and Dinsmore, P. (1985) Determination of biologically available phosphorus using a radiobioassay technique. *Freshwater Biology* 15, 597–603.

Jordan, C. and Smith, R.V. (1985) Factors affecting leaching of nutrients from an intensively managed grassland in County Antrim, Northern Ireland. *Journal of Environmental Management* 20, 1–15.

Kronvang, B. (1992) The export of particulate matter, particulate phosphorus and dissolved phosphorus from two agricultural river basins: implications on estimating the non-point phosphorus load. *Water Research* 10, 1347–1358.

Kronvang, B., Ærtebjerg, G., Grant, R., Kristensen, P., Hovmand, M. and Kirkegaard, J. (1993) Nationwide monitoring of nutrients and their ecological effects: state of the Danish aquatic environment. *Ambio* 22, 176–187.

Lund, J.W.G. and Moss, B. (1989) *Eutrophication in the United Kingdom.* A report to the Soap and Detergent Industry Association, Hayes, UK, 81 pp.

McGarrigle, M.L. (1993) Aspects of river eutrophication in Ireland. *Annales de Limnologie* 29, 355–364.

Meyer, D.H. and Harris, J. (1991) Prediction of phosphorus load from non-point sources to South African rivers. *Water South Africa* 17, 211–216.

Murphy, J. and Riley, J.B. (1962) A modified single solution method for the determination of phosphates in natural waters. *Analytica Chimica Acta* 27, 31–36.

Ng, H.Y.F, Mayer, T. and Marsalek J. (1993) Phosphorus transport in runoff from a small agricultural watershed. *Water Science and Technology* 28, 451–460.

OECD (1982) *Eutrophication of Waters – Monitoring Assessment and Control.* Organization for Economic Cooperation and Development, Paris, 154 pp.

Reddy, K.R., Khaleel, R., Overcash, M.R. and Westerman, P.W. (1980) Phosphorus – a potential nonpoint source pollution problem in the land areas receiving long-term application of wastes. In: Loehr, R., Haith, D.A., Walter, M.F. and Martin, C.S. (eds) *Best Management Practices for Agriculture and Silviculture.* Ann Arbor Science, Ann Arbor, pp. 193–211.

Rekolainen, S. (1991) Phosphorus and nitrogen load from forest and agricultural land in Finland. *Aqua Fennica* 19, 95–107.

Ryden, J.C., Syers, J.K. and Harris, R.F. (1973) Phosphorus in runoff and streams. *Advances in Agronomy* 25, 1–45.

Sharpley, A.N. and Withers, P.J.A. (1994) The environmentally sound management of agricultural phosphorus. *Fertilizer Research* 39, 133–146.

Sharpley, A.N., Chapra, S.C., Wedepohl, R.., Sims, J.T., Daniel, T.C. and Reddy, K.R. (1994) Managing agricultural phosphorus for protection of surface waters: issues and options. *Journal of Environmental Quality* 23, 437–451.

Smith, R.V. and Jordan, C. (1985) The fate of nitrogen and phosphorus in a rural ecosystem. In: *Annual Report on Research and Technical Work.* Department of Agriculture for Northern Ireland, HMSO, Belfast, 124 pp.

Smith, R.V. and Stewart, D.A. (1977) Statistical models of river loading of nitrogen and phosphorus in the Lough Neagh system. *Water Research* 11, 631–636.

Smith, R.V., Foy, R.H. and Lennox, S.D. (1994) Mathematical modelling techniques to evaluate the impact of phosphorus reduction on phosphate loads to Lough Neagh. In: Barnett, V. and Turkman, K.F. (eds) *Statistics for the Environment: Water Related Issues.* Wiley, London, pp. 271–284.

Smith, R.V., Lennox, S.D., Jordan, C., Foy, R.H. and McHale, E. (1995) Increase in soluble phosphorus transported in drainflow from a grassland catchment in response to soil phosphorus accumulation. *Soil Use and Management* 11, 204–209.

Sonzongoni, G., Chesters D.R., Jeffs, D.N., Konrad, J.C., Ostry, R.C. and Robinson, J.B. (1980) Pollution from land runoff. *Environmental Science and Technology* 14, 148–153.

Stevens, R.J. and Smith, R.V. (1978) A comparison of discrete and intensive sampling for measuring the loads of nitrogen and phosphorus in the River Main, County Antrim. *Water Research* 12, 823–830.

Stevens, R.J. and Stewart, D.A. (1981) The effect of sampling interval and method of calculation on the accuracy of estimated phosphorus and nitrogen loads in drainage water from two different sized catchment areas. *Record of Agricultural Research (Department of Agriculture, Northern Ireland)* 29, 29–38.

Stevens, R.J. and Stewart, B.M. (1982a) Concentration, fractionation and characterisation of soluble organic phosphorus in river water entering Lough Neagh. *Water Research* 16, 1507–1519.

Stevens, R.J. and Stewart, B.M. (1982b) Some components of particulate phosphorus entering Lough Neagh. *Water Research* 16, 1591–1596.

Stewart, B.M. (1991) Concentration, fractionation and identification of soluble organic phosphorus in agricultural runoff within the Lough Neagh catchment. MSc thesis, The Queen's University of Belfast, Northern Ireland.

Svendsen, L.M., Kronvang, B., Kristensen, P. and Græsbøl, P. (1995) Dynamics of phosphorus compounds in a lowland river system: importance of retention and non-point sources. *Hydrological Processes* 9, 119–142.

Tunney, H. (1990) A note on a balance sheet approach to estimating the phosphorus fertiliser needs of agriculture. *Irish Journal of Agricultural Research* 29, 149–154.

Uunk, E.J.B. (1991) Eutrophication of surface waters and the contribution of agriculture. *Proceedings of the Fertiliser Society* 303, 55 pp.

Vollenweider, R.A. (1975) Input–output models with special reference to the phosphorus loading concept in limnology. *Schweizerische Zeitschrift für Hydrologie* 37, 53–84.

3 Phosphorus Losses from Agriculture to Surface Waters in the Nordic Countries

S. Rekolainen,[1] P. Ekholm,[1] B. Ulén[2] and A. Gustafson[2]

[1]Finnish Environment Institute, Impact Research Division, PO Box 140, FIN-00251 Helsinki, Finland; [2]Swedish University of Agricultural Sciences, Division of Water Quality Management, PO Box 7072, S-750 07 Uppsala, Sweden

INTRODUCTION

Owing to geological development and climatic conditions, the Nordic countries in northern Europe are rich in freshwater lakes and rivers. In addition, the Nordic countries comprise most of the drainage basin of the brackish Baltic Sea. Although the countries are sparsely populated, eutrophication is considered a major environmental problem in Denmark, Sweden, Norway and Finland. This concerns both freshwater environments and the Baltic Sea. For example, in Finland about 20% of the lakes (total area) and 55% of the rivers (total length) show various degrees of degradation. Of these, a large fraction is affected mainly by nutrient inputs from agriculture (Finnish National Committee for UNCED, 1991). In most temperate fresh waters, phosphorus (P) is the limiting nutrient for algal growth. However, in the pelagian waters of the Baltic Sea, nitrogen plays the major role in eutrophication (Granéli et al., 1990; Kivi et al., 1993).

Despite the region's northerly location, approximately between 54° and 71° latitude in the northern hemisphere, agricultural activities are to a large extent favoured by the Gulf Stream and prevailing westerly winds causing mild winters. Owing to the comparatively mild climate, the fraction of agricultural land is high, not only in Denmark but also in large regions in southern and central Sweden, southern and western Finland and in certain parts of Norway. In the Nordic countries, the erosion of cultivated soils and leaching of nutrients are not considered a major threat to agriculture. However, in areas of intensive farming practices, increased P loads have often resulted in a gradual eutrophication of the receiving waters.

The objective of this chapter is to give an overview of the P losses from agriculture to surface waters in the Nordic countries and to discuss the possible reasons for differences in the loss estimates. Special emphasis is given to bioavailability of the P losses.

LOSSES OF TOTAL PHOSPHORUS IN THE NORDIC COUNTRIES

Estimates of agricultural P losses in the Nordic countries are based mainly on the monitoring of small drainage basins and rivers dominated by agricultural land use. The loss is calculated as a product of water flow and P concentration. In agricultural catchments, both the flow and the concentration fluctuate rapidly. Usually, the flow is measured continuously, but the concentration measurements are based on sampling. The high temporal variation causes difficulties in designing a suitable sampling strategy to produce reliable loss estimates. Using manual sampling, which is still often the method, the actual losses are underestimated (Rekolainen et al., 1991; Græsbøll et al., 1994). The degree of underestimation is site-specific, but in reported cases it is up to 40%, with monthly or twice-monthly sampling. There is also a high spatial variation in the P losses, but the available resources seldom allow a dense monitoring network covering a whole country. Usually, the loss estimates are reported as a range between different monitoring sites. It remains uncertain whether the reported range covers the true range between all watersheds.

The differences between the estimates from the Nordic countries are, however, so high that they probably cannot be explained by methodological factors (Table 3.1). When flow is measured continuously, the possible underestimation of the concentrations can hardly result in the observed variation; the P loss in Finland was four to six times higher than the P loss in Denmark (Table 3.1). The values from Sweden varied markedly but the P loss measured in small catchments approximately corresponded to the loss in Denmark. The P loss in Norway was slightly smaller than that estimated in Finland (Table 3.1).

In Sweden the national goal for total P concentration in drainage water is set at 0.05 mg l^{-1}. The annual runoff in agricultural areas in Sweden varies between 150 and 300 mm approximately. Consequently, the target losses vary from 0.08 to 0.15 kg ha^{-1} and are higher than the present losses, particularly on clay soils.

The contribution of agriculture to the total loading of P is estimated to be 39% in Denmark, 54% in Norway, 73% in Sweden and 79% in Finland (Löfgren and Olsson, 1990; Kronvang and Svendsen, 1991; Græsbøll et al., 1994; Rekolainen et al., 1995). The methods of calculating the agricultural contribution differ among the different countries and the estimates presented here represent different periods. Although the figures are not

Table 3.1. Estimates of annual phosphorus losses from agriculture to surface waters in the Nordic countries.

Country	Total P (kg ha^{-1})	Dissolved P (kg ha^{-1})	Study years	Reference
Denmark	0.23–0.34		1989–1992	Kronvang et al. (1995)
Small catchments		0.083*	1993	Græsbøll et al. (1994)
Finland	0.57		1965–1974	Kauppi (1984)
Small catchments	0.9–1.8		1981–1985	Rekolainen (1989a)
	0.8–1.7		1986–1990	Rekolainen et al. (1995)
		0.15–0.41*	1987–1989	Pietiläinen and Rekolainen (1991)
Sweden				
Small catchments	0.01–0.6	0.01–0.30†	1988–1994	B. Ulén, unpublished data
Experimental fields	0.01–2.7	0.01–2.5†	1988–1994	M. Hoffman and S. Wall-Ellström, unpublished data
Norway	0.7–1.4		1981–1989	Ulén et al. (1991)

* Orthophosphate (PO$_4$-P) analysed from a sample filtered using 0.45 µm membranes.
† PO$_4$-P analysed from a centrifuged sample.

comparable, they show that agriculture comprises the greatest single source of P to surface waters in these countries (this is also true for Denmark; the contribution from all other individual sources is less than 39%).

In Finland, the estimated P loss for the period 1981–1985 (0.9–1.8 kg ha^{-1} year^{-1}, Rekolainen, 1989a) was appreciably higher than the P loss for the period 1965–1974 (0.57 kg ha^{-1} year^{-1}, Kauppi, 1984). The higher P loss in the more recent period is partly caused by higher flow and more frequent sampling during that period. However, the absolute amounts of agricultural P losses have probably also increased due to the intensification of agricultural production.

Climatic variations cause difficulties when assessing the non-point load on an annual basis. This can be seen in the high differences in annual loss estimates for agriculture (Fig. 3.1). However, the main reason for the high contribution of agriculture to total P losses is the rapid decrease of P losses from point sources, particularly due to investments in municipal wastewater treatment in all the Nordic countries. This development is illustrated by an example from Denmark (Fig. 3.1), where a decrease in municipal P load has occurred in recent years (Kronvang et al., 1995). In Sweden and Finland the reduction efficiency of municipal treatment plants has exceeded 90% for several years.

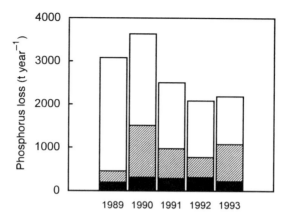

Fig. 3.1. Phosphorus load to the freshwater environment in Denmark from natural sources (lowest fraction of the bars), agriculture (middle fraction) and point sources (upper fraction). Point-source load includes sources from sewage plants, industry, rainwater constructions, freshwater fish farms and scattered dwellings. (Data adapted from Kronvang et al., 1995.)

In addition to diffuse sources, P is transported to surface waters from agricultural point sources, such as fodder silos and manure storages. However, there are only limited data on the amount of P losses from the point sources. In Sweden, the annual P load from agricultural point sources (milking sheds and manure pits) is estimated to be 266 t of total P. This corresponds to 4% of the total P load in 1982–1987 (Löfgren and Olsson, 1990). In Finland, the annual P losses from agricultural point sources have been estimated to be 300 t (Finnish Environment Agency, 1995).

LOSSES OF DISSOLVED PHOSPHORUS IN THE NORDIC COUNTRIES

The losses of dissolved P have not been measured as widely as the losses of total P. However, in all reported cases they are markedly lower than the losses of total P (Table 3.1). In Denmark, approximately 24–36% of the agricultural P loss is in the form of dissolved P. In Finland, the proportion of dissolved orthophosphate (PO_4-P) varied from 17 to 45% of total P in four small agricultural drainage basins (Pietiläinen and Rekolainen, 1991). In Sweden, 34% of the total P load, on average, is in dissolved form.

REASONS FOR THE DIFFERENCES IN THE NORDIC COUNTRIES

The Nordic countries are not situated in a uniform area; there is considerable variation in climatic conditions, soils, topography and the structure of agriculture. The differences in the reported loss estimates can be partly explained by the variability in natural conditions, but the relationships are often complex, non-linear and sometimes with apparent conflicting results.

Agriculture in the Nordic Countries

In Denmark, more than 60% of the land area is cultivated. Although the proportion of agricultural land is generally much lower in the other countries (< 10%), in certain relatively large regions it is almost as high as in Denmark.

Lack of space prevents a thorough overview of the structure and practice of agriculture in the Nordic countries in this chapter. As an example, we compare the P input to cultivated land during the 1980s (Fig. 3.2). The use of P has decreased recently in all these countries. The use is highest in Denmark and thus one would expect that the P losses would also be highest in Denmark, which, however, is not the case (Table 3.1). Different structures of agriculture can also be seen from Fig. 3.2, animal husbandry being much larger in Denmark than in Sweden and Finland.

Climatic Conditions

Temperature imposes limits for the cultivation of different crops and restricts the potential yield of each crop. The agricultural consequences of temperature differences can be illustrated by comparing growing degree-days in northern Europe (Fig. 3.3). As a result, yields per hectare for wheat, for example, are twice as high in Denmark as in Finland. Consequently, the P uptake in the yield is also much higher in Denmark and in southern Sweden than in more northern regions. This is one reason for lower P losses in Denmark and in southern Sweden than in Finland and Norway, for example.

In addition to temperature, precipitation and radiation are the most important climatic variables affecting runoff and P losses from agriculture. The differences in precipitation and evapotranspiration throughout the Nordic countries cause regional variations in annual runoff (Fig. 3.4). Although there is very little or no agriculture in regions with very high runoff volumes, the agricultural areas are situated in the regions with a wide range in runoff volumes (from 100 to 500 mm). The high variation in mean

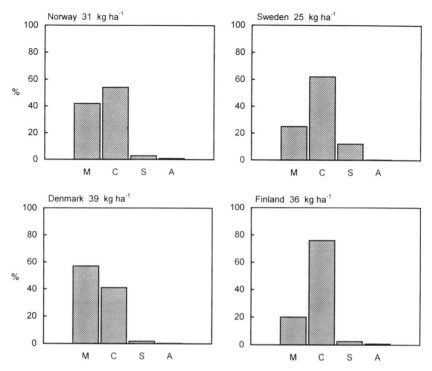

Fig. 3.2. Mean annual input of phosphorus to agricultural land in the Nordic countries in the 1980s. M, manure; C, chemical fertilizer; S, sewage sludge; A, atmospheric deposition.

annual precipitation among regions undoubtedly explains some observed differences in P loss estimates from different monitoring basins.

The seasonal and year-to-year variation in agricultural losses is mostly caused by temporal variations in precipitation and temperature. In Finland, the long-term averages show much higher monthly runoff in the spring snow-melt period and in autumn than in winter and summer. This is particularly true in agricultural areas and thus results in high peaks of losses during those seasons. In Denmark and in southern Sweden, maximum monthly runoff volumes decrease during the winter months.

The year-to-year variation in mean annual runoff in agricultural basins is considerable. For example, in Finland the annual runoff in wet years can be seven times higher than in dry years (Seuna, 1982). Consequently, the nutrient losses can vary this much even without any changes in agricultural practice. These high variations easily mask any trends in agricultural loss estimates.

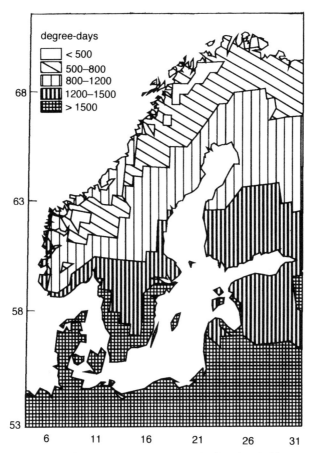

Fig. 3.3. Growing degree-days in the Nordic countries (5°C as threshold temperature). (From Carter *et al.*, 1991.)

Soil Types

Soil texture determines to a large extent the hydrological properties of a soil. Generally, the coarser the soil, the higher is its permeability and the lower its water-holding capacity. The soil composition in the Nordic countries is not homogeneous. In Finland, particularly in southern and western parts of the country, clay soils (proportion of clay > 30%, clay defined as particles with a diameter < 0.002 mm) are dominant, and in the whole country more than 35% of the agricultural soils fall into this class. Approximately 13% of the soils are heavy clays, in which the clay fraction exceeds 60%. Clay soils are also common in Sweden (about 39% of the total agricultural soils), whereas in Denmark there exist practically no soils with a clay fraction higher than 30%.

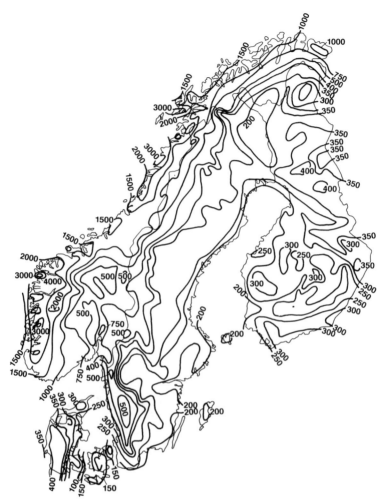

Fig. 3.4. Mean annual runoff in the Nordic countries in mm during 1931–1960. (Redrawn after Gottschalk, 1982.)

The differences in soil types between Finland and Denmark suggest that the surface runoff fraction (as opposed to the flow through the soil column) would be much higher in Finland than in Denmark. A high surface runoff volume results in high erosion rates, high particulate P losses and high dissolved P losses from the P-rich surface layers.

Special Nordic Features

Despite the climatic differences between the Nordic countries, winter snow cover is a common feature throughout the region. The fraction of the annual precipitation falling as snow varies from 5% in the south and in the coastal regions of the Atlantic Ocean up to 40% in the north (Fig. 3.5). The values presented here are based on calculations made by a snow-accumulation model (Rekolainen and Posch, 1993), based on temperature and precipitation in 1980–1990.

Snow cover and snow melting affect many biological and physical processes that govern P transport. Snow-melt water has much lower kinetic energy than raindrops. Thus, its capacity to detach soil particles is assumed to be low. Consequently, lower erosion rates would be expected during snow-melt. However, observations from Finland show that the snow-melt fraction of annual erosion is higher than the snow fraction of the annual precipitation (Rekolainen and Posch, 1993). This is probably due to the high transport capacity of snow-melt water. Usually snow in field lands melts rapidly, producing a high runoff volume, whereas rainstorms are rarely as intense in the Nordic countries as they are, for example, in the USA (Posch and Rekolainen, 1993).

The length of the ground-frost period in the Nordic countries varies from a couple of weeks to 5 months. Independently of the length of the frozen period, the soil is usually still frozen and almost impermeable when snow melting takes place. This prevents infiltration and thus the runoff consists almost totally of surface runoff. This increases the erosion capacity of the snow-melt water.

Freezing and thawing of plant residues may cause leaching of dissolved P, which is transported to surface waters with the snow-melt. Another reason for the observed high dissolved P losses during the snow-melt (e.g.

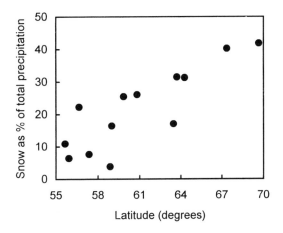

Fig. 3.5. Snow (calculated as mm of water equivalent) as percentage of the total precipitation shown as a function of the latitude of some weather stations in the Nordic countries.

Rekolainen, 1989b) might be the extraction of P-rich surface soil with dilute snow-melt water.

Phosphorus losses are usually associated with surface runoff and erosion. However, in certain conditions large amounts of P can be transported with preferential flow. This is particularly true for clay soils, in which cracks can be formed during dry periods and frosts. Remarkably rapid P losses have been reported in such conditions (e.g. Ulén, 1995). This means that some erosion control measures do not necessarily have a positive effect in all conditions.

BIOAVAILABILITY OF AGRICULTURAL PHOSPHORUS LOSS

Agricultural P losses consist of several different P compounds. In the receiving waters, algae and other primary producers readily use a part of the P, whereas some P forms may become available only slowly or they may be entirely inert. The bioavailability of P depends on the rate at which it is transformed into dissolved PO_4, the only P compound directly used by algae.

In some agricultural point sources, such as leakage from fodder silos and manure storage, the concentration of dissolved P may be high. However, most of the bioavailability studies have concentrated on particulate P, which is often the dominant fraction in the surface runoff from cultivated fields. The potential bioavailability of P can be experimentally determined most reliably by bioassays. In the most common bioassay technique, the test algae are incubated in a P-free nutrient medium, so that the particles are the sole source of P. During the test, which may last several weeks, the cell number of the algae is counted. The amount of bioavailable P is calculated from a standard curve determined between the cell number and the concentration of PO_4 in the particular experimental set-up.

The test algae and particles may also be separated from each other by a permeable membrane that allows the transport of dissolved substances between the two compartments. The desorption of P is assumed to be the main mechanism by which P is released from particles. Because the particles are the only P source, the increase in the algal P content equals the amount of bioavailable P in the samples.

Independently of the technique employed, the bioassays are laborious and efforts have been made to substitute them with chemical methods. These methods either mimic bioassays (e.g. extractions with iron (Fe)-hydroxide) or aim to determine directly those P forms (e.g. P adsorbed by aluminium (Al)- and Fe-hydroxides) that are bioavailable. However, according to a review by Hegemann and Keenan (1985), no single chemical fraction represents bioavailable P satisfactorily.

The Nordic studies, in which bioassays are used, give conflicting results on the potential bioavailability of particulate P (Table 3.2). In rivers draining agricultural land in Sweden, an average of 41% of the particulate P was available to the green alga *Chlamydomonas* (Persson, 1990). Even higher values were obtained when the results were presented as the maximum bioavailability of combined *Chlamydomonas* and *Aphanizomenon* assays. Berge and Källqvist (1990) found that 23–30%, on average, of the particulate P was bioavailable in water samples from cultivated areas in Norway. Krogstad and Løvstad (1991) reported higher values (~20–70%) for Norwegian soil samples assayed with planktonic blue-green algae. The lowest bioavailability was presented for particulate P in Finnish rivers from agricultural catchments (mean 5.1%; Ekholm, 1994). All the studies cited above dealt with the southern parts of the countries. In studies conducted outside the Nordic countries (USA and Hungary), the mean bioavailability values have varied from 11–20% (DePinto *et al.*, 1981; Zlinszky and Herodek, 1990) to about 30% (Dorich *et al.*, 1980; Williams *et al.*, 1980).

The variation among the reported values may arise from differences in the soil type and in agricultural management practices, such as the use of fertilizers. It seems, however, that the bioassay method also affects the results. Contrary to the other Nordic studies, the test algae and particles were separated from each other in the Finnish study. It has been stated that the algae, when in direct contact with particles, can liberate 'extra' P by their surface-bound enzymes (Williams *et al.*, 1980). This would partly explain the low bioavailability values obtained in the Finnish study. It must be noted, however, that when algae and particles are mixed, the algal P uptake may be overestimated, for example owing to various sample pretreatments (e.g. Källqvist and Berge, 1990). Bioavailability values exceeding 100% have been obtained from this type of bioassay (Persson, 1990).

Other experimental factors may also affect the results. In bioassays, pH often increases owing to photosynthesis. This enhances the desorption of P in non-calcareous conditions. In one of the Nordic studies, Persson (1990) reported increased pH values. The assay suspension used by Krogstad and Løvstad (1991) was intentionally made alkaline to mimic the conditions in the recipient eutrophic lake (Table 3.2). In contrast, Ekholm (1994) buffered the pH of the assay suspension at approximately pH 7. Thus, the differences in pH may explain some discrepancy in the results. Berge and Källqvist (1990) did not report pH.

The bioavailability studies in which chemical methods have been employed support the results obtained by bioassays. In Finnish river-water samples, the amount of P extracted by Fe-hydroxide-impregnated filter-paper strips was even lower than the amount used by the test algae (Ekholm and Yli-Halla, 1992). Correspondingly, in Norwegian water samples, the proportion of reactive P often varied between 25 and 75% of the total P (Krogstad and Løvstad, 1989). Reactive P, measured by a molybdate method on unfiltered samples, was assumed to equal bioavailable P.

Table 3.2. Potential bioavailability (BAP) of particulate phosphorus (PP) in Nordic agricultural catchments and in cultivated soils, as estimated by bioassays.

Country and sample type	BAP as % of PP		Characteristics of the bioassay method				Reference
	Mean	Range	Type	Sample pretreatment	pH in bioassay	Test organism	
Finland							
Rivers, cultivated areas	5.1	0–13.2	b	Particles concentrated by various methods or used without any pretreatment	Buffered at ~7	Green algae *Selenastrum* sp.	Ekholm (1994)
Norway							
Surface soil, cultivated fields	?	Often 20–70	a	Soil samples dried and sieved	Initially 9.2–9.9	Planktonic blue-green algae	Krogstad and Løvstad (1991)
Subsoil, cultivated fields	?	Often > 50	a	"	"	"	"
Surface runoff, cultivated area	30	22–46	a	Samples sterilized by γ-radiation	Not presented	Green algae *Selenastrum* sp.	Recalculated from Berge and Källqvist (1990)
Drainage water, cultivated area	28	0–58	a	"	"	"	"
Ditch-water, cultivated area	23	0–45	a	"	"	"	"
Sweden							
Runoff, agricultural area	41	?	a	Particles concentrated and fractionated, concentrates frozen	7.3–11.0	Green algae *Chlamydomonas*	Persson (1990)
River, agricultural and forested areas, or waste-water loading	41	?	a	"	"	"	"

a, Algae and particles mixed, algal P uptake estimated from cell yield; b, algae and particles separated from each other, P uptake determined directly.

The bioavailability of the dissolved P in water from agricultural catchment has seldom been assessed. Usually it is simply assumed that dissolved PO_4 (PO_4-P analysed from a filtered sample) is equal to immediately available P and that dissolved organic P (measured as the difference between total P and PO_4-P, both analysed from a filtered sample) is largely inert (Lee et al., 1980; Sonzogni et al., 1982). The results of Ekholm (1994) support this assumption, although in samples with a high amount of humus, dissolved PO_4 may overestimate the biologically available P (P. Ekholm, unpublished data).

Only one Nordic study reported the bioavailability of P in agricultural point sources. In leakage from fodder silos and manure storage, the proportion of dissolved P is greater than in runoff water from cultivated fields (Källqvist and Berge, 1990). Correspondingly, the bioavailability of total P is also higher. The P in the runoff water from a cattle-production area was also more available than P in the runoff water from a field with grain production (Källqvist and Berge, 1990).

As in the case of agriculture, the reported bioavailability values for other non-point and point sources vary widely. The results from the Nordic countries and the USA suggest that the bioavailability of particulate P is generally higher in point sources than in diffuse sources. In the diffuse sources, the mean bioavailability of particulate P varied from 0 to 6% in natural erosion material (Young et al., 1982; Ellis and Stanford, 1988) and from 6 to 55% in forest runoff (Ellis and Stanford, 1988; Persson, 1990) and was 30% in urban runoff (Cowen and Lee, 1976). In the point sources, the mean bioavailability of total P in purified waste water has been reported to vary from 36 to over 90% (Young et al., 1982; Shannon, 1983; Berge and Källqvist, 1990; Persson, 1990; P. Ekholm, unpublished data) and in the effluent of a pulp and paper mill unit it was about 80% (Priha, 1994).

Finally, it must be noted that the conditions in the bioassays are often very favourable for the release of P from particles. In receiving waters the actual release of the potentially bioavailable P depends on the physically determined transport and sedimentation of the particles and on the chemical and biological conditions (Sonzogni et al., 1982; Boström et al., 1988). The concentration of PO_4 and oxygen, in particular, as well as pH, affects the release of P (e.g. Ekholm, 1994). Furthermore, the capacity of eroded material to release P may markedly differ between brackish or marine waters and freshwater environments.

CONCLUSIONS

Agriculture comprises the greatest single source of P to surface waters in Denmark, Sweden, Norway and Finland. Thus, the focus of environmental policy has been shifted from point-source pollution to agriculture and other diffuse sources when considering eutrophication. Several programmes and information campaigns are pursued in these countries by both governmental agencies and non-governmental organizations (Nordic Council of Ministers, 1993). Within the European Union the recent adoption of a new agroenvironmental support scheme (EEC, 1992) has brought new measures to the European Union (EU) member states in the Nordic region. However, it can be foreseen that the contribution of the agricultural P load in eutrophication will also be high in the future, because the losses from point sources will most probably be further reduced, whereas the reduction programmes for agricultural losses will take effect more slowly. The implementation of agricultural water-protection measures will probably take several years. Furthermore, the impact of these measures, for example on the P status of cultivated soil, is slow (Sharpley and Rekolainen, Chapter 1, this volume).

Within the Nordic region, the reported P losses differ from country to country and from region to region. The differences are caused by variation in natural conditions, such as climate, soils and topography, but also by different agricultural structures and practices. This is important to recognize when integrating agricultural and environmental policies and harmonizing them in large regions such as the EU. All legislative measures and monetary incentives should be designed in such a way that each country and region could select an approach which is most suitable for the local conditions. In order do this, the impact of natural variability on agricultural losses should be known and recognized as a basis for an abatement policy.

There exists no general agreement on the bioavailability of P from agricultural sources. To design effective measures against eutrophication, it would be essential to know whether reported variability in the bioavailability is real or caused by methodological differences. If the bioavailability of particulate P is assumed to be low, the measures should be directed towards reduction of dissolved P. However, if the opposite is true, the prevention of erosion is also important. A Nordic intercalibration would reveal the actual differences in the bioavailability.

Even if a large proportion of particulate P is available to algae, the dissolved P is the most harmful P fraction in terms of eutrophication of surface waters. Several studies have shown that the leaching of dissolved P from cultivated areas is related to the soil P status (e.g. Sharpley *et al.*, 1981; Yli-Halla *et al.*, 1995). To reduce the loss of dissolved P, either the content of labile P in the soil or the amount of surface runoff should be decreased. In addition, buffer zones which can also retain dissolved P should be developed.

ACKNOWLEDGEMENTS

The authors are grateful to Brian Kronvang and Lars Svendsen for providing the Danish data. Michael Bailey revised the English.

REFERENCES

Berge, D. and Källqvist, T. (1990) *Biotilgjengelighet av fosfor i jordbruksavrenning. Sammenliknet med andre forurensningskilder, Sluttrapport.* NIVA, Oslo, 130 pp.
Boström, B., Persson G. and Broberg, B. (1988) Bioavailability of different phosphorus forms in freshwater systems. *Hydrobiologia* 170, 133–155.
Carter, T.R., Parry, M.L. and Porter, J.H. (1991) Climatic change and future agroclimatic potential in Europe. *International Journal of Climatology* 11, 251–269.
Cowen, W.F. and Lee, G.F. (1976) *Algal Nutrient Availability and Limitation in Lake Ontario during IFYGL.* Part 1. *Available Phosphorus in Urban Runoff and Lake Ontario Tributary Waters.* EPA-600/3-76-94a, US Environmental Protection Agency, Environmental Research Laboratory, Duluth, Minnesota.
DePinto, J.V., Young T.C. and Martin, S.C. (1981) Algal-available phosphorus in suspended sediments from Lower Great Lakes tributaries. *Journal of Great Lakes Research* 7, 311–325.
Dorich, R.A., Nelson, D.W. and Sommers, L.E. (1980) Algal availability of sediment phosphorus in drainage water of the Black Creek watershed. *Journal of Environmental Quality* 9, 557–563.
EEC (1992) Council Regulation (EEC) No. 2078/92 of 30 June 1992 on agricultural production methods compatible with the requirements of the protection of the environment and the maintenance of the countryside. *Official Journal of the European Communities* L215, 85–90.
Ekholm, P. (1994) Bioavailability of phosphorus in agriculturally loaded rivers in southern Finland. *Hydrobiologia* 287, 179–194.
Ekholm, P. and Yli-Halla, M. (1992) Reversibly adsorbed phosphorus in agriculturally loaded rivers in southern Finland. *Aqua Fennica* 22, 35–41.
Ellis, B.K. and Stanford, J.A. (1988) Phosphorus bioavailability of fluvial sediments determined by algal assays. *Hydrobiologia* 160, 9–18.
Finnish Environment Agency (1995) Vesiensuojelun tavoiteohjelma vuoteen 2005. A manuscript.
Finnish National Committee for UNCED (1991) *National Report to UNCED 1992.* Publication 13/91, Ministry for Foreign Affairs, Helsinki.
Gottschalk, L. (1982) *Hydrologi.* Kompendium. Institut für Teknisk Vattenresurslära. Lunds Teknisk Högskola Lunds Universitet, 351 pp.
Græsbøll, P., Erfurt, J., Hansen, H.O., Kronvang, B., Larsen, S.E., Rebsdorf, A. and Svendsen, L.M. (1994) Vandmiljøplanens Overvågningsprogram 1993. Ferske vandområder, vandløb og kilder. *Faglig Rapport fra Danmarks Miljøundersøgelser* 119, 186 pp.
Granéli, E., Wallström, K., Larsson, U., Granéli, W. and Elmgren, R. (1990) Nutrient limitation and primary production in the Baltic Sea area. *Ambio* 19, 142–151.
Hegemann, D.A. and Keenan, J.D. (1985) Measurement of watershed phosphorus:

a review. *Toxicological and Environmental Chemistry* 9, 265–289.

Källqvist, T. and Berge, D. (1990) Biological availability of phosphorus in agricultural runoff compared to other phosphorus sources. *Verhandlungen der Internationale Vereinigung für Theoretische und angewandte Limnologie* 24, 214–217.

Kauppi, L. (1984) The contribution of agricultural loading to eutrophication in Finnish lakes. *Water Science and Technology* 17, 1133–1140.

Kivi, K., Kaitala, S., Kuosa, H., Kuparinen, J., Leskinen, E., Lignell, R., Marcussen, B. and Tamminen, T. (1993) Nutrient limitation and grazing control of the Baltic plankton community during annual succession. *Limnology and Oceanography* 38, 893–905.

Krogstad, T. and Løvstad, Ø. (1989) Erosion, phosphorus and phytoplankton response in rivers of south-eastern Norway. *Hydrobiologia* 183, 33–41.

Krogstad, T. and Løvstad, Ø. (1991) Available soil phosphorus for planktonic blue-green algae in eutrophic lake water samples. *Archiv für Hydrobiologie* 122, 117–128.

Kronvang, B. and Svendsen, L.M. (1991) Phosphorus supply to the soil – freshwater environment in the Nordic countries. In: Svendsen, L.M. and Kronvang, B. (eds) *Phosphorus in the Nordic Countries – Methods, Bioavailability, Effects and Measures*. Nordisk Ministerråd, København Nord 47, 101–108. (In Danish, with English summary, legends and headings.)

Kronvang, B., Grant, R., Larsen, S.E., Svendsen, L.M. and Kristensen, P. (1995) Non-point-source nutrient losses to the aquatic environment in Denmark: impact of agriculture. *Marine Freshwater Research* 46, 167–177.

Lee, G.F., Jones, R.A. and Rast, W. (1980) Availability of phosphorus to phytoplankton and its implications for phosphorus management strategies. In: Loehr, R.C., Martin, C.S. and Rast, W. (eds) *Phosphorus Management Strategies for Lakes*. Ann Arbor Science, Ann Arbor, pp. 259–308.

Löfgren, S. and Olsson, H. (1990) Tillförsel av kväve ach fosfor till vattendrag i Sveriges inland – Underlagsrapport till Hav -90, Aktionsprogram mot havsföroreningen. *Naturvårdsverkets rapport* 3692. (With English summary.)

Nordic Council of Ministers (1993) Mod strømmen ... en vurdering af styringsmidler overfor afstrømmingen af næringssalte fra landbruget i de nordiske lande. *Nordiske Seminar-og Arbejdsrapporter* 565, 165 pp.

Persson, P. (1990) Utilization of phosphorus in suspended particulate matter as tested by algal bioassays. *Verhandlungen der Internationale Vereinigung für Theoretische und angewandte Limnologie* 24, 242–246.

Pietiläinen, O.-P. and Rekolainen, S. (1991) Dissolved reactive and total phosphorus load from agricultural and forested basins to surface waters in Finland. *Aqua Fennica* 21, 127–136.

Posch, M. and Rekolainen, S. (1993) Estimation of rainfall erosivity in Finland. *Agricultural Science in Finland* 2, 271–279.

Priha, M. (1994) Bioavailability of pulp and paper mill effluent phosphorus. *Water Science and Technology* 29, 93–103.

Rekolainen, S. (1989a) Phosphorus and nitrogen load from forest and agricultural areas in Finland. *Aqua Fennica* 19, 95–107.

Rekolainen, S. (1989b) Effect of snow and soil frost melting on the concentrations of suspended solids and phosphorus in two rural watersheds in Western Finland. *Aquatic Sciences* 51, 211–223.

Rekolainen, S. and Posch, M. (1993) Adapting the CREAMS model for Finnish conditions. *Nordic Hydrology* 24, 309–322.

Rekolainen, S., Posch, M., Kämäri, J. and Ekholm, P. (1991) Evaluation of the accuracy and precision of annual phosphorus load estimates from two agricultural basins in Finland. *Journal of Hydrology* 128, 237–255.

Rekolainen, S., Pitkänen, H., Bleeker, A. and Felix, S. (1995) Nitrogen and phosphorus fluxes from Finnish agricultural areas to the Baltic Sea. *Nordic Hydrology* 26, 55–72.

Seuna, P. (1982) *Frequency Analysis of Runoff of Small Basins.* Publications of the Water Research Institute 48, National Board of Waters, Finland, 77 pp.

Shannon, E.E. (1983) *Investigation of Bioavailable Phosphorus (BAP) in Municipal Wastewaters and the Effects of Wastewater Treatment on BAP.* Final report submitted to Environment Canada by Canviro Consultants, Ltd., Kitchener, Ontario.

Sharpley, A.N., Ahuja, L.R., Yamamoto, M. and Menzel, R.G. (1981) The kinetics of phosphorus desorption from soil. *Soil Science of America Journal* 45, 493–496.

Sonzogni, W.C., Chapra, S.C., Armstrong, D.E. and Logan, T.J. (1982) Bioavailability of phosphorus inputs to lakes. *Journal of Environmental Quality* 11, 555–563.

Ulén, B. (1995) Episodic precipitation and discharge events and their influence on losses of phosphorus and nitrogen from tiledrained arable fields. *Swedish Journal of Agricultural Research* 25, 25–31.

Ulén, B., Kronvang, B. and Svendsen, L.M. (1991) Loss of phosphorus from woodland, natural land and agricultural land. In: Svendsen, L.M. and Kronvang, B. (eds) *Phosphorus in the Nordic Countries – Methods, Bioavailability, Effects and Measures.* Nordisk Ministerråd, København Nord 47, 83–100. (In Swedish, with English summary, legends and headings.)

Williams, J.D.H., Shear, H. and Thomas, R.L. (1980) Availability to *Scenedesmus quadricauda* of different forms of phosphorus in sedimentary materials from the Great Lakes. *Limnology and Oceanography* 25, 1–11.

Yli-Halla, M., Hartikainen, H., Ekholm, P., Turtola, E., Puustinen, M. and Kallio, K. (1995) Assessment of soluble phosphorus load in surface runoff by soil analyses. *Agriculture, Ecosystems, and Environment* 56, 53–62.

Young, T.C., DePinto, J.V., Flint, S.E., Switzenbaum, M.S. and Edzwald, J.K. (1982) Algal availability of phosphorus in municipal wastewater. *Journal of the Water Pollution Control Federation* 54, 1505–1516.

Zlinszky, J. and Herodek, S. (1990) Biologically available phosphorus retention by the Kis-Balaton reservoir. *Archiv für Hydrobiologie Beiheft Ergebnisse der Limnologie* 33, 703–707.

4 Reconstructing Historical Phosphorus Concentrations in Rural Lakes Using Diatom Models

N.J. Anderson
Division of Environmental History and Climate,
Geological Survey of Denmark and Greenland (GEUS),
Thoravej 8, DK-2400 Copenhagen NV, Denmark

INTRODUCTION

The nutrient enrichment of lowland lakes continues to be a problem in many areas, despite major reductions in point-source inputs associated with improved sewage treatment (Cullen and Forsberg, 1988; Sas, 1989; Sharpley *et al.*, 1994; Foy *et al.*, 1995). Regional water-quality surveys indicate that there are many lakes in the industrial and intensive-agricultural areas of Western Europe and North America that have elevated phosphorus (P) concentrations when compared with what are assumed to be natural levels or those derived from eutrophication guidelines (e.g. Jeppesen *et al.*, 1991; Gibson *et al.*, 1995). The extent to which such lakes have undergone enrichment, however, is unclear, primarily because of the absence of long-term water-quality data. Also, some lakes with elevated nutrient concentrations are assumed to be naturally eutrophic, which in some cases may overlook their disturbance over long time periods (Anderson, 1995a).

Questions concerned with the extent of the eutrophication problem in rural lakes require some measure of past conditions with which to compare the present situation. There are very few lakes anywhere with more than 10 years' continuous monitoring data and for the majority of rural lakes little or nothing is known. It is difficult, therefore, to ascertain what the natural background status was, or when the lake began to change and at what rate (Ford, 1988). Such information is, of course, important for lake management – for example, for designing restoration scenarios and determining present rates of change, together with future trajectories (Smol, 1992). Moreover, as lake chemistry and lake biota exhibit variability at a variety of time-scales, there is a clear need for a longer-term ecological perspective (Schoonmaker and Foster, 1991; Anderson, 1995a, b).

DETERMINING BACKGROUND CONDITIONS IN LAKES

Traditionally, lake sediment records have been used to infer past changes in lake ecosystems (Battarbee, 1991). There are, however, other approaches (see Ford, 1988, for a review); these are long-term monitoring data, nutrient export coefficient models that utilize documentary catchment data, and dynamic models, which can be hindcast as well as used for forecasting. Regional eutrophication regression-type models can be used to determine apparent past concentrations but they make assumptions about past P loading and P retention and have no temporal component (OECD, 1982; Ahl, 1994).

The export-coefficient approach has a limited temporal perspective compared with sediment records because it is dependent on available documentary data (for primary data input into the model). It does, however, offer a complementary approach for about the last 50 years. For example, for some lakes it is possible to infer historical changes in P and nitrogen, using export-coefficient models (e.g. O'Sullivan, 1992) or a combination of this approach with sediment records and geochemical methods (Heathwaite, 1994). Dynamic models can be hindcast over longer periods (Schelske *et al.*, 1988) but are subject to problems of parameterization and often do not translate readily to new study sites, or at least, not without considerable recalibration (Ahlgren *et al.*, 1988; Anderson, 1995c).

Lake Sediment Records

Lake sediments are natural environmental archives, which collect a variety of biotic and abiotic information over a range of time-scales. Most northwest European and North American temperate lakes were formed at the end of the last glaciation and as a result have sediment sequences covering ~10,000 years. There are a number of detailed reviews which provide introductions to the wide range of information preserved in lake sediments, as well as the recent progress in the reliability of the interpretations made (e.g. Battarbee, 1991; Anderson, 1993; Anderson and Battarbee, 1994). The time-scales and temporal resolution of sediment cores vary enormously (Anderson, 1995b) but, importantly, high-resolution records of decadal trends can be readily obtained from eutrophic lakes, where sediment accumulation rates are often greater than 0.5 cm year^{-1}. In this review, emphasis is placed on changes that have occurred over the last 100–150 years. In part, this time-scale is a reflection of the substantial anthropogenic impacts on both terrestrial and aquatic ecosystems in this period, but also this time period can be reliably dated using the lead isotope ^{210}Pb (Oldfield and Appleby, 1984). Longer sediment sequences are dated using a variety of techniques but most commonly carbon-14 (^{14}C) and the use of annually laminated sediments (Battarbee, 1991). Reliable chronologies are fundamental to palaeolimnological investigations.

Geochemical Phosphorus in Lake Sediments

Lake sediments contain, on average, between 0.01 and 10 mg P g^{-1} dry weight (DW) (Holtan et al., 1988). It is therefore reasonable to assume that direct geochemical determination of P in sediments would be a measure of the P loading on the lake. There have been numerous such geochemical studies of lake eutrophication (e.g. Williams et al., 1976; Bengtsson and Persson, 1978), but few have attempted to determine the historical P loading from such data or in-lake epilimnetic total P (TP) concentrations. Historical loading estimates require a multicore approach to overcome the problem of spatial variability in the sediment record (Evans and Rigler, 1980). For example, Moss (1980) used multiple cores to calculate the historical P load to Barton Broad (Norfolk, England) and also changes in the lakewater P concentration. However, calculating both the loading and the lake-water TP concentration from geochemical analyses makes assumptions about the permanent retention of P within the sediments, which in reality may not remain constant and can decrease with increasing eutrophication (Fig. 4.1).

Increased organic loading to sediments accompanies eutrophication with the result that there is increased anoxia in both the hypolimnion and the sediments themselves. The lowered redox potential results in increased mobility of P (along with iron (Fe) and manganese (Mn)) in the sediments (Carignan and Flett, 1981). Ultimately, release from the sediments creates the internal P-loading problem (Marsden, 1989), which means that even substantial reductions in external loading may not produce similar reductions in the in-lake TP concentration. From a historical perspective, the alteration of the P cycle with progressive eutrophication means that the amount of P retained in lake sediments may not correlate with the loading (Engstrom and Wright, 1984; Fig. 4.1). It is against this background that the recent statistical developments in palaeolimnology (Battarbee, 1991; Anderson, 1993; Birks, 1993) have permitted the development of diatom–P models, the means of estimating background TP concentrations and their change over time.

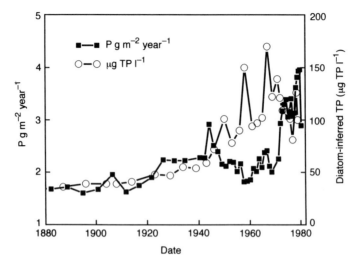

Fig. 4.1. A comparison of geochemical P (■) and diatom-inferred TP (o) for Lough Augher (Northern Ireland), which illustrates the impact of eutrophication on sediment and hypolimnetic redox and hence enhanced release of P from the sediments. As the diatom-inferred TP values peaked in the late 1960s, the geochemical phosphorus flux was at a minimum, due to the reduced sediment retention. (Redrawn from data in Anderson and Rippey, 1994.)

DIATOMS AND LAKE EUTROPHICATION

Diatoms are an ecologically sensitive group of algae, widely distributed over a range of environmental conditions and habitats (Round, 1981). Because of their siliceous cell wall, diatoms generally preserve well in lake sediments and can be very abundant, 10^6–10^8 frustules g^{-1} DW (Battarbee, 1986a). An introduction to diatom analysis is given by Battarbee (1986a). The ecology of the dominant freshwater planktonic diatom genera is well understood in terms of their responses to nutrients, light and temperature, and is derived from contemporary experimental and lake-survey data (Tilman *et al.*, 1982; Reynolds, 1984; Willén, 1991). Species-specific growth-rate information is available for a few common taxa, e.g. *Asterionella formosa* and *Aulacoseira subarctica*, but for many of the rarer planktonic taxa and the great majority of the benthic forms (which live attached to stone or plant substrates) relatively little is known.

Diatoms preserved in lake sediments have been widely used to provide records of changing nutrient enrichment of lakes and similar species sequences have been found throughout the world in response to deforestation, agricultural development and point-source pollution (Haworth, 1972; Bradbury, 1975; Battarbee, 1978, 1986a, b; Gaillard *et al.*, 1991). These sequences are interpreted qualitatively in terms of nutrient enrichment, using the contemporary relationships between diatoms and water chemistry that have existed for some time (Stoermer, 1984). These qualitative interpretations are valid and ecologically sound.

DIATOM–PHOSPHORUS MODELS

Now, however, it is possible to quantify the relationships between diatom and P using weighted-averaging (WA) regression and calibration models (ter Braak and Prentice, 1988; ter Braak and van Dam, 1989). This approach utilizes the developments made during lake-acidification research in the 1980s, where diatoms were used to reconstruct the changes over time of pH in lakes (Birks *et al.*, 1990). As a result, it is possible to infer TP quantitatively from diatoms preserved in lake sediments (Anderson *et al.*, 1993).

Establishing a diatom–lake–P model is a two-step process: first, regression – the calculation of the optima; and second, calibration – the use of the species' optima to infer a lake's TP concentration. Inherent to all the diatom–water-chemistry models is the construction of a training set to determine the distribution of the individual species along the environmental gradient in question. This is normally done by taking surface-sediment samples from a number of lakes ($n \geq 40$) for which the water chemistry is known. For each of the surface-sediment samples, the relative abundance of each diatom species is determined (Fig. 4.2). The species found in the surficial sediment (0.5–1 cm) are assumed to represent the last year's growth and therefore are presumably related to the water chemistry for that period. Planktonic diatoms respond very quickly to nutrient additions (Reynolds, 1984) and these changes in the water column are transferred without any significant delay into sediment assemblages, where they can be used for estimating P. The rapidity of these responses means that diatom-inferred TP can be used to follow lake TP concentrations in response to nutrient reduction projects and with a high resolution (Anderson and Rippey, 1994). Moreover, diatoms respond to the within-lake TP concentration and so, even if this P is primarily derived from the sediments (i.e. internal loading), they will still reflect the increased internal-loading component (Fig. 4.1).

The first diatom–nutrient models were used to reconstruct lake trophic-status categories rather than TP, but, more importantly, these models were of a linear type and utilized simple linear regression of diatom species and/or categories against TP (Whitmore, 1989; Brenner *et al.*, 1993). In this approach, diatoms are placed a priori into trophic-status categories. In these methods, a linear response between the diatoms and the environmental variable is assumed. However, it has been shown that diatom species distributions along water-chemical gradients (pH, TP, salinity) are essentially non-linear (Birks *et al.*, 1990) and are generally better described by unimodal or Gaussian distributions (Fig. 4.2; ter Braak and Prentice, 1988). In unimodal models, it is assumed that the optimum of a species occurs at or close to its maximum abundance (ter Braak and van Dam, 1989). It has been shown that this optimum can be reliably estimated by a simple abundance-weighted average of the species abundance along the gradient in question (ter Braak and Looman, 1986).

In WA models, the estimate of a species optimum ($\hat{u}_K$) (WA regression) is:

$$\hat{u}_k = \sum_{i=1}^{n} y_{ik} x_i \bigg/ \sum_{i=1}^{n} y_{ik}$$

where y_{ik} is the abundance of the species k at site i, x_i is the observed TP concentration at site i, m is the number of species at the site and n is the number of sites.

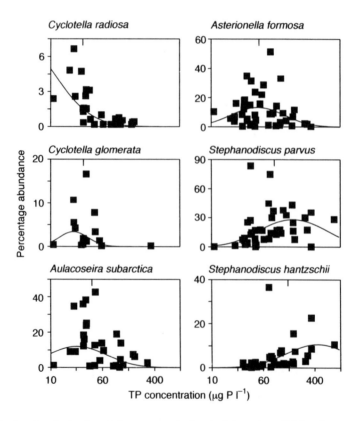

Fig. 4.2. Species distribution of six common planktonic diatoms in surficial sediments of Northern Irish lakes, which illustrates their changing abundance along a phosphorus gradient. The vertical axes are relative abundances (of total diatom count); the horizontal axis is TP concentration on a log scale; only positive occurrences are shown. The weighted-averaging optimum for each species is indicated by the vertical dash on the upper vertical axis and the Gaussian logit curves fitted to the abundance are indicated (original, from N.J. Anderson, unpublished data).

These optima can then be used to infer the TP concentration of a sample ($\hat{x}_i$) (WA calibration):

$$\hat{x}_i = \sum_{k=1}^{m} y_{ik}\, \hat{u}_k \Big/ \sum_{k=1}^{m} y_{ik}$$

where y_{ik} is the abundance of the species k in the sample i, $\hat{u}_k$ is the optimum of the species k, and m is the number of species at the site. As averages are taken twice within WA, once within each of the regression and calibration calculations, the original gradient length is reduced and has to be corrected. This correction is done by 'deshrinking' using either classical or inverse regression (Birks et al., 1990). Inverse deshrinking results in lower errors of prediction, whereas classical deshrinking results in a better fit of estimated versus observed over the whole range of the environmental variable being modelled. It is a subjective decision as to the type of deshrinking used.

Error statistics for diatom models are usually given as the root-mean-square of the error (RMSE) (measured – predicted TP) and the correlation between measured and estimated TP (Birks et al., 1990). However, as the RMSE is based solely on the training set, it is an underestimate of the true error and is referred to as the apparent RMSE. The majority of diatom-TP training sets are not sufficiently large for splitting into training and test-data sets to determine the RMSE independently (Birks et al., 1990). It has become common, therefore, to use resampling techniques (bootstrapping and jackknifing) to estimate the true error (Birks et al., 1990). Both summary statistics are reported here, together with the apparent r^2 value for measured versus estimated TP. These calculations of WA regression and calibration are made using the program WACALIB (Line and Birks, 1990), the latest version of which (WACALIB 3.3; Line et al., 1994) also includes an option for calculating reconstruction errors for each individual fossil sample by bootstrapping.

As an example of the predictive ability of diatom models, the north-west European training set is presented in Fig. 4.3. This training set is the combination of five regional data sets (Table 4.1) and covers a TP gradient of 10 to ~1000 µg TP l^{-1} (Bennion et al., 1996a). The predictive ability of this WA partial least squares (WA-PLS) model (see below) is marginally better than a simple WA model, as measured by the RMSE$_{jack}$ – 0.21 versus 0.22 log$_{10}$ TP units, respectively (Bennion et al., 1996a). Simple WA models utilize only a small amount of information in the relationship between the taxa and the variable being modelled. This recent statistical development, WA-PLS (ter Braak and Juggins, 1993), utilizes secondary relationships to improve the predictive ability of the models.

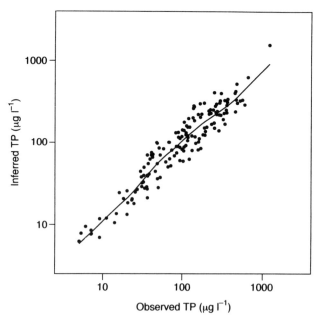

Fig. 4.3. Diatom-inferred TP derived from the north-west European training set (see Bennion *et al.*, 1996a for details) plotted against observed TP. The diatom-inferred TP values are derived from a two-component, weighted-averaging, partial least-squares model (see ter Braak and Juggins, 1993, for a full description of the methodology). (Redrawn from data in Bennion *et al.*, 1996b.)

Error Sources

There are primarily four main error sources in diatom–water-chemistry models: first, sediment sampling/representativity; second, water-chemistry sampling/seasonality effects; third, errors in species' optima; and, finally, the no-analogue problem. These are essentially different kinds of sampling problems. Sampling problems of both surface sediments and cores have been discussed elsewhere (Charles *et al.*, 1991; Anderson, 1997).

Analogue problems, where diatom species are present in the sediment cores but not in the training set, create errors in the final reconstruction (inference) of historical TP concentrations. Analogue problems tend to be small when the lake being used for TP reconstructions is located within the same geographical region and the sediment core covers only short (~100-year) time periods. Over longer time-scales, and as a result of inappropriate application of the training set, the resultant reconstructions must, for the moment, be treated with some caution.

The final accuracy of the diatom models is heavily dependent on the estimated species' optima, which in turn are influenced by the length of the sampled TP gradient (ter Braak and Looman, 1986). To overcome this, larger data sets are required covering a longer TP gradient (5–1000 µg TP l^{-1}). Such a large data set is now available for north-western Europe (Bennion *et al.*, 1996a). Although the ecological basis for diatom–P models is very sound – the impact of P enrichment on algae has been well known for some time (Reynolds, 1984) – the actual physiological basis of how the models work is less apparent. However, for applied studies, that is not so relevant as the ability of the model to reconstruct accurately past TP concentrations. An important ecological benefit of this approach, however, is the derivation of information about the majority of diatom species encountered in a particular region, in terms of their ranking along a P gradient, and the calculation of a P optimum for each species (see Fig. 4.2).

Independent validation has become an important aspect of the study recently (Bennion *et al.*, 1995). For the moment, models have not been compared with any of the long-term P records, such as those available for Esthwaite Water (Talling and Heaney, 1988), but a comparison of diatom-inferred TP and monitored records at White Lough (Co. Tyrone, Northern Ireland; B. Rippey *et al.*, 1997) indicates that the diatom models pick up-trends accurately, although there was a slight discrepancy between the monitored and diatom-inferred TP values.

Application of Weighted-Averaging Models to Diatom Stratigraphy

Applying WA models to diatoms preserved in lake sediments turns the changing abundances of the different species, which can be interpreted ecologically and qualitatively (see above), into a quantitative estimate of past TP concentrations. At Lough Patrick, for example (Fig. 4.4a), the replacement of a number of small *Cyclotella* species and *Achnanthes minutissima* by *Aulacoseira subarctica, Aulacoseira ambigua* and *Stephanodiscus parvus* is clearly indicative of enrichment. Such changes have been found at many sites throughout north-western Europe, Scandinavia and North America (Bradbury, 1975; Battarbee, 1986a). At Lough Patrick, the diatom-inferred TP profile suggests that prior to initial disturbance, TP concentrations were ~ 10 µg TP l^{-1} but increased steadily to 35 µg TP l^{-1} at the core surface, which agrees well with the present-day measured TP value, 36 µg TP l^{-1} (Fig. 4.4b).

Regional Diatom–Phosphorus Models

Table 4.1 lists the currently available diatom–P training sets and their associated error statistics. Models are at present available for only North America (six) and Europe (seven). The predictive ability of the models is generally similar, although it is difficult to compare directly the RMSE values between training sets because of the different data transformations

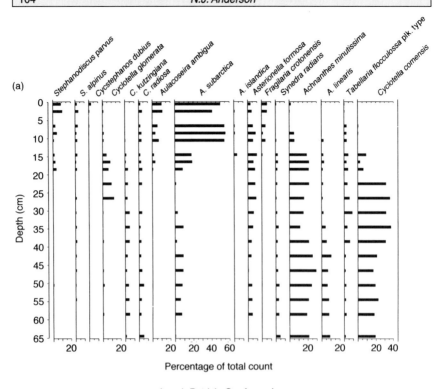

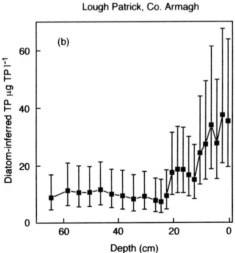

Fig. 4.4. Conversion of diatom assemblages in sediment cores to diatom-inferred TP. (a) Summary diatom stratigraphy (percentage abundance) from Lough Patrick, Co. Armagh, Northern Ireland. (b) Diatom-inferred TP (μg TP l^{-1}) for Lough Patrick derived from application of the Northern Irish training set, 95% confidence limits were derived from bootstrapping of individual samples (see Line *et al.*, 1994). (From N.J. Anderson, unpublished.)

that were used in the original construction of the models (Table 4.1). However, a clear difference is the much shorter gradient covered by the North American models compared with some of the European models. The former generally cover a range of 5–60 µg TP l^{-1}, compared with 20 to ~1000 µg TP l^{-1} for the training sets from Denmark, Northern Ireland, English meres and south-west England. The training-set TP ranges, which primarily reflect extant lake-water chemistry, compare well with the differences in soil-test P values between North America and those in Europe (Sharpley et al., 1994). The Austrian Alpine models are more comparable to the North American data sets in terms of their P range.

The number of water-chemistry measurements used in the calculation of the mean annual TP value is variable among the different data sets, ranging from one (Michigan) to 19 (Denmark). The implications of this variably defined 'mean' TP value for the diatom models have yet to be critically evaluated (Anderson, 1995d; Gibson et al., 1996). The amount of variance in the unconstrained species data explained by TP in these training sets is normally between 8 and 12%. Although this is low, it is typical for predictive models derived from noisy biological data sets and compares well with that observed for diatoms and pH (Birks et al., 1990).

In some of the training sets, TP is not the dominant explanatory variable in a constrained ordination of diatoms and water-chemical variables (e.g. environmental monitoring and assessment program – surface waters (EMAP-SW), Dixit and Smol, 1994; south-central Ontario, Hall and Smol, 1996); pH, for example, explains more of the variance in the diatom-species matrix. This can result in a poorer fit between predicted versus observed TP (lower r^2, higher RMSE). In the south-central Ontario data set, a partial canonical correspond analysis (CCA) indicated that there is no statistically significant independent variance explicable by TP (Hall and Smol, 1996). It remains a matter of debate whether diatom–P relationships should be modelled in these instances. A similar problem occurs when two chemical variables explain similar amounts of variance. For example, Christie and Smol (1993) chose to model total nitrogen rather than TP, although there was no difference between the two variables in terms of the ability to be modelled using WA.

Comparisons of species' optima in the different regional training sets is beyond the scope of this chapter, but they have important implications for the biogeography and autecology of the different species/forms found in the training sets. For example, *Stephandiscus parvus* in British Columbia has an optimum of 18 µg TP l^{-1}, compared to 136 µg TP l^{-1} in Northern Ireland and 142 µg TP l^{-1} in Denmark. Clearly, if these are the same species, its true optimum will be somewhere in between.

Table 4.1. Summary details of regional weighted-averaging diatom–phosphorus models.

Geographical area	Number of lakes	TP range ($\mu g\ TP\ l^{-1}$)	r^2 (apparent)	RMSE (apparent)	RMSE (boot.)	Reference
Alpine (Austria, Bavaria and Italy)	62	2–266	0.57	0.32	0.35	Wunsam and Schmidt, 1995
British Columbia (enlarged)	59	5–85	0.73	0.33	0.44	Reavie et al., 1995a
British Columbia (original)	37	5–28	0.73	0.25	n/k	Hall and Smol, 1992
Denmark	28	27–1190	0.80	0.15	0.30	Anderson and Odgaard, 1994, and unpublished
English meres	33	68–1161	0.61	0.17	0.25	Bennion, 1995
Michigan (USA)	41	1–51	0.73	0.41	n/k	Fritz et al., 1993
North-eastern USA (EMAP-SW)	66	1–154	0.62	0.66	n/k	Dixit and Smol 1994
Northern Ireland	49	11–800	0.75	0.15	0.23	Anderson et al., 1993
North-western Europe (combined) (WA-PLS; see text)	164	5–1190	0.91	0.22	0.21	Bennion et al., 1996b
Ontario	28	2–63	0.80	0.15	n/k	Agbeti, 1992
Ontario (south-central)	54	3–24	0.62	3.5	0.41	Hall and Smol, 1996
South-eastern England	31	25–646	0.79	0.16	0.28	Bennion, 1994
Switzerland	72	5–200	0.79	n/k	0.19	A. Lotter, personal communication

boot., Bootstrapping; n/k, not known; EMAP-SW, Environmental Monitoring and Assessment Program – Surface Waters.

RECONSTRUCTION OF LAKE EUTROPHICATION HISTORIES USING DIATOM–PHOSPHORUS MODELS

Most of the training sets listed in Table 4.1 have been applied to diatom stratigraphies in sediment cores, but only some of the lakes are rural. The total number of TP reconstructions is still relatively limited. Detailed reconstructed TP histories have been made for point-source lakes, such as Lough Augher (Anderson and Rippey, 1994) and Marsworth Pond (Bennion, 1994). Fritz *et al.* (1993) applied the Michigan data set to four lakes in extensively forested catchments in the lower peninsula of Michigan. All the lakes were dominated by small, oligotrophic *Cyclotella* species and, as a result, the reconstructed TP values are all low (< 10 µg TP l^{-1}) and comparatively stable over hundreds of years. At three of the lakes, TP values increased slightly at the time of settlement and logging of the catchments but subsequently declined.

Anderson *et al.* (1993) applied the Northern Irish training set to two small rural lakes and compared the results with geochemical P profiles in undated sediment cores. This study has been followed by a reanalysis of one of the lakes, White Lough, where the massive increase in diatom-inferred TP concentrations (~200 µg TP l^{-1}) proved to be relatively recent (*c.* 1976) and very short-lived (2–4 years). Originally thought to be a result of agricultural activity, Rippey *et al.* (1997) concluded that this increase was largely driven by increased P recycling associated with the general longer-term eutrophication of the lake. It was the drier winters of the mid-1970s that resulted in a reduced flushing rate and caused increased retention of the sediment-derived P within the lake, hence increasing the in-lake concentration (Rippey *et al.*, 1997).

The Northern Irish training set has also been applied to ^{210}Pb-dated sediment cores from six rural lakes in Northern Ireland (N.J. Anderson, unpublished data). The lakes (Loughs Brantry, Creeve, Heron, Patrick, Ballywillin and Corbet) are all small (< 20 ha) with rural catchments and have present-day TP concentrations ranging from ~30 to > 300 µg TP l^{-1}. The cores cover variable time periods, which in part explains the large range in TP concentrations inferred for the oldest samples in each core. Currently eutrophic lakes, such as Loughs Corbet (measured TP = 285 µg TP l^{-1}) and Ballywillin (measured TP = 310 µg TP l^{-1}), have basal TP concentrations of 80 and 60 µg TP l^{-1}, respectively, but the higher sediment accumulation rates in these lakes means that these samples are only ~100 years old (Fig. 4.5a). In contrast, the basal TP estimate at Lough Patrick is only 10 µg TP l^{-1}; the age of this sample is unknown, but it is probably hundreds of years old (see Fig. 4.4b). Four of the six lakes (Brantry, Creeve, Corbet and Ballywillin) have unambiguously higher rates of TP increase for the period since 1940–1960, while at Lough Heron the increase of TP has been approximately linear since ~1850 (Fig. 4.5a; N.J.

Anderson, unpublished). At Lough Patrick, ambiguities in the ^{210}Pb chronology did not permit post-1950 rates of TP increase to be calculated. Battarbee (1978, 1986b) reported increases in rate of eutrophication in the postwar period for Loughs Erne and Neagh in Northern Ireland. At three of the smaller rural lakes studied by N.J. Anderson (unpublished) (Ballywillin, Brantry and Creeve), there have been recent decreases in diatom-inferred TP concentrations, indicating some lake recovery but also the catchment-specific nature of enrichment.

Bennion et al. (1996b) applied the north-west European data set to three lakes of conservation importance on Anglesey (North Wales). The lakes have primarily agricultural catchments. Two of the lakes (Llyn Dinam and Llyn Coron) show large increases in diatom-inferred TP (to more than 100 µg TP l^{-1}) and have background TP concentrations of around 60 µg TP l^{-1}. Unfortunately, no dates are available for these cores to determine when the increases occurred. At the third site (Llyn Penrhyn), there is a large discrepancy between the reconstructed diatom-inferred TP value and the

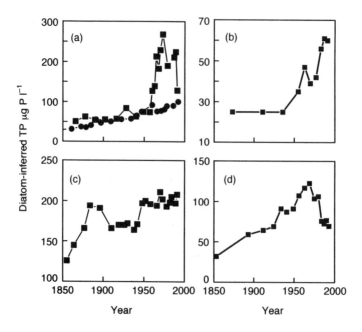

Fig. 4.5. Regional applications of diatom–phosphorus weighted-averaging calibration data sets (Table 4.1). (a) Two Northern Irish lakes (Ballywillin Lough (■–■), Lough Heron (●–●)) (from N.J. Anderson, unpublished). (b) Charlie Lake, British Columbia, Canada (redrawn from data in Reavie et al., 1995a). (c) Lange Sø, Denmark (redrawn from Anderson and Odgaard, 1994). (d) Nørre Sø, Denmark (from B. Odgaard, J.A. Wolin and N.J. Anderson, unpublished). For all lakes, the diatom-inferred TP values are plotted against sediment age derived from ^{210}Pb.

measured value, suggesting that use of the model is inappropriate at this site. Bennion *et al.* (1996b) attributed this misfit, in part, to the very poor diatom preservation in this shallow lake.

Reavie *et al.* (1995b) applied the British Columbia data set to six lakes in the province. Three of the lakes showed significant enrichment since European settlement (*c.* 1850), while there was little change at the other sites, despite anthropogenic activity in the catchments. Background TP concentrations were above 10 µg TP l^{-1} at all sites. At Charlie Lake, diatom-inferred TP increased from ~20 to 60 µg TP l^{-1} since 1936 (Fig. 4.5b). Initial enrichment was caused by deforestation for agriculture but the later increases relate both to agricultural inputs and possibly to the contamination of groundwater inputs, which are being enriched by domestic sewage disposal.

One of the British Columbia lakes, Pinantan, was dominated by *S. parvus* throughout the sediment core, including the period which predates European settlement (> 200 years ago). Reavie *et al.* (1995) interpret this lake as being naturally eutrophic. Other lakes also have natural meso-eutrophic diatom assemblages prior to settlement. In contrast, the long-term uniformity of the diatom assemblages and the often high diatom-inferred TP values at some European sites probably attest to the length of cultural activity in Europe (Anderson, 1995b).

At sites such as Lange Sø, on the Danish island of Fuen, the diatom-inferred TP record indicates that TP concentrations were around 100 µg TP l^{-1} over 100 years ago (Anderson and Odgaard, 1994). Even allowing for substantial errors in the inference model, the constancy of the plankton diatom stratigraphy throughout the length of the core suggests that there have been only limited changes in water chemistry at this site (Fig. 4.5c). However, even at this site there was a clear increase in TP concentrations after 1950. At Nørre Sø, also on Fuen, concentrations again increased more rapidly in the immediate postwar period, but only until ~1970, since when they have declined steadily (Fig. 4.5d). This decline has also been observed in the limited monitoring data available, and the surface-inferred value agrees very well with the measured annual mean (69 versus 65 µg TP l^{-1}). Although it has an agricultural and forest catchment, Nørre Sø is instructive because of the large groundwater component of its nutrient budget. It is difficult to determine to what extent changing groundwater pollution has influenced the in-lake TP concentration over historical time.

Time-scales of Phosphorus Increases

There have been only a few lakes to which the diatom models have been applied to date and there is also the problem of local between-lake differences, which are a reflection of natural factors (soil type, geology, relief, rainfall; Dillon and Kirchner, 1975), as well as the variable disturbance histories. However, a number of generalizations are possible. Initial lake disturbance in North America is commonly associated with the European settlement (Bradbury, 1975; Brugam and Speziale, 1983). In contrast, European time-scales of disturbance are more variable and, generally, much longer (Fritz, 1989). High-resolution studies of laminated lake sediments in northern Sweden indicate the rapidity with which forest clearances changed the diatom communities, and by implication the water chemistry (Anderson et al., 1995). Unfortunately, no P reconstruction of this site has been made.

In fact, there have been very few applications of diatom–P models to long-term sediment records, the reconstruction for Diss Mere (England), which covers 6000 years of agricultural and urban expansion, being the best example to date (Birks et al., 1995). At this lake, the shift from slash-and-burn agriculture during the neolithic period to more sedentary cereal production during the early Bronze Age resulted in TP increasing from 33 to 221 µg TP l^{-1}. The later part of this record, from the postmedieval period onwards, is irrelevant to this discussion because of the growth of the town of Diss as P source for the lake (Fritz, 1989). Wunsam and Schmidt (1995) applied the Alpine model to a long core from Längsee (Carinthia, Austria). There was a dramatic increase in diatom-inferred TP concentrations (from < 10 to > 100 µg TP l^{-1}) at the time of the Bronze Age deforestation, as indicated by an increase in grass pollen.

While lowland lakes have probably always been more productive than low-alkalinity upland lakes (e.g. Foy and Withers, 1995), lowland areas in north-western Europe have histories of cultural impacts which largely preclude use of the term 'natural' (Anderson, 1995b). However, an important consideration for these older sediments is the errors in the reconstructions associated with ill-defined species optima, caused in part by analogue problems in the training set. Many of the European training sets lack low-P, high-alkalinity sites which would lower the optima of a number of key species (e.g. *Cyclotella ocellata*), which in turn would lower the reconstructed value.

Given that most studies to date have been of cores dated by ^{210}Pb (i.e. covering only the last 150 years), it is possible to reach two main conclusions. First, changes during the twentieth century result in higher rates of TP concentration increase, primarily after ~1930 or later. Second, the relatively high background TP concentrations are probably the result of long-term agricultural practices within the catchments, the nature and intensity of which are locally variable. These background levels (Fig. 4.5a, c, d) are high in comparison with, for example, Organization for Economic Cooperation and Development (OECD) guidelines, and suggest that the lakes either are naturally eutrophic or have been enriched over long periods of time by agricultural activity within the catchments (Anderson, 1995b).

The post-1950 increases in diatom-inferred TP observed at some sites in Northern Ireland (Fig. 4.5a), for example, probably relate to changed agricultural practice. These predominantly pastoral catchments are quite stable and have low surface-soil losses. Since 1940, there has been little change in the input of P fertilizer, but field drainage has increased substantially. Animal slurry inputs have also to be considered. Agricultural point sources (such as uncontrolled slurry disposal, runoff from paved farmyards) are also responsible at some sites, and may account for the very rapid increases observed at Ballywillin Lough (Fig. 4.5a). Foy and Withers (1995) suggested that rapid increases in P concentrations indicate point-source inputs. For the other lakes, the implication is that better soil drainage combined with increasing P saturation of the soils has resulted in the observed TP increases (Fig. 4.5a; N.J. Anderson, unpublished). While such sources are difficult to identify in the sediment record, a combination of documentary studies of catchment change and geochemical markers may be useful (Krug, 1993; Heathewaite, 1994).

Foy and Withers (1995) calculated that soil P in Northern Ireland has increased by 10 kg ha^{-1} year^{-1} since 1945 as a result of P inputs (as fertilizer and animal foodstuffs) being greater than the amount exported. N.J. Anderson's (unpublished) estimates of rates of TP increase in rural Northern Irish lakes (1–2 µg TP l^{-1} year^{-1} since 1950; determined by regressing diatom-inferred TP on ^{210}Pb age) compare very well with those reported by Foy and Withers (1995), derived from long-term monitoring programmes: dissolved reactive phosphorus has been increasing at an average of 1.5 µg TP l^{-1} year^{-1} since 1974.

It is reasonable to conclude that, despite long periods of enrichment, the twentieth century has seen major increases in P concentrations, probably the greatest since the initial clearances for agriculture (Birks *et al.*, 1995). It would be interesting to determine what the impact of the initial deforestation was on the P budget for an agricultural catchment and compare these changes with the post-1950 increases. Types and sources of P could be addressed by using a variety of geochemical analyses and diatom–P models.

FUTURE WORK

Lake responses to disturbance can be individualistic, due to a variety of factors – e.g. importance of local conditions; geology, degree and length of the perturbation – but, importantly, sediment records provide this site-specific record. However, there is undoubtedly a need for better regional coverage, with more lakes being studied. The core top-and-bottom approach, which works so well in North America, with its readily defined pre-European baseline (Brugam, 1988), is more difficult to apply in Europe, with its longer cultural history (Anderson, 1993). The top-and-bottom method consists of comparing surface sediments with a single sample from the base of a short core (< 40 cm) which is assumed to predate cultural impacts. Widely used in acidification studies (Cumming et al., 1992), this approach, to date, has been applied to TP reconstructions in the USA and Canada (Dixit and Smol, 1994; Hall and Smol, 1996). The approach, while useful in that it provides regional patterns, is limited by its lack of a temporal framework. However, it is possible, using a structured approach to the subsampling of cores, to work with reduced numbers of analyses and samples per core, to increase the number of lakes studied and thereby to permit generalizations on the regional scale. It is also possible to provide coarse dating using only a few samples per core. This approach is also much cheaper and is being pioneered by H. Bennion and P.G. Appleby (personal communication) to derive time-controlled estimates of changing TP concentrations in English lakes.

Another area for development is the combination of diatom models and more orthodox modelling (Anderson, 1995c), for both model validation and inputs to other model types (parameterization). Combining geochemical determinations of P in sediments and diatom-inferred P permits the calculation of past loading to lakes, as the diatom-inferred estimate provides a measure of in-lake P concentration, which reflects the enhanced release of P from the sediments as enrichment progresses (Rippey and Anderson, 1996; see Fig. 4.1).

CONCLUSION

The diatom–P models that have been developed provide an important tool in the present debate about the rate and timing of lake enrichment due to agriculture. That lake enrichment is widespread in rural lakes in northwestern Europe is suggested by the high frequency of lakes with TP concentrations above 30 µg TP l^{-1} in both predominantly grazing areas, such as Northern Ireland (Foy and Withers, 1995; Gibson et al., 1995), and intensive agricultural areas, such as Denmark (Jeppesen et al., 1990). The few palaeolimnological studies available suggest that this enrichment is probably largely anthropogenic in origin (Anderson and Odgaard, 1994; Anderson, 1995b).

Given the increased relevance of palaeolimnological studies to contemporary limnology and management (Smol, 1992; Anderson, 1993), greater attention is now being given to organisms other than diatoms, in an effort to broaden the reconstructions to other trophic levels and communities (Anderson and Battarbee, 1994). Holistic approaches to the responses of lake biota to disturbances, such as eutrophication, are now being attempted. Multidisciplinary approaches (i.e. analyses of diatoms, pigments, zooplankton, benthic invertebrates, macrophytes (via plant macrofossils), geochemistry) used in conjunction with diatom-inferred TP estimates offer a broader perspective of the response of lakes to disturbance, including the time-scale of these changes.

Clearly, rural lakes are changing both chemically and biologically; the question is, what are we to do? What type of agriculture do we want and what environmental price are we prepared to pay for it? Many lakes have changed over long time periods and their natural state is probably irretrievable, but the rate of change for many lakes has increased rapidly in the postwar period. These rates of change, site-specific and user-defined baselines and restoration targets can be determined from palaeolimnological studies and used within the socio-economic debate that will surround the future of agriculture at the start of the twenty-first century.

ABSTRACT

The debate about the impact of agriculture on phosphorus (P) enrichment on rural lakes demands that the extent of the problem be defined. Determining whether a lake is enriched and by how much is particularly difficult for rural lakes because of the lack of background data. Lake sediments are natural archives that provide one method of determining change in lakes. Diatoms preserved in sediments have been used for some time to infer the qualitatively changing nutrient status of lakes. However, the development of weighted averaging (WA) regression and calibration has permitted statistically robust and ecologically sound methods of inferring past P concentrations. This review provides an introduction to diatom–P models, the main error sources and the predictive ability of the training sets currently available. Applications to rural lakes are summarized and the implications of the data for background conditions and rates of enrichment discussed. In general, north-west European lakes have long-term disturbance histories but there is evidence of increased rates of P enrichment in the post-1950 period in those lakes studied. North American lakes have strong disturbance signals associated with the start of European settlement. Rates of P increase over the last 40–50 years calculated for some Northern Irish lakes from the diatom-inferred total (TP) profiles are very similar to those derived from long-term monitoring exercises (1–2 µg TP l^{-1} year^{-1}).

REFERENCES

Agbeti, M.D. (1992) Relationship between diatom assemblages and trophic variables: a comparison of old and new approaches. *Canadian Journal of Fisheries and Aquatic Sciences* 49, 1171–1175.

Ahl, T. (1994) Regression statistics as a tool to evaluate excess (anthropogenic) phosphorus, nitrogen, and organic matter in classification of Swedish fresh water quality. *Water Air Soil Pollution* 74, 169–187.

Ahlgren, I., Frisk, T. and Kamp-Nielsen, L. (1988) Empirical and theoretical models of phosphorus loading, retention and concentration vs. lake trophic state. *Hydrobiologia* 170, 285–303.

Anderson, N.J. (1993) Natural versus anthropogenic change in lakes: the role of the sediment record. *Trends in Ecology and Evolution* 8, 356–361.

Anderson, N.J. (1995a) Using the past to predict the future: lake sediments and the modelling of limnological disturbance. *Ecological Modelling* 78, 149–172

Anderson, N.J. (1995b) Naturally eutrophic lakes: reality, myth or myopia. *Trends in Ecology and Evolution* 10, 137–138.

Anderson, N.J. (1995c) Temporal scale, phytoplankton ecology and palaeolimnology. *Freshwater Biology* 34, 367–378

Anderson, N.J. (1995d) Diatom-based phosphorus transfer functions – errors and validation. In: Patrick, S.T. and Anderson, N.J. (eds) *Ecology and Palaeoecology of Lake Eutrophication*. Service Report 7, Geological Survey of Denmark, Copenhagen, pp, 39–40.

Anderson, N.J. (1997) Variability of diatom stratigraphy in a small lake basin. II: Ordination and weighted averaging-inferred phosphorus. *Journal of Paleolimnology* (in press).

Anderson, N.J. and Battarbee, R.W. (1994) Aquatic community persistence and variability: a palaeolimnological perspective. In Giller, P.S., Hildrew, A.G. and Raffaelli, D.G. (eds) *Aquatic Ecology: Scale, Pattern and Process*. Blackwell Scientific Publications, Oxford, pp. 233–259.

Anderson, N.J. and Odgaard, B. (1994) Recent palaeolimnology of three shallow Danish lakes. *Hydrobiologia* 275/276, 411–422.

Anderson, N.J. and Rippey, B. (1994) Monitoring lake recovery from point-source eutrophication: the use of diatom-inferred epilimnetic total phosphorus and sediment chemistry. *Freshwater Biology* 32, 625–639.

Anderson, N.J., Rippey, B. and Gibson, C.E. (1993) A comparison of sedimentary and diatom-inferred phosphorus profiles: implications for defining pre-disturbance nutrient conditions. *Hydrobiologia* 253, 357–366.

Anderson, N.J., Segerström, U. and Renberg, I. (1995) Diatom production responses to the development of early agriculture in a boreal forest lake-catchment (Kassjön, northern Sweden). *Journal of Ecology* 83, 809–822.

Battarbee, R.W. (1978) Observations on the recent history of Lough Neagh and its drainage basin. *Philosophical Transactions of the Royal Society London, Series B* 281, 303–345.

Battarbee, R.W. (1986a) Diatom analysis. In: Berglund, B.E. (ed.) *Handbook of Holocene Palaeoecology and Palaeohydrology*. Wiley, Chichester, pp. 527–570.

Battarbee, R.W. (1986b) The eutrophication of Lough Erne inferred from changes in the diatom assemblages of ^{210}Pb- and ^{137}Cs-dated sediment cores. *Proceedings of the Royal Irish Academy* 86 (Section B), 141–168.

Battarbee, R.W. (1991) Recent paleolimnology and diatom-based environmental reconstruction. In: Shane, L.C.K. and Cushing, E.J. (eds) *Quaternary Landscapes*. University of Minnesota Press, Minneapolis, pp. 129–174.

Bengtsson, L. and Persson, T. (1978) Sediment changes in a lake used for sewage reception. *Polskie Archiv für Hydrobiologie* 25, 17–33.

Bennion, H. (1994) A diatom–phosphorus transfer-function for shallow, eutrophic ponds in southeast England. *Hydrobiologia* 275/276, 391–410.

Bennion, H. (1995) A summary of diatom–phosphorus training sets for the ponds of south-east England, and the meres of Cheshire and Shropshire, north-west

England. In: Patrick, S.T. and Anderson, N.J. (eds) *Ecology and Palaeoecology of Lake Eutrophication.* Service Report 7, Geological Survey of Denmark, Copenhagen, pp. 33–35.

Bennion, H., Wunsam, S. and Schmidt, R. (1995) The validation of diatom–phosphorus transfer-functions: an example from Mondsee, Austria. *Freshwater Biology* 34, 271–283.

Bennion, H., Juggins, S. and Anderson, N.J. (1996a) Predicting epilimnetic phosphorus concentrations using an improved diatom-based transfer function, and its application to lake eutrophication management. *Environmental Science and Technology* 30, 2004–2007.

Bennion, H., Duigan, C.A., Haworth, E.Y., Allott, T.E.H., Anderson, N.J., Juggins, S. and Monteith, D.T. (1996b) The Anglesey Lakes, Wales, U.K.: changes in trophic status of three standing waters as inferred from diatom transfer functions and their implications for conservation. *Aquatic Conservation: Marine and Freshwater Ecosystems* 6, 81–92.

Birks, H.J.B. (1993) Quaternary palaeoecology and vegetation science – current contributions and possible future developments. *Review of Palaeobotany and Palynology* 79, 153–177.

Birks, H.J.B., Line, J.M., Juggins, S., Stevenson, A.C. and ter Braak, C.J.F. (1990) Diatoms and pH reconstruction. *Philosophical Transactions of the Royal Society London, Series B* 327, 263–278.

Birks, H.J.B., Anderson, N.J. and Fritz, S.C. (1995) Post-glacial changes in total phosphorus at Diss Mere, Norfolk inferred from fossil diatom assemblages. In: Patrick, S.T. and Anderson, N.J. (eds) *Ecology and Palaeoecology of Lake Eutrophication.* Service Report 7, Geological Survey of Denmark, Copenhagen, pp. 48–49.

Bradbury, J.P. (1975) Diatom stratigraphy and human settlement in Minnesota. *Geological Society of America Special Paper* 171, 1–74.

Brenner, M., Whitmore, T.J., Flannery, M.S. and Binford, M.W. (1993) Paleolimnological methods for defining target conditions in lake restoration: Florida case studies. *Lake and Reservoir Management* 7, 209–217.

Brugam, R.B. (1988) Long-term history of eutrophication in Washington lakes. *Aquatic Toxicology and Hazard Assessment* 10, 63–70.

Brugam, R.B. and Speziale, B.J. (1983) Human disturbance and the paleolimnological record of change in the zooplankton community of Lake Harriet, Minnesota. *Ecology* 64, 578–591.

Carignan, R. and Flett, R.J. (1981) Post depositional mobility of phosphorus in lake sediments. *Limnology and Oceanography* 26, 361–366.

Charles, D.F., Dixit, S.S., Cumming, B.F. and Smol, J.P. (1991) Variability in diatom and chrysophyte assemblages and inferred pH: paleolimnological studies of Big Moose Lake, New York, USA. *Journal of Paleolimnology* 5, 267–284.

Christie, C.E. and Smol, J.P. (1993) Diatom assemblages as indicators of lake trophic status in southeastern Ontario lakes. *Journal of Phycology* 29, 575–586.

Cullen, P. and Forsberg, C. (1988) Experiences with reducing point sources of phosphorus to lakes. *Hydrobiologia* 170, 321–336.

Cumming, B.F., Smol, J.P., Kingston, J.C., Charles, D.F., Birks, H.J.B., Camburn, K.E., Dixit, S.S., Uutala, A.J. and Selle, A.R. (1992) How much acidification has occurred in Adirondack lakes (New York, USA) since pre-industrial times? *Canadian Journal of Fisheries and Aquatic Sciences* 49, 128–141.

Dillon, P.J. and Kirchner, W.B. (1975) The effects of geology and land-use on the

export of phosphorus from watersheds. *Water Research* 9, 135–148.
Dixit, S.S. and Smol, J.P. (1994) Diatoms as indicators in the environmental monitoring and assessment program-surface waters (EMAP-SW). *Environmental Monitoring and Assessment* 31, 275–306.
Engstrom, D.R. and Wright, Jr, H.E. (1984) Chemical stratigraphy of lake sediments as a record of environmental change. In: Lund, J.W.G. and Haworth, E.Y. (eds) *Lake Sediments and Environmental History*. University of Leicester Press, Leicester, pp. 11–67
Evans, R.D. and Rigler, F.H. (1980) Measurement of whole lake sediment accumulation and phosphorus retention using lead-210 dating. *Canadian Journal of Fisheries and Aquatic Science* 37, 817–822.
Ford, J. (1988) The effects of chemical stress on aquatic species composition and community structure. In: Lewin, S.A., Harwell, M.A., Kelly, J.R. and Kimball, K.D. (eds) *Ecotoxicology: Problems and Approaches*. Springer-Verlag, New York, pp. 99–114.
Foy, R.H. and Withers, P.J.A. (1995) The contribution of agricultural phosphorus to eutrophication. *Proceedings of the Fertilizer Society* 365, 1–32.
Foy, R.H., Smith, R.V., Jordan, C. and Lennox, S.D. (1995) Upward trend in soluble phosphorus loadings to Lough Neagh despite phosphorus reduction at sewage treatment works. *Water Research* 29, 1051–1063.
Fritz, S.C. (1989) Lake development and limnological response to prehistoric and historic land-use in Diss, Norfolk, U.K. *Journal of Ecology* 77, 182–202.
Fritz, S.C., Kingston, J.C. and Engstrom, D.R. (1993) Quantitative trophic reconstruction from sedimentary diatom assemblages: a cautionary tale. *Freshwater Biology* 30, 1–23.
Gaillard, M.-J., Dearing, J.A., El-Daoushy F., Enell, M. and Håkansson, H. (1991) A late Holocene record of land-use history, soil erosion, lake trophy and lake-level fluctuations at Bjäresjösjön (South Sweden). *Journal of Paleolimnology* 6, 51–81.
Gibson, C.E., Wu, Y., Smith, S.J. and Murphy-Wolfe, S.A. (1995) Synoptic limnology of a diverse geological region: catchment and water chemistry. *Hydrobiologia* 306, 213–277.
Gibson, C.E., Bailey-Watts, A.E. and Foy, R.H. (1996) An analysis of the total phosphorus cycle in some temperate lakes: the response to enrichment. *Freshwater Biology* 35, 525–532.
Hall, R.I. and Smol, J.P. (1992) A weighted-averaging regression and calibration model for inferring total phosphorus concentration from diatoms in British Columbia (Canada) lakes. *Freshwater Biology* 27, 417–434.
Hall, R.I. and Smol, J.P. (1996) Paleolimnological assessment of long-term water quality changes in south-central Ontario lakes affected by cottage development and acidification. *Canadian Journal of Fisheries and Aquatic Sciences* 53, 1–17.
Haworth, E.Y. (1972) The recent diatom history of Loch Leven, Kinross. *Freshwater Biology* 2, 131–141.
Heathwaite, A.L. (1994) Chemical fractionation of lake sediments to determine the effects of land-use change on nutrient loading. *Journal of Hydrology* 159, 395–421.
Holtan, H., Kamp-Nielsen, L. and Stuanes, A.O. (1988) Phosphorus in soil, water and sediment: an overview. *Hydrobiologia* 170, 19–34.
Jeppesen, E., Kristensen, P., Jensen, J.P., Søndergaard, M., Mortensen, E. and Lauridsen, T. (1991) Recovery resilience following a reduction in external

phosphorus loading of shallow, eutrophic Danish lakes: duration, regulating factors and methods for overcoming resilience. *Memorie dell' Istituto Italiano di Idrobiologia* 48, 127–148.

Krug, A. (1993) Drainage history and land use pattern of a Swedish river system – their importance for understanding nitrogen and phosphorus load. *Hydrobiologia* 251, 285–296.

Line, J.M. and Birks, H.J.B. (1990) WACALIB version 2.1 – a computer program to reconstruct environmental variables from fossil assemblages by weighted averaging. *Journal of Paleolimnology* 3, 170–173.

Line, J.M., ter Braak, C.J.F. and Birks, H.J.B. (1994) WACALIB version 3.3 – a computer program to reconstruct environmental variables from fossil assemblages by weighted averaging and to derive sample-specific errors of prediction. *Journal of Paleolimnology* 10, 147–152.

Marsden, M.W. (1989) Lake restoration by reducing external phosphorus loading: the influence of sediment phosphorus release. *Freshwater Biology* 21, 139–162.

Moss B. (1980) Further studies on the palaeolimnology and changes in the phosphorus budget of Barton Broad, Norfolk. *Freshwater Biology* 10, 261–279.

OECD (1982) *Eutrophication of Waters, Monitoring Assessment, Control.* OECD, Paris, 154 pp.

Oldfield, F. and Appleby, P.G. (1984) Empirical testing of ^{210}Pb-dating models for lake sediments. In: Haworth, E.Y. and Lund, J.W.G. (eds) *Lake Sediments and Environmental History.* Leicester University Press, Leicester, pp. 93–124.

O'Sullivan, P.E. (1992) The eutrophication of shallow coastal lakes in southwest England – understanding and recommendations for restoration, based on palaeolimnology, historical records, and the modelling of changing phosphorus loads. *Hydrobiologia* 243/244, 421–434.

Reavie, E., Hall, R.I. and Smol, J.P. (1995a) An expanded weighted-averaging model for inferring past total phosphorus concentrations from diatom assemblages in eutrophic British Columbia (Canada) lakes. *Journal of Paleolimnology* 14, 49–67.

Reavie, E., Smol, J.P. and Carmichael, N.B. (1995b) Post-settlement eutrophication histories of six British Columbia (Canada) lakes. *Canadian Journal of Fisheries and Aquatic Science* 52, 2388–2401.

Reynolds, C.S. (1984) *The Ecology of Freshwater Phytoplankton.* Cambridge University Press, Cambridge.

Rippey, B. and Anderson, N.J. (1996) The reconstruction of lake phosphorus loading and dynamics using the sedimentary record. *Environmental Science and Technology* 30, 1786–1788.

Rippey, B.H., Anderson, N.J. and Foy, R.H. (1997) Accuracy of diatom-inferred total phosphorus concentrations and the accelerated eutrophication of a lake due to reduced flushing. *Canadian Journal of Fisheries and Aquatic Science* (in press).

Round, F.E. (1981) *The Ecology of the Algae.* Cambridge University Press, Cambridge.

Sas, H. (ed.) (1989) *Lake Restoration by Reduction of Nutrient Loading: Expectations, Experiences, Extrapolations.* Academia Verlag, St Augustin, 497 + XXL pp.

Schelske C.L., Robbins, J.A., Gardner, W.S., Conley, D.J. and Bourbonniere, R.A. (1988) Sediment record of biogeochemical responses to anthropogenic perturbations of nutrient cycles in Lake Ontario. *Canadian Journal of Fisheries and Aquatic Sciences* 45, 1291–1303.

Schoonmaker, P.K. and Foster, D.R. (1991) Some implications of paleoecology for contemporary ecology. *Botanical Review* 57, 204–245.

Sharpley, A.N., Chapra, S.C., Wedephol, R., Sims, J.T., Daniel, T.C. and Reddy, K.R. (1994) Managing agricultural phosphorus for protection of surface waters: issues and options. *Journal of Environmental Quality* 23, 437–451.

Smol, J.P. (1992) Paleolimnology: an important tool for effective ecosystem management. *Journal of Ecosystem Health* 1, 49–58.

Stoermer, E.F. (1984) Qualitative characteristics of phytoplankton assemblages. In: Shubert, L.E. (ed.) *Algae as Ecological Indicators.* Academic Press, London, pp. 49–67.

Talling, J.F. and Heaney, S.I. (1988) Long-term changes in some English (Cumbrian) lakes subjected to increased nutrient inputs. In: Round, F.E. (ed.) *Algae and the Aquatic Environment.* Biopress, Bristol, pp. 1–29.

ter Braak, C.J.F. and Juggins, S. (1993) Weighted averaging partial least squares regression (WA-PLS): an improved method for reconstructing environmental variables from species assemblages. *Hydrobiologia* 269/270, 483–502.

ter Braak, C.J.F. and Looman, W.N. (1986) Weighted averaging, logistics regression and the Gaussian response model. *Vegetatio* 65, 3–11.

ter Braak, C.J.F. and Prentice, I.C. (1988) A theory of gradient analysis. *Advances in Ecological Research* 18, 271–317.

ter Braak, C.J.F. and van Dam, H. (1989) Inferring pH from diatoms: a comparison of old and new calibration methods. *Hydrobiologia* 178, 209–223.

Tilman, D., Kilham, S.S. and Kilham, P. (1982) Phytoplankton community ecology: the role of limiting nutrients. *Annual Review of Ecology and Systematics* 13, 349–372.

Whitmore, T.J. (1989) Florida diatom assemblages as indicators of trophic state and pH. *Limnology and Oceanography* 34, 882–895.

Willén, E. (1991) Planktonic diatoms – an ecological overview. *Algological Studies* 62, 69–106.

Williams J.D.H., Murphy, T.P. and Mayer, T. (1976) Rates of accumulation of phosphorus forms in Lake Erie sediments. *Journal of the Fisheries Research Board of Canada* 33, 430–439.

Wunsam, S. and Schmidt, R. (1995) A diatom–phosphorus transfer function for Alpine and pre-alpine lakes. *Memorie dell'Istituto di Idrobiologia* 53, 85–99.

5 The Dynamics of Phosphorus in Freshwater and Marine Environments

C.E. Gibson
*Department of Agriculture for Northern Ireland,
Agricultural and Environmental Science Division,
Newforge Lane, Belfast BT9 5PX, UK*

INTRODUCTION

Phosphorus (P) is a potent agent of change in aquatic ecosystems. Although its name derives from the Greek 'bearer of light', the discussion on its effects and sources in the environment has often been heated and acrimonious, rather than illuminating. In the late 1960s and early 1970s an intense debate took place about the enrichment of fresh water, the appearance of dense algal scums and the factors responsible for promoting the fertility (perhaps overfertility) which was observed worldwide. This debate, although occasioned by an essentially applied environmental problem, stimulated much fundamental science, and significant symposia were held to air the topic (Rohlich, 1969; Likens, 1972). The enrichment of lakes was considered by some to be an ecological disaster equivalent to the dust bowl of the American prairies (Vallentine, 1974). Interventive experimentation with whole lakes (Schindler *et al.*, 1971; Schindler and Fee, 1974) showed that, in some cases at least, the addition of P alone was sufficient to stimulate cyanobacterial blooms. Such striking ocular evidence, skilfully presented and backed up by numerous other, less dramatic studies, effectively ended the debate over the role of P in lake enrichment. Remedial measures were instituted in many lake catchments, often with striking success, but sometimes without result (see, *inter alia*, Cullen and Forsberg, 1988; Sas, 1989, for reviews). In general, the enrichment debate dwindled to be overtaken by new environmental concerns, such as acid rain and, latterly, global warming. Nevertheless, the problem of enrichment has not gone away and scientific papers on P in the aquatic environment continue to appear at a steady rate. The key words 'phosphorus' and 'freshwater' produced 1179 hits on a reference database originating in 1978, of which 629 have appeared since 1988. The P problem is still with us and is now assuming a renewed prominence. Nevertheless, the debate has moved on.

It is now axiomatic that a lake and its catchment are one unit. Early remedial measures focused on easily achievable goals, and point sources of P were both exactly quantifiable and relatively easily managed, at least technically. Now, attention is turning to the subtler and less tractable problem of diffuse sources of P. To manage P in the environment effectively, it is important to understand its main pathways through the ecosystem; the aim of this chapter is to produce a framework against which to assess the problem and its possible solution. The focus of this chapter is mainly on the importance of P in standing fresh water, which is in itself a very large topic. Some aspects of P in the marine field are also addressed briefly, although P generally plays a less crucial role in the enrichment of the sea than in fresh waters. Several extensive reviews have appeared in recent years, covering most aspects of the enrichment of fresh waters (Persson and Janson, 1988; Harper, 1992; Van Liere and Gulati, 1992; Mortensen *et al.*, 1994), and, taken together, they provide a well-balanced introduction to the topic.

WHY PHOSPHORUS?

There are several reasons why P is of particular significance in freshwater ecology. Firstly, it is commonly thought to be limiting for the production of biomass. 'Limiting' here means that P is the resource in shortest supply, so that increases or decreases in P have a direct effect on the phytoplankton biomass. There have been numerous data sets published (e.g. Sakamoto, 1966; Dillon and Rigler, 1974; Straskraba, 1980; Smith, 1982; Ahlgren *et al.*, 1988; Prairie *et al.*, 1989) modelling the relationship between total P and phytoplankton biomass, measured by chlorophyll *a*. It is not wise to claim too much for the predictive power of models based on a univariate relationship between total P and chlorophyll *a*. One of the obvious difficulties is that neither total P nor chlorophyll is an unambiguous measure of either the resource or the ecological response. Phosphorus may be held by non-algal particulates to a variable extent, whereas the implicit assumption of the models is that, at maximum biomass, all the P is contained in the phytoplankton. Furthermore, the chlorophyll *a* and P contents of phytoplankton are quite variable, so that the apparently good fit of a log:log relationship between chlorophyll *a* and total P should be seen for what it is – a useful generalization. There is agreement that, as total P in the water increases, the maximum biomass rises. Below 100 µg P l^{-1}, the relationship tends to be linear; but additional P can only stimulate additional biomass production within limits and, when total P increases above 100 μg P l^{-1}, other factors, principally light availability, become increasingly important. Figure 5.1 shows the relationship between total P and biomass suggested by Prairie *et al.* (1989) and plots on it some Irish lakes (Foy and Withers, 1995). In general, the lakes obey the proposed relationship, with the obvious exception of Lough Erne, where the maximum biomass allowed by the P resource is never realized. This is because Lough Erne water is deep, well mixed and peat-stained, so that light limitation overrides P limitation in the open waters. A further point of interest is that, although the Lough Neagh data fall on the predicted region of the curve, there is considerable scatter and the points lie close to the point of inflection at 100 μg P l^{-1} and, in fact, P limitation only occurs for a brief period at the end of the spring bloom. Unlike Lough Erne, where light limitation is caused by the properties of the lake water, the main cause of light limitation in greatly enriched lakes is the phytoplankton itself, so-called 'self-shading'.

A second reason for the concentration on P is that so little is required for biomass production. Redfield (1934) first pointed out that marine biomass tends to a constant atomic ratio of 105:15:1 carbon (C):nitrogen (N):P, and this was extended to fresh waters by Uhlmann and Albrecht (1968). The implication of this fact is that one atom of P can support the production of as much biomass as 15 atoms of N or 105 atoms of C. Hence, adding P has a much greater effect than adding either N or C, even assuming that they were limiting. Conversely, controlling the P supply is

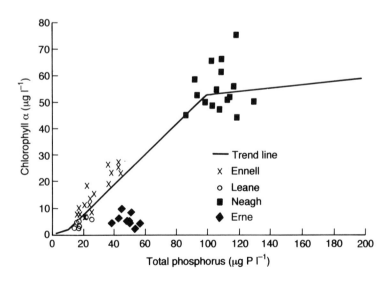

Fig. 5.1. Lake total phosphorus vs. lake chlorophyll *a* (annual mean values) for Irish lakes. The fitted line is according to the model of Prairie *et al.* (1989) and does not include the Irish data. Reproduced from Foy and Withers (1995).

the most effective and generally the most achievable management option. Figure. 5.1 shows that in the model 100 µg P l^{-1} is predicted to support an annual mean chlorophyll *a* of 50 µg l^{-1}, which implies a much larger maximum value, since the biomass may vary several-fold from maximum to minimum through the year. On average, 1 µg P supports 1 g chlorophyll *a*. To grasp the scale of the relationship, it is useful to consider the effects on the clarity of the water. According to published equations (Sas, 1989), in a clear-water lake, an increase in chlorophyll from 1 µg l^{-1} to 10 µg l^{-1} reduces the water clarity from approximately 9 m to 3 m, as measured by a Secchi disc. In other words, 10 µg P l^{-1} is sufficient to make the water noticeably cloudy. To scale the quantity of P involved, 10 µg P l^{-1} is equivalent to only 5.0 kg P in a 10-hectare lake of average depth 5 m. Further increases in P to greater than 50 µg P l^{-1} tend to produce many consequent problems, such as deoxygenation of the bottom water, loss of amenity value and greatly increased treatment costs for potable water supplies.

The tendency for enrichment to increase the incidence of cyanobacterial blooms, sometimes toxic, provides a compelling third motive for the focus on P. Why P enrichment favours cyanobacteria is still not entirely resolved, and there has been no definitive study that defines the exact point on the enrichment gradient at which blooms may be expected. Many of the most objectionable bloom-forming genera, such as *Anabaena* and *Aphanizomenon*, are capable of fixing atmospheric N to satisfy their N requirements and hence are able to avoid N limitation. Increasing the P supply makes N limitation more likely for species incapable of N fixation and so gives some nuisance blooms a competitive advantage.

CYCLES OF PHOSPHORUS IN LAKES

Input–output budgets of P in lakes, although they do little to explain the mechanisms involved, are useful tools to evaluate the main pathways of P through a lake ecosystem. The evaluation of input, standing stock, sedimentation and loss down the outflow form the framework of such budgets. Each of the compartments can be subdivided to quantify the importance of different inputs, the location of P in the lake itself, the interaction between the lake water and the bottom sediment and the overall retention of P by the sediment. The following account of the P cycle in lakes will be structured in that way.

Sources of Phosphorus

Natural sources
Some P originates naturally by rock weathering, and Dillon and Kirchner (1975) showed that, in virgin Canadian catchments, the export of P was greater from catchments on sedimentary than from those on igneous rocks. In these virtually uninhabited regions, geology can play an important part in determining the P economy. However, even low-intensity agriculture and small-scale settlements disturb this primeval state. Schelske et al. (1985) found evidence in a core from the bottom sediment of Lake Ontario that an increase in P accumulation rate was already evident in 1669, coinciding with the earliest European colonization.

Atmospheric inputs
In some situations, atmospheric inputs of P may be significant. Figure 5.2 shows the range of atmospheric P inputs to be found in the literature, which varies from 5 to over 100 kg P km^{-2} year^{-1}. It is likely that humans have a direct influence on the magnitude of atmospheric inputs of P, and the lowest values are found in areas remote from civilization. There is evidence that the rate of deposition has increased historically; for example, a study of a Swedish raised bog (Ahl, 1988) suggests that, in that area, the atmospheric P deposition was 2.1 kg P km^{-2} year^{-1} in AD 900, 2.7 kg P km^{-2} year^{-1} in AD 1400 and has increased to 13.3 kg P km^{-2} year^{-1} at the present time. A slightly greater figure is currently found in Ireland. Jordan (1987) found an average value of 19 kg P km^{-2} year^{-1} in three rural sites in Northern Ireland, and Gibson and Wu (1995) measured an atmospheric input of 22 kg P km^{-2} year^{-1} at Loughgarve, a relatively remote site on the Antrim plateau. The importance of this input depends on the ratio of lake surface to catchment area, the retention of P in the catchment and the scale of other P sources. In an upland catchment, such as Loughgarve, P may be almost entirely supplied by rainfall. In that lake, the ratio of land catchment to water surface is 4.5:1 and the catchment is acidic peat, uninhabited and uncultivated except for very light grazing by sheep. As a result, very little P was retained by the soil in the catchment and the entire P budget could be accounted for by the atmospheric deposition. At the other end of the scale, atmospheric P inputs to Lough Neagh are trivial. Direct atmospheric inputs to the lake surface account for only 1.5% of the total P budget and the deposition on the rest of the catchment is swamped by the influence of humans – the mean export of P from the Lough Neagh catchment is approximately 110 kg P km^{-2} year^{-1}.

Anthropogenic sources
In many lakes, anthropogenic sources dominate the P budget. There have been numerous reviews of P loss rates from catchments (Vollenweider, 1968; Weibel, 1970; Gächter and Furrer, 1972; Dillon and Kirchner, 1975;

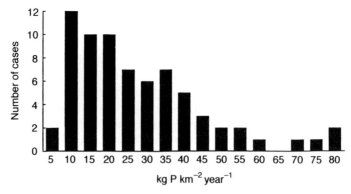

Fig. 5.2. Reported values for precipitation inputs of phosphorus in the literature. Based on Gibson and Wu (1995), with some additional data. Values are categorized by the upper boundary; hence 5 equals 0 to 5, etc.

Smith, 1977), which vary widely depending on population density and agricultural intensity. The assessment of P loading from urban centres is potentially straightforward, even if it cannot be measured directly. Human populations can be weighted by a contemporary per capita P contribution and due allowance made for possible industrial contributions. There are more uncertainties in assessing the rural human contribution, because, unlike the urban population, the pathway from the point of discharge to the water body is undefined and there are several mechanisms (particle adsorption, precipitation, plant uptake) by which P can be removed. It is nevertheless possible to put an upper bound on the P loading from the rural population by calculating the number of people, weighted by a likely per capita P contribution. This first estimate, seen to be too great, may nevertheless be a useful figure for management purposes to indicate whether or not the rural human population is a significant contributor to the P loading. To attempt a similar approach for agricultural losses is much more difficult. The problem is that the potential P losses from agriculture are very large, whereas the actual losses, unlike the human contribution, are a small fraction of that potential. A per capita calculation of the P contribution from livestock, for example, will often completely dominate the P budget for a catchment, whereas this may not actually be the case for the realized contributions from the two sources. It is difficult to appraise the per capita approach to agriculture, because a small difference in the assumed fractional loss will make a very large difference to the calculated agricultural loss. It has been calculated (Foy and Withers, 1995) that 1000 kg P km^{-2} year^{-1} accumulates in the soils of Northern Ireland because of the mismatch between application and harvest of P. This has led to a progressive increase in the diffuse P loss in the Lough Neagh catchment. On average, the P entering the lake from this source is increasing by 4.5

tonnes every year. While this is of considerable significance to the P economy of the lake, it accounts for only 0.5% of the annual agricultural P surplus in the catchment.

Nevertheless, models based on animal populations and crop type have been developed to predict nutrient losses from farmland (e.g. Johnes and O'Sullivan, 1989), and these can be helpful in a historical context to estimate the magnitude of P losses in the recent past. Because one is trying to estimate a small loss from a large input and because the relationship between P loss and P application rate is variable, such an approach may be open to significant and unquantifiable errors.

Location of Phosphorus in the Lake

The phytoplankton are only one part of the complex web of organisms that participate in the functioning of lake ecosystems. As the primary producers, they are responsible for much of the C input to the lake, by photosynthesis, and the demonstrated relationship between P and phytoplankton biomass (e.g. as in Fig. 5.1) implies that much of the lake P is in the phytoplankton and this is probably generally true. Figure 5.3 shows the annual cycle of dissolved P and phytoplankton in Lough Neagh. During the winter, most of the P is dissolved, but, during the spring, the increased light drives photosynthesis, which exerts a P demand greater than the rate of supply, so that the soluble P concentration drops as the particulate P (phytoplankton) increases. This is the common pattern.

There have been few published studies of complete P inventories in lakes, but it is likely that, in many lakes, soluble and phytoplankton P account for most of the P in the water column. In very shallow lakes, however, this may not be so. In a Dutch lake formed as a result of peat

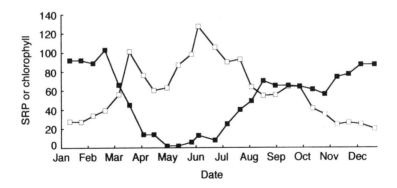

Fig. 5.3. The annual cycle of phosphorus (filled symbols) and phytoplankton chlorophyll *a* (empty symbols) concentrations in Lough Neagh, Northern Ireland, 1994. Both variables are µg l^{-1}. Values are categorized by the upper boundary; hence 5 means 0 to 5, etc.

cutting in the Middle Ages (Van Liere and Janse, 1992), the amount of P in the fish populations was equal to that in the phytoplankton, and in other cases zooplankton (Salonen *et al.*, 1994) or rooted macrophyte vegetation (Adams and Prentki, 1982) may also contain a high proportion of the total P stock. Finally, by far the largest quantity of P is likely always to be contained in the bottom sediment. In all lakes, at least some P will have been accumulating on the bottom since the formation of the lake, so that the total quantity held there may be very large. In the case of Lough Neagh, for example, approximately 30% of the annual P input is retained in the lake sediment. Hence, in the last 20 years alone, the sediment has accumulated 3360 tonnes of P, compared with an average annual input of 489 tonnes. The interaction between the sediment and the lake water can therefore have an important influence on the P cycle in any lake.

Phosphorus and Lake Sediments

The material that settles on the lake bed is a mixture of in-washed (allochthonous) inorganic and organic particulates and the remains of organisms, predominantly phytoplankton, which have grown in the lake (autochthonous). With low P loadings and low productivity, the sediments generally contain less than 1 mg P g^{-1} dry sediment, but, as the P loading increases, so does autochthonous sedimentation and with it both the degradable C and P contents of the surface sediment. The increased C input creates a greater oxygen demand, hence generally promoting chemically reducing conditions in the sediment and this, combined with the greater P content, causes a progressive tendency for more P to be released from the mud. A large survey of lakes in which the P inputs had been reduced (Sas, 1989) suggested that there was a threshold at approximately 1 mg P g^{-1} dry sediment below which sediments tended not to release P back into the water column.

Much of the feedback mechanism between sediment and water P is chemically driven – either by the oxidation–reduction state of the iron–P complexes, or in calcareous sediments by the pH/calcium/P reactions. The disturbance of the bottom by animals may also be important, by turning over sediment and resuspending it in the water column or by irrigating deeper layers with water currents. Bottom-feeding fish (Lamarra, 1975) or chironomid midge larvae (Gallepp, 1979) can be significant factors in this regard. Similarly, the growth of rooted macrophytes ('pondweed') may act as a nutrient pump, taking up P from the sediment and releasing it to the water column, primarily when the plants die back in the autumn (Granéli and Solander,1988). In a shallow Norfolk Broad (Moss *et al.*, 1986), isolation of the lake from its inflows at first restricted the nutrient input and the sediment P release. However, greater water clarity due to smaller phytoplankton crops allowed rooted macrophytes to become re-established and P release began again, perhaps due to the input of organic matter from the plant remains. This illustrates the importance of the hydrological cycle: a lake with little hydrological flushing is likely to accumulate nutrients and the regime promotes closed nutrient cycles in which sediment–water fluxes dominate the annual budget.

A more straightforward example is given by Lough Ennel, County Westmeath (Foy *et al.*, 1996), where the P budget was dominated by the input from Mullingar sewage works. There was a close connection between the input of P, the turbidity of the water due to phytoplankton and the success of the *Chara* (stonewort) beds for which the lake is noted. Changes in the management regime of the sewage works, which reduced the P loading in two phases, were closely tracked by the lake response (Fig. 5.4). Increases in P loading brought about cloudier water and loss of *Chara*, but the relatively rapid water-flushing rate and the fact that the sediment played little role in the P cycle meant that reduction of the P input was quickly followed by an improvement in water quality.

Some Case-Study Lakes

Northern Ireland offers an excellent natural laboratory for observing different P economies, because it possesses a wide range of lake types with different catchment geology, hydrology and P loadings (Gibson *et al.*, 1995). Contrasting examples, some already referred to, are discussed below and summarized in Table 5.1. Loughgarve (Gibson and Wu, 1995) is a shallow upland lake on a peaty catchment whose P economy relies entirely on atmospheric inputs. Water turnover time is rapid and measurable

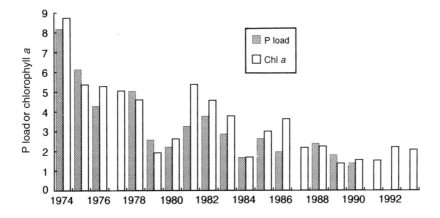

Fig. 5.4. Mullingar sewage phosphorus loading (t year^{-1}) and Lough Ennell annual mean chlorophyll *a* concentrations (μg l^{-1}). For display purposes, the chlorophyll *a* values have been divided by 5. Data are from Foy *et al.* (1995).

quantities of P are not retained in the catchment. Lough Erne (Hayward *et al.*, 1993) is a very much larger lake, where atmospheric inputs are not important. Water is exchanged about every 6 months, but about 35% of the P input settles to the lake bed, so that large quantities of P have accumulated there. However, due to the well-oxygenated water column, no P is released from the mud. In Lough Neagh (Gibson *et al.*, 1988), much of the P loading is sedimented to the bottom by the spring diatom bloom. Although the lake is generally well mixed and there is no sustained deoxygenation of a bottom water layer, there is nevertheless a rapid release of P to the water column, almost equivalent to the total annual catchment input. During the autumn, equilibration occurs and a percentage of the released P re-enters the sediment, but the net release is still sufficient to significantly increase the P concentration in the lake water. Finally, White Lough (Foy, 1985) is a small lake set in rural County Tyrone; the water column stratifies strongly in summer and there is complete deoxygenation of the bottom water, accompanied by a massive release of P. In 1979 (Table 5.1), the release of P from the sediment was greater than the catchment input and the lake was actually exporting P, but in 1982, in spite of P release from the sediment, the lake retained 28% of the load. It is likely (Rippey and Anderson, 1996) that, at the time of the study, the lake was in a state of change, and palaeolimnological evidence suggests that P concentration in the lake had been much greater in the years immediately prior to the study and the lake was re-equilibrating with its external load. White Lough contrasts strongly with Lough Neagh, because of its stratified water column. In White Lough, although P was released from the sediment, it was largely

Table 5.1. Phosphorus budgets and some characteristics of some contrasting Northern Irish lakes.

	Loughgarve	Lower Erne	Neagh	White 1979	White 1982
Area (km^2)	0.03	109.5	387	0.074	0.074
Catchment area (km^2)	0.16	4212	4453	1.076	1.076
Catchment/lake area	5.5	38.5	11.5	14.5	14.5
Mean depth (m)	0.55	11.9	8.9	6.1	6.1
Flushing time (year^{-1})	0.07	0.51	1.24	0.73	0.73
P budget (tonnes P year^{-1})					
Atmospheric	0.003	1.6	7.0	0.001	0.001
Catchment	0.0	213	460	0.04	0.034
Lake content	0.0002	53	397	0.06	0.04
Sedimentation	0	75	148	−0.044	0.010
Sedimentation %	0	35	32	−105	32
Outflow	0.003	140	312	0.085	0.024

held in the bottom water during the summer and only became available to the lighted upper layers by gradual entrainment through the thermocline – the region of temperature change. When the surface water cooled sufficiently to allow top-to-bottom mixing, the P was distributed throughout the lake.

PHOSPHORUS IN RIVERS

The ecology of rivers is greatly affected by the flow regime. In many rivers, the residence time of the water is too short to allow phytoplankton to develop, but in long lowland reaches or where the water is impounded for reservoirs, a distinctive phytoplankton may develop. Such rivers behave in a very similar fashion to lakes and the response to P is the same. Hence, if the residence time of the water is sufficiently long, enrichment with P may produce water blooms. Such conditions are relatively uncommon in the British Isles. Perhaps a more pressing concern is the effect of enrichment on smaller, more torrential rivers. While there is much anecdotal evidence (Whitton, 1975) to suggest that enrichment causes greater growth of weed – either algae such as *Cladophora* (blanket weed), or flowering plants, e.g. *Ranunculus* (water crowfoot) – there are rather few studies (e.g. Wharfe *et al.*, 1984; Dodds, 1991) which directly relate to the factors responsible for the perceived increase in weed growth. One problem is that baseline studies do not always exist to describe the situation before enrichment. A model study of the effect of P enrichment on a tundra stream (Peterson *et al.*, 1993) suggested that the effects were quite complex. Although initially there was 'bottom-up' stimulation of the riverine food web, based on greater autotrophic production, in later years there was a strong 'top-down' feedback of insects grazing on the epilithic algae. Further complications are the relative importance of the bottom sediment and the river water as sources of nutrient (Bushong and Bachmann, 1989), the difficulty of accounting for differences in flow regime, due to climatic variation or changes in drainage pattern and other changes in management practice. Low flows or the removal of shading bank-side vegetation may have as much influence on weed growth as the nutrient regime.

PHOSPHORUS IN THE MARINE ENVIRONMENT

Nutrient cycles in the sea are dominated by recycling in the upper layers and, in oceanic water, by the entrainment of nutrient-rich deep water (upwelling), which counterbalances losses from settlement of particles downwards. In the marine terminology, 'new' production (Eppley and Peterson, 1979) means nutrient gained from the vast store held in the deep water layers. This situation is quite unlike that in most fresh waters, except the very large deep lakes of the world (e.g. Baikal, Lake Tanganyika), where water turnover time is very long and the bottom water, as in the sea, contains a vast reserve of nutrients. However, in some coastal areas and in estuaries, land-based inputs may be a source of completely 'new' nutrients, as in most lake catchment systems. In these regions, local enrichment due to land runoff can be expected. Hecky and Kilham (1988) have made an extensive comparison of nutrient limitation in marine and fresh waters and have cast doubt on the general dogma (e.g. Ryther and Dunstan, 1971) that, in the sea, it is N rather than P that is generally limiting to phytoplankton production. The situation in the open oceans is difficult to assess, because the ecosystem is finely balanced and nutrients are rapidly recycled. This fuelled the recent 'iron-limitation' controversy (Chisholm and Morel, 1991). There are areas of the open ocean where there are apparently excess nutrients, but here it may be zooplankton grazing that limits the attainment of maximum biomass (Cullen, 1991). In coastal situations, there is a clear annual cycle of uptake and regeneration. In the Irish Sea, there is little evidence of enrichment by P, except for the area off the Mersey (Foster, 1984) and in other much more localized areas in sea loughs or near the mouths of large rivers (Gillooly et al., 1992). A recent unpublished synoptic survey by the Department of Agriculture for Northern Ireland shows that the phytoplankton in the open Irish Sea is unlikely to be P-limited. Nitrogen and silica are depleted before P and detectable concentrations of P remain throughout the year. This is prima facie evidence that P is not limiting.

There are some areas of Europe where P enrichment may play a role in stimulating phytoplankton production. For example, in the low-salinity environment of the Baltic Sea (Granéli et al., 1990), P may be a factor in promoting cyanobacterial blooms. It is also reported (Schaub and Gieskes, 1990) that the phytoplankton biomass in the North Sea near the Dutch coast is linked to the discharge from the Rhine. Flushing of nutrients from agricultural land in the Dutch part of the Rhine catchment after heavy rain counteracts dilution. However, Anon. (1993) suggests that at least at the Dutch–German border, river phosphate concentrations have declined markedly since the mid-1980s. Worldwide, what has been described as a 'global epidemic of novel phytoplankton blooms' (Smayda, 1990; see also Cosper et al., 1989) has been tentatively linked to P enrichment. It is suggested (Smayda, 1990) that P enrichment has lowered the silicon (Si):P ratio and favoured non-diatom blooms. In very shallow waters, so-called 'green tides' of macroalgae, e.g. *Cladophora* (Lapointe and O'Connell, 1989), may be stimulated by localized P enrichment.

CONCLUSION

This review has tried to put in context the importance of P in aquatic ecosystems, the different sources of P to the environment and its pathways through it. Lakes rarely grow richer in P without the intervention of humans. Lakes which had existed for thousands of years (e.g. Lough Neagh, Battarbee and Flower, 1993; Lake Ontario, Schelske *et al.*, 1985) showed no sign of enrichment until humans made a serious impact on the catchment. The driving force behind the P cycle is almost always the input from the catchment and the key to managing a lake is to manage the catchment. There are some instances where the intervention of the sediment P store has delayed the recovery of a managed lake, but this is not to be expected in most lakes. Unless the lake is shallow and hydrologically isolated, backflow of P from the sediment will eventually be flushed out of the lake if the incoming P concentration can be reduced sufficiently. This is not always easily achieved. If diffuse inputs are dominant, remedial measures may be difficult to institute and, because of the reservoir of soil P, the response may be slow (Lijklema, 1994), and this remains a challenge to be faced. Our scientific knowledge on the topic is still incomplete, and although it is clear that the more intensive the agriculture, the greater the leakage of P into the aquatic environment is likely to be, the mechanisms controlling the loss of P to the drainage water are imperfectly understood. Nevertheless, the ecological imperative to reduce P inputs to fresh waters remains clear.

ACKNOWLEDGEMENTS

I would like to acknowledge with thanks Dr Bob Foy for his careful reading of the final draft of this chapter and Mrs April McKinney for considerable help with checking the text and with the diagrams.

REFERENCES

Adams, M.S. and Prentki, R.T. (1982) Biology, metabolism and function of littoral submerged weedbeds of Lake Wingra, Wisconsin, USA: a summary and review. *Archiv für Hydrobiologie Supplement* 62, 333–409.

Ahl, T. (1988) Background yield of phosphorus from drainage area and atmosphere: an empirical approach. *Hydrobiologia* 170, 35–44.

Ahlgren, I., Frisk, T. and Kamp-Nielsen, L. (1988) Empirical and theoretical models of phosphorus loading, retention and concentration vs lake trophic state. *Hydrobiologia* 170, 285–303.

Anon. (1993) *Nutrients in the Convention Area*. Report of the Oslo and Paris Commissions, London, 86 pp.

Battarbee, R.W. and Flower, R.J. (1993) The recent sediments of Lough Neagh, Part

A: structure, stratigraphy and geochronology. In: Wood, R.B. and Smith, R.V. (eds) *Lough Neagh*. Kluwer, Dordrecht, pp. 113–132.

Bushong, S.J. and Bachmann, R.W. (1989) *In situ* nutrient enrichment experiments with periphyton in agricultural streams. *Hydrobiologia* 170, 1–10.

Chisholm, S.W. and Morel, F.M.M. (eds) (1991) What controls phytoplankton production in nutrient-rich areas of the open sea? *Limnology and Oceanography* 36, 1507–1965.

Cosper, E.M., Bricelj, V.M. and Carpenter, E.J. (eds) (1989) *Novel Phytoplankton Blooms*. Springer Verlag, Berlin, 460 pp.

Cullen, J.J. (1991) Hypotheses to explain high-nutrient conditions in the open sea. *Limnology and Oceanography* 36, 1578–1599.

Cullen, P. and Forsberg, C. (1988) Experiences with reducing point sources of phosphorus to lakes. *Hydrobiologia* 170, 321–336.

Dillon, P.J. and Kirchner, W.B. (1975) The effects of geology and land use on the export of phosphorus from watersheds. *Water Research* 9, 135–148.

Dillon, P.J. and Rigler, F.H. (1974) The phosphorus–chlorophyll relationship in lakes. *Limnology and Oceanography* 19, 767–773.

Dodds, W.K. (1991) Factors associated with the dominance of the filamentous green alga *Cladophora glomerata*. *Water Research* 25, 1325–1332.

Eppley, R.W. and Peterson, B.J. (1979) Particulate organic matter flux and planktonic new production in the deep ocean. *Nature* 282, 677–680.

Foster, P. (1984) Nutrient distribution in the winter regime of the northern Irish Sea. *Marine Environmental Research* 13, 81–95.

Foy, R.H. (1985) Phosphorus inactivation in a eutrophic lake by the direct addition of ferric aluminum sulphate: impact on iron and phosphorus. *Freshwater Biology* 15, 613–629.

Foy, R.H. and Withers, P.J.A. (1995) The contribution of agricultural phosphorus to eutrophication. *Proceedings of the Fertilizer Society* 365, 1–32.

Foy, R.H, Champ, W.S.T. and Gibson, C.E. (1996) The effectiveness of restricting phosphorus loadings from sewage treatment works as a means of controlling eutrophication in Irish lakes. In: Giller, P.S. and Myers, A.A. (eds) *Disturbance and Recovery in Ecological Systems*. Royal Irish Academy, Dublin, pp. 134–152.

Gächter, R. and Furrer, O.J. (1972) Der Beitrag der Landwirtschaft zur Eutrophierung der Gewasser in der Schweiz, I. Ergebnisse von direkten Messungen in Einzugsgebiet verschiedener Vorfluter. *Schweizerischer Zeitschrift für Hydrologie* 34, 41–70.

Gallepp, G.W. (1979) Chironomid influence on phosphorus release in sediment – water microcosms. *Ecology* 60, 547–566.

Gibson, C.E. and Wu, Y. (1995) Substance budgets of an upland catchment: the significance of atmospheric phosphorus inputs. *Freshwater Biology* 33, 385–392.

Gibson, C.E., Smith, R.V. and Stewart, D.A. (1988) A long term study of the phosphorus cycle in Lough Neagh, Northern Ireland. *Internationale Revue der gesamten Hydrobiologie* 73, 249–257.

Gibson, C.E., Wu, Y., Smith, S.J. and Wolfe-Murphy, S.A. (1995) Synoptic limnology of a diverse geological region: catchment and water chemistry. *Hydrobiologia* 306, 213–227.

Gillooly, M., O'Sullivan, G., Kirkwood, D. and Aminot, A. (1992) *The Establishment of a Database for Trend Monitoring of Nutrients in the Irish Sea*. Report of EC NORSAP Contract, B6618-89-03, 70 pp.

Granéli, W. and Solander, D. (1988) Influences of aquatic macrophytes on phosphorus cycling in lakes. *Hydrobiologia* 170, 245–266.

Granéli, E., Wallstroem, K., Larsson, U., Granéli, W. and Elmgren, R. (1990) Nutrient limitation of primary production in the Baltic Sea area. *Ambio* 19, 142–151.

Harper, D. (1992) *Eutrophication of Freshwaters: Principles, Problems and Restoration.* Chapman and Hall, London, 327 pp.

Hayward, J., Foy, R.H. and Gibson, C.E. (1993) Nitrogen and phosphorus budgets in the Erne system 1974–89. *Biology and Environment* 93B, 33–44.

Hecky, R.E. and Kilham, P. (1988) Nutrient limitation of phytoplankton in freshwater and marine environments: a review of recent evidence on the effects of enrichment. *Limnology and Oceanography* 33, 796–822.

Johnes, P.J. and O'Sullivan, P.E. (1989) The natural history of Slapton Ley Nature Reserve. XVIII Nitrogen and phosphorus losses from the catchment – an export coefficient approach. *Field Studies* 7, 285–309.

Jordan, C. (1987) Precipitation chemistry at rural sites in Northern Ireland. *Record of Agricultural Research (Department of Agriculture, Northern Ireland)* 35, 53–66.

Lamarra, V.A. (1975) Digestive activities of carp as a major contributor to the nutrient loading of lakes. *Verhandlungen der internationale Vereinigung für theoretische and angewandte Limnologie* 19, 2461–2468.

Lapointe, B.E. and O'Connell, J. (1989) Nutrient-enhanced growth of *Cladophora prolifera* in Harrington Sound, Bermuda: eutrophication of a confined phosphorus-limited marine ecosystem. *Estuarine, Coastal and Shelf Science* 28, 347–360.

Lijklema, L. (1994) Nutrient dynamics in shallow lakes: effects of changes in loading and role of sediment–water interactions. *Hydrobiologia* 275, 335–348.

Likens, G.E. (ed.) (1972) *Nutrients and Eutrophication: the Limiting Nutrient Controversy.* American Society of Limnology and Oceanography, Lawrence, Kansas, 328 pp.

Mortensen, E., Jeppesen, E., Søndergaard, M. and Kamp-Nielsen, L. (eds) (1994). *Nutrient Dynamics and Biological Structure in Shallow Freshwater and Brackish Lakes. Hydrobiologia* 275/276, 507 pp.

Moss, B., Balls, H., Irvine, K. and Stansfield, J. (1986) Restoration of two lowland lakes by isolation from nutrient-rich water sources with and without removal of sediment. *Journal of Applied Ecology* 23, 391–414.

Persson, G. and Jansson, M. (eds) 1988. *Phosphorus in Freshwater Ecosystems. Hydrobiologia* 170, 340 pp.

Peterson, B.J., Deegan, L., Helfrich, J., Hobbie, J.E., Hullar, M., Moller, B., Ford, T.E., Hershey, A. and Hilter, A. (1993). Biological responses of a tundra river to fertilization. *Ecology* 74, 653–672.

Prairie, Y.T., Duarte, C.M. and Kalff, J. (1989) Unifying nutrient chlorophyll relationship in lakes. *Canadian Journal of Fisheries and Aquatic Science* 46, 1176–1182.

Redfield, A.C. (1934) On the proportions of organic derivatives in sea water and their relationship to the composition of plankton. In: *James Johnston Memorial Volume.* Liverpool University Press, Liverpool, pp. 176–192.

Rippey, B. and Anderson, N.J. (1996) Accuracy of diatom-inferred total phosphorus concentrations and the accelerated eutrophication of a lake due to reduced flushing. *Environmental Science and Technology* 30, 1786–1788.

Rohlich, G.A. (1969) (ed.) *Eutrophication: Causes, Consequences, Correctives.* National Academy of Sciences, Washington, DC, 661 pp.

Ryther, J.H. and Dunstan, W.M. (1971) Nitrogen, phosphorus and eutrophication in the coastal marine environment. *Science* 171, 1008–1013.

Sakamoto, M. (1966) Primary production by phytoplankton community in some Japanese lakes and its dependence on lake depth. *Archiv für Hydrobiologie* 62, 1–28.

Salonen, K., Jones, R.I., De Haan, H. and James, M. (1994) Radiotracer study of phosphorus uptake by plankton and redistribution in the water column of a small humic lake. *Limnology and Oceanography* 39, 69–83.

Sas, H. (1989) *Lake Restoration by Reduction of Nutrient Loading.* Academia Verlag, Sankt Augustin, 497 pp.

Schaub, B.E.M. and Gieskes, W.W.C. (1990) The eutrophication of the North Sea: the relation between Rhine river discharge and chlorophyll-a concentration in Dutch coastal waters. *Hydrobiologia* 195, 85–90.

Schelske, C.L., Conley, D.J. and Warwick, W.F. (1985) Historical relationships between phosphorus loading and biogenic silica accumulation in Bay of Quinte sediments. *Canadian Journal of Fisheries and Aquatic Science* 42, 1401–1409.

Schindler, D.W. and Fee, E.J. (1974) Experimental Lakes Area: whole lake experiments in eutrophication. *Journal of the Fisheries Research Board of Canada* 31, 937–953.

Schindler, D.W., Armstrong, F.A.J., Holmgren, S.K. and Brunskill, C.J. (1971) Eutrophication of Lake 227, Experimental Lakes Area, North-western Ontario, by addition of phosphate and nitrate. *Journal of the Fisheries Research Board of Canada* 28, 1763–1782.

Smayda, T.J. (1990) Novel and nuisance phytoplankton blooms in the sea: evidence for a global epidemic. In: Granéli, E., Sundstroem, B., Edler, L. and Anderson, D.M. (eds) *Toxic Marine Phytoplankton.* Elsevier, New York, pp. 29–40.

Smith, R.V. (1977) Domestic and agricultural contributions to the inputs of phosphorus to Lough Neagh. *Water Research* 11, 453–459.

Smith, V.H. (1982) The nitrogen and phosphorus dependence of algal biomass in lakes: an empirical and theoretical overview. *Limnology and Oceanography* 27, 1101–1112.

Straskraba, M. (1980) The effects of physical variables on freshwater production: analyses based on models. In: Le Cren, C.D. and Lowe-McConnel, R.H. (eds) *The Functioning of Freshwater Ecosystems.* Cambridge University Press, Cambridge, pp. 13–84.

Uhlmann, D. and Albrecht, E. (1968) Biogeochemische Faktoren der Eutrophierung von Trinkwassen-Talsperren. *Limnologica (Berlin)* 6, 225–245.

Vallentine, J.R. (1974) *The Algal Bowl, Lakes and Man.* Canadian Department of the Environment, Fisheries and Marine Service, Ottawa, 185 pp.

Van Liere, L. and Gulati, R. (eds) (1992) *Restoration and Recovery of Shallow Eutrophic Lake Ecosystems in the Netherlands. Hydrobiologia* 233, 287 pp.

Van Liere, L. and Janse, J.H. (1992) Restoration and resilience to recovery of the Lake Loosdrecht ecoystem in relation to its phosphorus flow. *Hydrobiologia* 233, 95–104.

Vollenweider, R.A. (1968) *Water Management Research: Scientific Fundamentals of Lakes and Flowing Waters, with Particular Reference to Nitrogen and Phosphorus as Factors in Eutrophication.* Technical Report DAS/C31/68.27, OECD, Paris.

Weibel, S.R. (1970) Urban drainage as a factor in eutrophication. In: Rohlich, G.A. (ed.) *Eutrophication: Causes, Consequences, Correctives.* National Academy of Science, Washington, DC, pp. 383–403.

Wharfe, J.R., Taylor, K.S. and Montgomery, H.A.C. (1984) The growth of *Cladophora glomerata* in a river receiving sewage effluent. *Water Research* 18, 971–979.

Whitton, B.A. (1975) Algae. In: Whitton, B.A. (ed.) *River Ecology.* Blackwell Scientific Publications, Oxford, pp. 81–105.

6 The Behaviour of Soil and Fertilizer Phosphorus

M.A. Morgan
Department of Environmental Resource Management (Soil Science), Faculty of Agriculture, University College, Dublin 4, Ireland

INTRODUCTION

The role of phosphorus (P) in the occurrence of accelerated eutrophication in surface waters has long been established. In Ireland, as in many other countries, much attention has focused on the contribution of P from rural/agricultural sources and on the management steps required to minimize such contributions. In particular, there has been increasing awareness of the need for sensitive management of animal manures on farms, including the question of adequate storage capacity so as to optimize, as far as possible, the times at which land-spreading can take place. In general, the same comments apply to farmyard wastes, such as silage effluent and dirty water. It is not clear, however, that the same general awareness applies to the use of P fertilizers and to the possibility that P from fertilizers could be lost to watercourses after application to land.

Build-up of available (extractable) P in soils in a period when P fertilizer use remained relatively static, as well as a positive P balance with respect to total P input and off-take for agricultural production, has been interpreted as being significant for the P status of rivers and lakes in Ireland (Tunney, 1990). Such a proposition implies that P from a current fertilizer application, or residual P from previous applications, is transported from land to water. It does not, however, imply a particular mechanism of loss.

Loss of P from land can occur in three ways (Ryden *et al.*, 1973).

1. As water-soluble and/or particulate P in surface runoff (overland flow), referring to P picked up by rainwater which flows over land surfaces to streams or rivers.
2. As water-soluble and/or particulate P in subsurface runoff (leaching), referring to P picked up by water which enters the soil profile and moves

through the soil to streams or rivers without ever reaching the main water-table.

3. As water-soluble and/or particulate P in flow to groundwater, referring to P picked up by water that passes to the water-table and which is subsequently discharged to streams, rivers or lakes as seepage.

A general account of native soil P compounds, the likely reactions which soluble inorganic P fertilizers undergo in mineral soils and the agronomic value of P fertilizer residues is presented here. Sample data for P concentrations in extracted soil solutions, P concentrations in solution required for growth and P loss by surface and subsurface runoff are presented and discussed. A marked dependence of soil and fertilizer P behaviour on site characteristics is indicated.

PHOSPHORUS COMPOUNDS IN MINERAL SOILS

In comparison with other essential major nutrients for growth, the P content of the earth's crust (1100–1200 mg kg^{-1}) is low (Jackson, 1964; Tisdale *et al.*, 1985). Likewise, the total P content of mineral topsoils is relatively low (50–1100 mg kg^{-1}; Tisdale *et al.*, 1985; Brady, 1990). For Irish grassland soils, Hanley and Murphy (1973) reported that the range of total P values was 493–936 mg kg^{-1}, with an overall mean value of 733 mg kg^{-1}. Varying quantities (20–80%) of the total P in soils occur in organic form (Hanley and Murphy, 1973; Tisdale *et al.*, 1985; Brady, 1990).

Inorganic P constituents in mineral soils have generally been classified into two groups, namely calcium phosphates and iron and aluminium phosphates. Among the calcium phosphates, $Ca_{10}F_2(PO_4)_6$ (fluorapatite), $Ca_{10}(OH)_2(PO_4)_6$ (hydroxyapatite), $Ca_{10}O(PO_4)_6$ (oxyapatite) and $Ca_{10}CO_3(PO_4)_6$ (carbonate apatite) are most abundant, while $FePO_4.2H_2O$ (strengite) and $AlPO_4.2H_2O$ (variscite) are the other main P-bearing minerals. Inositol phosphates, of which phytic acid is the most significant component, phospholipids, nucleic acids, nucleotides and unidentified sugar phosphates are the principal compounds in the organic P fraction. In general, native inorganic sources of soil P are highly stable and of little short-term consequence for commercial crop production. Although part of the organic P pool may undergo mineralization (Dormaar, 1972) or occur as dissolved materials in the soil water (Ron Vaz *et al.*, 1993), the same comment generally applies to the organic fraction. Historically, the major issues relating to P nutrition in agricultural systems were poor crop yields, commonly associated with P deficiency symptoms, and inadequate levels of P in herbage for animal requirements.

PHOSPHORUS FERTILIZATION

The need to supplement soils with water-soluble or potentially water-soluble P fertilizers arises from inability of the relatively small pool of native soil P to supply and maintain adequate amounts of soluble orthophosphate ($H_2PO_4^-$ and HPO_4^{2-}) to the soil solution for satisfactory crop growth and animal performance. In principle, this approach is not different from supplying nitrogen (N) or potassium (K) fertilizers. However, the reactions which P fertilizers undergo in mineral soils are quite different from those undergone by N and K fertilizers and result in poor (25% or less) efficiency of recovery of an annual application in the growing crop (Barrow, 1980; Haynes, 1984). In contrast, recovery of fertilizer N or K by a crop in the season of application may be as high as 80%. In essence, then, whereas N and K fertilizers in soils are relatively accessible to crop roots, this is not so with P fertilizers, which, after dissolution in the soil water, are quickly immobilized by reactions with various soil constituents. As a result, P nutrition of field crops is largely dependent on the subsequent release of P from these reaction products to the soil water.

PHOSPHORUS RETENTION – GENERAL

The mechanisms by which P fertilizers in mineral soils are converted from a soluble state to a less soluble state are referred to as P retention/P fixation. While the precise sequence of events that follows from an application of soluble fertilizer to soils is unclear, there is a general consensus as to the principal components of the retention process. These are: (i) events associated with the dissolution of the P fertilizer particle; (ii) the occurrence of precipitation reactions involving the orthophosphate anion, in which both reactants are initially dissolved in the soil solution; and (iii) the occurrence of adsorption reactions, in which only one of the reactants (orthophosphate) is initially in solution. Part of the terminology related to P retention includes the terms chemisorption, anion exchange, sorption, physical adsorption, specific adsorption and non-specific adsorption, each referring to particular aspects of the P retention mechanism. In the present discussion, the terms adsorption and precipitation (only) are used to characterize the components of the retention process. No attempt is made to assess the relative significance of these components, about which, in any event, there is rather little general agreement (Sample *et al.*, 1980). Many accounts of the P retention mechanism are available in the literature (e.g. Parfitt, 1978; Sample *et al.*, 1980; White, 1980; Haynes, 1984; Tisdale *et al.*, 1985; Brady, 1990), and the following represents a short distillation of them.

PHOSPHORUS RETENTION – A MACRO VIEW

Regardless of pH, all mineral soils contain aluminium and iron oxides and hydrous oxides, which occur as discrete particles or as coatings on other soil particles, especially clay. In addition, amorphous aluminium hydroxy compounds may be present in interlayer locations of expandable aluminium silicates. Such materials are highly efficient in adsorbing $H_2PO_4^-$ ions that may be present in the soil solution, the general view being that retention occurs as a result of exchange between the $H_2PO_4^-$ anion and hydroxyl (OH^-) ions associated with the iron and/or aluminium. Specifically, under alkaline conditions in the presence of free calcium carbonate ($CaCO_3$), adsorption of $H_2PO_4^-/HPO_4^{2-}$ on to calcite can also occur by replacement of water, bicarbonate (HCO_3^-) or OH^- ions present on the calcite particles. Concurrent with these adsorption reactions, $H_2PO_4^-$ ions in solution may undergo precipitation reactions, the nature of which vary with the pH of the soil. Under acid (pH < 5.0) conditions, presence of active aluminium, iron or manganese may result in the formation of poorly soluble hydroxy metal phosphates (e.g. $Al(OH)_2H_2PO_4$). In contrast, under alkaline conditions, presence of active calcium causes precipitation of dicalcium phosphate anhydrous.

PHOSPHORUS RETENTION – A MICRO VIEW

Details of the types of reactions that occur/that could occur in the immediate vicinity of a dissolving granule of monocalcium phosphate monohydrate ($Ca(H_2PO_4)_2.H_2O$) fertilizer were described more than 30 years ago (Lindsay and Stephenson, 1959a, b, c; Lindsay et al., 1959). On initial contact with the soil, water moves to the fertilizer particle and a highly acidic (pH ~1.0) solution containing ~4.5 mol l^{-1} P and ~1.3 mol l^{-1} calcium (Ca) is produced. During dissolution, the reaction zone is characterized by the presence of undissolved monocalcium phosphate as well as dicalcium phosphate dihydrate ($CaHPO_4.2H_2O$). At the end of dissolution, when the driving forces for water movement to the granule site and solution movement from the site have diminished, ~25% of the P fertilizer remains as $CaHPO_4.2H_2O$ and/or $CaHPO_4$. The following reactions have been used to describe the dissolution process:

$$Ca(H_2PO_4)_2.H_2O \rightarrow CaHPO_4 + H_3PO_4 + H_2O$$
(Lindsay and Stephenson, 1959a)
$$Ca(H_2PO_4)_2.H_2O + H_2O \rightarrow CaHPO_4.2H_2O + H_3PO_4$$
(Brady, 1990)

Fertilizers based on monammonium phosphate ($NH_4H_2PO_4$) or diammonium phosphate (($NH_4)_2HPO_4$) also form solutions that are highly concentrated in P (~2.9 and ~3.8 mol l^{-1}, respectively), although they differ markedly in their acidity (pH 3.5 and 8.0, respectively). Like monocalcium phosphate monohydrate, these fertilizers are highly water soluble, and so it can be assumed that movement of water to, and of the concentrated P solution from, the dissolving granules of these materials is also quite rapid.

The P reaction products/probable reaction products that form as the P-rich solution moves away from fertilizer particles have been studied in two ways. In one (Lindsay et al., 1962), soils were shaken for various periods of time with saturated solutions of P fertilizers and the soil–solution suspensions filtered. Aliquots of the filtrates were then stored for up to several months and the solid-phase products that subsequently appeared in the filtrates identified. In the case of the very acidic solution of $Ca(H_2PO_4)_2.H_2O$, Lindsay et al. (1962) noted that iron and aluminium (especially) were dissolved from the soil materials and that a colloidal ferric aluminium phosphate of variable composition $((FeAlX)PO_4.nH_2O)$ was precipitated after ~3 days. Amount of precipitate formed increased with time and with further dilution of the aqueous phase with water. With monammonium phosphate, the initial product identified after reaction with an acid (pH 4.9) sandy loam was taranakite $((NH_4)_3Al_5H_6(PO_4)_8.18H_2O)$, while a calcareous loam (pH 8.5) initially (15 min) yielded a precipitate of $CaHPO_4.2H_2O$ and subsequently (3 days) $MgNH_4PO_4.6H_2O$ (struvite). With diammonium phosphate, unidentified colloidal materials, $MgNH_4PO_4.6H_2O$ and basic aluminium phosphate $(NH_4Al_2(PO_4)_2OH.8H_2O)$ were the initial products found. In all, more than 30 crystalline products containing P, as well as colloidal materials of unknown composition, were identified shortly after reaction of soils, or common soil constituents, with saturated solutions of the commonly used water-soluble P fertilizers (Lindsay et al., 1962). A more expansive listing of the likely products of P fertilizer reactions with soils is given by Sample et al. (1980).

The procedure used by Bell and Black (1970) to study reactions of P fertilizers in soils involved contact of a thin layer of solid P fertilizer with columns of soil for 4, 16 and 48 weeks. The soil columns were then cut into sections at short (cm) distances from the fertilizer layer and newly formed P compounds in the different sections identified. With monocalcium phosphate and monammonium phosphate, $CaHPO_4.2H_2O$ was the main product identified, with small quantities of $CaHPO_4$ also present where monocalcium phosphate was used. With diammonium phosphate, the products identified were $Ca(NH_4)_2(HPO_4)_2.H_2O$, $Ca_8H_2(PO_4)_6.5H_2O$, $MgNH_4PO_4.6H_2O$ and $CaHPO_4.2H_2O$, especially where the soils were high in calcium content. With time (> 4 weeks), $Ca(NH_4)_2(HPO_4)_2.H_2O$ altered, leaving a residue of $CaHPO_4.2H_2O$, while $Ca_8H_2(PO_4)_6.5H_2O$ also reverted to $CaHPO_4.2H_2O$ where soil pH was 6.3 or more. It was further noted that movement of P from the fertilized zone was greatest with diammonium phosphate (9.7 cm) and least (2.6 cm) with monocalcium phosphate after 48 weeks of contact. Generally, transport of P from the fertilized layer decreased as clay content and amounts of exchangeable Ca and magnesium increased.

In summary, the ongoing reactions between fertilizer P and soils can be envisaged to result in the P occurring in three different states, namely P in soil solution, and P associated with soil particles and in 'equilibrium' with solution P (Barrow, 1980). The third P pool is regarded as being more firmly retained than the latter, is not in direct equilibrium with the soil solution and is also less accessible to plants/growing crops. In this general scenario, a clear distinction between adsorption and precipitation processes cannot be made. As to the location of these processes relative to the site of fertilizer dissolution, Sample *et al.* (1980) suggest that as the P solution moves away from the granule site, the P concentration decreases (by precipitation and dilution by soil water) and a location is reached at which adsorption becomes dominant and precipitation negligible. The point at which the transition from one to the other process occurs is not known.

Implicit in the foregoing view of P fertilizer behaviour is that the early-formed products of the retention mechanism are only 'relatively' stable and are therefore able to supply adequate amounts of soluble P for early/speedy utilization by growing crops. Over time, however, the initial reaction products are altered, and, while the manner of alteration is not known, there seems to be general agreement as to the composition of the ultimate products of the fixation mechanism (Sample *et al.*, 1980). Available evidence indicates that the 'stable' products of the fixation reaction sequence are octacalcium phosphate, hydroxyapatites and fluorapatites under alkaline/calcareous conditions, and strengite and variscite under acid/neutral conditions. Apparently these materials can appear within a matter of months (Sample *et al.*, 1980), their rate of production being especially dependent on the prevailing pH.

In the transition from the 'relatively' stable to the 'stable' state, it is not thought that there is quantitative conversion from one type of P compound to another. Rather, it is more likely that orthophosphate ions which appear during transition are exposed to a number of competing reaction possibilities, including absorption by roots and transport to new sites of reaction by water or cultivation practices. Additional stability of reaction products may also arise from occlusion phenomena or from growth in crystal size of precipitated P compounds.

SINGLE-FACTOR EFFECTS ON PHOSPHORUS RETENTION

Retention/fixation of fertilizer P does not occur to the same degree in all soils. Since the adsorption component of the fixation process is associated with the clay (< 2 µm) and hydrous oxide fractions, it follows that P retention will be greater in soils of higher clay content (Brady, 1990). In contrast, soils in which inorganic colloids are absent/essentially absent retain little P fertilizer (Fox and Kamprath, 1971). The same applies to soils that are predominantly sandy in nature (Mattingly, 1970; Humphreys and Pritchett, 1971). Elevated soil temperatures (Beaton *et al.*, 1965) and presence of kaolinite and calcite (Tisdale *et al.*, 1985) each increase P retention, the latter more so when it contains a high proportion of iron oxide impurities (Kuo and Lotse, 1972; Holford and Mattingly, 1975; Ryan *et al.*, 1985). In general, P retention is minimal where soil pH is ~6.0–6.5.

The net effect of the presence of organic matter is difficult to assess. Whereas iron and aluminium hydroxy humic complexes have been shown capable of adsorbing P (Wild, 1950), organic acid-induced desorption of P from kaolinite and iron/aluminium oxides has also been reported (Nagarajah *et al.*, 1970). However, more subtle effects of organic constituents on fertilizer P behaviour have been suggested (Sample *et al.*, 1980). Arising from observations that the advancing front of the concentrated solution developed from a dissolving fertilizer particle can dissolve a portion of soil organic matter, Sample *et al.* (1980) hypothesized: (i) that displaced Ca, Fe or Al ions from metal–organic complexes could cause P to be precipitated; (ii) that dissolution of organic-matter coatings from clay particles could expose new clay surfaces for participation in adsorption/desorption reactions; and (iii) that, as solubilized organic matter is removed from one surface, it may be deposited on another, thereby protecting the latter from interacting with P in the fertilizer solution.

RESIDUAL FERTILIZER BEHAVIOUR

As noted previously, there is general acceptance that only a small proportion of applied fertilizer P is recovered by growing crops in the year of application and that the unused portion remains in the soil as part of the total soil P pool. From an agronomic viewpoint, availability of P in these residues to succeeding crops is an important issue, where 'availability' refers not only to the reactivity of the residues *per se*, but also to their distribution within the profile. The significance of the latter concept (i.e. 'positional' availability) relates to the small effective diffusion coefficient of the orthophosphate ion and to the small proportion of total soil volume occupied by roots (~1–2%). Whereas some studies indicate that movement of fertilizer P away from the initial reaction site is limited to a few centimetres or so (Khasawneh *et al.*, 1974), others show deeper penetration of fertilizer P in the profile (Cooke and Williams, 1970). Presumably, over time, P can diffuse outwards and downwards from the uppermost soil layers, resulting in more uniform distribution in the profile as a whole. On the reasonable assumption that part of the redistribution process occurs in solution phase, downward diffusive flux of P raises the question of risk of soluble P loss from soils in subsurface runoff. To some degree, at least, risk of such loss would be moderated by exposure of the diffusing P to soil particles that had not previously interacted with the fertilizer or with any of its reaction products. Redistribution of P may also occur as a result of normal tillage operations and possibly too by recycling through dead or decaying plant roots (Read, 1982).

There is clear evidence from the literature that fertilizer P which is not used by crops in the year of application remains in the soil and is capable of being used by crops in succeeding years. In some cases, the contribution of residual P to growth may last for 8–10 years (Kamprath, 1967; Spratt, 1978; Halvorson and Black, 1981–1982), at the end of which essentially all of the original application may have been recovered in the growing crops. However, such observations may not be generally applicable and, in particular situations, benefit of residual P for growth will depend on a number of factors, including original rate of fertilizer application, amount of P already removed in growth, soil buffer capacity and soil pH. Generally, experimental results show that the agronomic effectiveness of water-soluble P fertilizers decreases with time of contact with soils (Leamer, 1963; Devine et al., 1968). From a farming standpoint, the critical question here is knowing the amount of available P in the residual pool below which crop yield is not acceptable. Even where soil test values for P are 'high', however, additional yield response may follow from small 'starter' applications of P fertilizer, possibly as a compensatory factor for low soil temperatures (Alessi and Power, 1980). Among the P fertilizer reaction products referred to earlier, potassium and ammonium taranakites and amorphous and crystalline aluminium phosphate (Taylor et al., 1963) as well as octacalcium phosphate (Tisdale et al., 1985), are all capable of supplying P for crop utilization.

PHOSPHORUS REMOVAL FROM SOIL: CROP REQUIREMENT AND LOSSES TO WATER

The source of P for growth is that which is present as orthophosphate in the soil water (soil solution). As the P concentration in solution decreases through uptake, there must be a mechanism whereby the soil solution is replenished in order that P uptake be sustained over time. As noted earlier, replenishment occurs through a combination of dissolution and desorption reactions, involving reaction products from the most recent P fertilizer application, and soil-P compounds, resulting from historical P fertilizer applications.

It is not known with any certainty what concentration of P in solution is necessary to sustain optimal crop growth. An index of the probable requirement is provided by Asher and Loneragan (1967), who showed that 5 μM phosphate (~0.16 $\mu g\ ml^{-1}$ P) was generally satisfactory for the plants studied. Other plants, however, required as much as 25 μM phosphate (~0.8 μg ml^{-1} P). Generally, these early data correspond well with values from a number of sources summarized by Kamprath and Watson (1980) and Olsen and Khasawneh (1980). Sample data for soluble P concentrations in extracted soil solutions range from 0.02 μg ml^{-1} (Curtin and Smillie, 1983) to > 200 μM (> 6 μg ml^{-1}) in heavily fertilized soils (Adams et al., 1982). It is apparent from these data, then, that the concentrations of P likely to be found in soil solutions may not always be sufficient for optimal crop growth. Under commercial farming conditions, this problem is overcome by use of P fertilizers. Adjusting the level of P in soil solution to that required for sustained crop growth, however, may not be compatible with current water-quality parameters, because the concentrations of soluble P apparently required for crop growth (see above) are always likely to exceed the value (0.01 μg ml^{-1}) above which biological growth in water bodies may be stimulated (Sharpley and Menzel, 1987). Indeed, even where soils have not received P fertilizer, the concentrations of P in soil solutions may sometimes exceed the critical level for biological production (Sharpley and Menzel, 1987).

From a management perspective, it is important that the possible routes of P loss from soil be differentiated from each other. It is generally held that transport of soluble inorganic P from soils in subsurface runoff is small because of P retention by soil particles in the profile as a whole. This may not always be true, however, as borne out in the data summarized by Sharpley and Menzel (1987), which show that concentration (μg ml^{-1}) and annual loss (kg ha^{-1}) of soluble P in subsurface drainage from cropped soils fertilized with P were 0–0.21 and 0–0.44, respectively. Highest concentrations and highest total loss of P occurred under tiled-drainage conditions, which would be expected to increase infiltration and percolation, thereby reducing time of contact between soluble P and soil particles in the profile. Significantly, though, the data (Sharpley and Menzel, 1987) also reveal situations in which soluble P (0.02–0.18 μg ml^{-1}) was detected in water draining land that had not been fertilized, giving total soluble P losses of up to 0.12 kg ha^{-1} year^{-1}. Similar data for dissolved P in subsurface runoff have been published by Ryden et al. (1973). Although Ryden et al. (1973) refer to the 'reasonable proportion' of P loss that can occur by subsurface runoff from arable watersheds, Sharpley and Menzel (1987) take the view that subsurface losses of soluble P are 'small', with applications of P fertilizer at recommended rates having no significant effect on P losses. Diminished retention/fixation capacity, more speedy movement of P and therefore greater susceptibility to loss in subsurface drainage flow can be expected in predominantly sandy (Sawhney, 1977) or organic (Fox and Kamprath, 1971) soils.

It is clear from the literature that losses of P by surface runoff are likely to be much larger than those by subsurface runoff, and that such losses are likely to be related to P fertilizer use. In the data referred to above

(Sharpley and Menzel, 1987), soluble P concentrations in runoff water from unfertilized cropped land ranged from 0.01 to 0.30 µg ml^{-1} (0.01 to 0.50 kg ha^{-1} year^{-1} P), with particulate P concentrations at 0.06–1.8 µg ml^{-1} (0.02–1.4 kg ha^{-1} year^{-1} P). Where P fertilization was carried out, however, soluble P concentrations were 0.03–3.7 µg ml^{-1} (0.04–4.3 kg ha^{-1} year^{-1}), and particulate P concentrations were 0.14–9.61 µg ml^{-1} (0.02–18.19 kg ha^{-1} year^{-1}). Amounts of surface runoff losses of total P and of various fractions of the total P from a selection of sloping sites are also available in Ryden et al. (1973). Part of the soluble P component of surface drainage may be envisaged as deriving from undissolved fertilizer particles lying on the soil surface at the time a runoff event commences. More of the soluble P may arise from dissolution/desorption of P associated with soil particles in the upper few millimetres of the profile (Sharpley et al., 1981). Finally, aqueous leachate from living and dead vegetation (Sharpley, 1981) may also contribute to P runoff, especially following a freeze–thawing cycle (White, 1973). Overall, though, it seems as if the bulk of P loss by the surface-runoff mechanism is P associated with particulate matter.

CONCLUSIONS

All soils, whether fertilized or not, give up finite quantities of nutrients to percolating water and, on the face of it, increasing nutrient losses would appear to be related to increasing levels of fertilization. In the case of fertilizers containing P, however, there is clear evidence that the water-soluble P fraction interacts with mineral soil constituents to give a range of P compounds of lesser solubility and reactivity than the original fertilizer material, implying diminished mobility of the P. The few data quoted in this chapter relating to P losses from soils do not provide a solid basis for making unqualified statements as to the general occurrence of P losses by the surface and subsurface runoff processes. Rather, they indicate that high P losses may occur under some, but not all, circumstances and that losses are more likely to occur as particulate and soluble P in surface runoff than as soluble P in subsurface drainage. Perhaps the most obvious aspect of the issue of P loss from land is that it is a highly site-specific process. Accordingly, risk of P loss will be determined by a variety of soil properties (plant-available and total P content, organic-matter content, P buffer capacity, depth, slope, texture, structure, hydrological features, pH, temperature, etc.), by cropping features (extent of ground cover, type of crop, rate of growth and degree of root proliferation), by fertilization programme (type and amount of fertilizer, and manner and timing of applications relative to previous and future weather conditions) and by soil management. Doubtless, interactive effects are also likely to occur. It is apparent, therefore, that application of localized observations/events to a regional, much less national, level is likely to give a distorted view of the real picture. This situation may be further aggravated by the use of occasional/ periodic measurements of P loss, which cannot account for the spasmodic and highly variable nature of the loss processes over time. Especially in respect of P, accurate definition of actual loss from land, as well as origin of loss, is best achieved by long-term monitoring of P transport in specific areas/watersheds, leading to an assessment of the steps that need to be put in place to reduce or prevent the leakage.

REFERENCES

Adams, J.F., Adams, F. and Odom, J.W. (1982) Interaction of phosphorus rates and soil pH on soybean yield and soil solution composition of two phosphorus-sufficient ultisols. *Soil Science Society of America Journal* 46, 323–328.

Alessi, J. and Power, J.F. (1980) Effects of banded and residual fertilizer phosphorus on dryland spring wheat yield in the northern plains. *Soil Science Society of America Journal* 44, 792–796.

Asher, C.J. and Loneragan, J.F. (1967) Response of plants to phosphate concentration in solution culture: I. Growth and phosphorus content. *Soil Science* 103, 225–233.

Barrow, N.J. (1980) Evaluation and utilization of residual phosphorus in soils. In: Khasawneh, F.E., Sample, E.C. and Kamprath, E.J. (eds) *The Role of Phosphorus in Agriculture.* American Society of Agronomy, Madison, Wisconsin, pp. 333–359.

Beaton, J.D., Speer, R.C. and Brown, G. (1965) Effect of soil temperature and length of reaction period on water solubility of phosphorus in soil fertilizer reaction zones. *Soil Science Society of America Proceedings* 29, 194–198.

Bell, L.C. and Black, C.A. (1970) Crystalline phosphates produced by interaction of orthophosphate fertilizers with slightly acid and alkaline soils. *Soil Science Society of America Proceedings* 34, 735–740.

Brady, N.C. (1990) *The Nature and Properties of Soils*, 10th edn. Macmillan Publishing Company, New York, 621 pp.

Cooke, G.W. and Williams, R.J.B. (1970) Losses of nitrogen and phosphorus from agricultural land. *Journal of the Society for Water Treatment and Examination* 19, 253–276.

Curtin, D. and Smillie, G.W. (1983) Soil solution composition as affected by liming and incubation. *Soil Science Society of America Journal* 47, 701–707.

Devine, J.R., Gunary, D. and Larsen, S. (1968) Availability of phosphate as affected by duration of fertilizer contact with soil. *Journal of Agricultural Science* 71, 359–364.

Dormaar, J.F. (1972) Seasonal pattern of soil organic phosphorus. *Canadian Journal of Soil Science* 52, 107–112.

Fox, R.L. and Kamprath, E.J. (1971) Adsorption and leaching of P in acid organic soils and high organic matter sand. *Soil Science Society of America Proceedings* 35, 154–156.

Halvorson, A.D. and Black, A.L. (1981–1982) Long-term benefits from a single application of P. *Better Crops Plant Food* 66, 33–35.

Hanley, P.K. and Murphy, M.D. (1973) Soil and fertilizer phosphorus in the Irish ecosystem. *Water Research* 7, 197–210.

Haynes, R.J. (1984) Lime and phosphate in the soil–plant system. *Advances in Agronomy* 37, 249–315.

Holford, I.C.R. and Mattingly, G.E.G. (1975) The high- and low-energy phosphate adsorbing surfaces in calcareous soils. *Journal of Soil Science* 26, 407–417.

Humphreys, F.R. and Pritchett, W.L. (1971) Phosphorus adsorption and movement in some sandy forest soils. *Soil Science Society of America Proceedings* 35, 495–500.

Jackson, M.L. (1964) Chemical composition of soils. In: Bear, F.E. (ed.) *Chemistry of the Soil.* ACS Monograph No. 160, Reinhold Publishing Corporation, New York, pp. 71–141.

Kamprath, E.J. (1967) Residual effect of large applications of phosphorus on high phosphorus fixing soils. *Agronomy Journal* 59, 25–27.

Kamprath, E.J. and Watson, M.E. (1980) Conventional soil and tissue tests for assessing the phosphorus status of soils. In: Khasawneh, F.E., Sample, E.C. and Kamprath, E.J. (eds) *The Role of Phosphorus in Agriculture.* American Society of Agronomy, Madison, Wisconsin, pp. 433–469.

Khasawneh, F.E., Sample, E.C. and Hashimoto, I. (1974) Reactions of ammonium ortho- and polyphosphate fertilizers in soil: I. Mobility of phosphorus. *Soil Science Society of America Proceedings* 38, 446–451.

Kuo, S. and Lotse, E.G. (1972) Kinetics of phosphate adsorption by calcium carbonate and Ca-kaolinite. *Soil Science Society of America Proceedings* 36, 725–729.

Leamer, R.W. (1963) Residual effects of phosphorus fertilizer in an irrigated rotation in the southwest. *Soil Science Society of America Proceedings* 27, 65–68.

Lindsay, W.L. and Stephenson, H.F. (1959a) Nature of the reactions of monocalcium phosphate monohydrate in soils: I. The solution that reacts with the soil. *Soil Science Society of America Proceedings* 23, 12–18.

Lindsay, W.L. and Stephenson, H.F. (1959b) Nature of the reactions of monocalcium phosphate monohydrate in soils: II. Dissolution and precipitation reactions involving iron, aluminium, manganese, and calcium. *Soil Science Society of America Proceedings* 23, 18–22.

Lindsay, W.L. and Stephenson, H.F. (1959c) Nature of the reactions of monocalcium phosphate monohydrate in soils: IV. Repeated reactions with metastable triple-point solution. *Soil Science Society of America Proceedings* 23, 440–445.

Lindsay, W.L., Lehr, J.R. and Stephenson, H.F. (1959) Nature of the reactions of monocalcium phosphate monohydrate in soils: III. Studies with metastable triple-point solution. *Soil Science Society of America Proceedings* 23, 342–345.

Lindsay, W.L., Frazier, A.W. and Stephenson, H.F. (1962) Identification of reaction products from phosphate fertilizers in soils. *Soil Science Society of America Proceedings* 26, 446–452.

Mattingly, G.E.G. (1970) Residual value of basic slag, Gafsa rock phosphate and superphosphate in a sandy podzol. *Journal of Agricultural Science* 75, 413–418.

Nagarajah, S., Posner, A.M. and Quirk, J.P. (1970) Competitive adsorption of phosphate with polygalacturonate and other organic anions on kaolinite and oxide surfaces. *Nature* 228, 83–85.

Olsen, S.R. and Khasawneh, F.E. (1980) Use and limitations of physical–chemical criteria for assessing the status of phosphorus in soils. In: Khasawneh, F.E., Sample, E.C. and Kamprath, E.J. (eds) *The Role of Phosphorus in Agriculture.*

American Society of Agronomy, Madison, Wisconsin, pp. 361–410.

Parfitt, R.L. (1978) Anion adsorption by soils and soil materials. *Advances in Agronomy* 30, 1–50.

Read, D.W.L. (1982) Bio-cycling of phosphorus in the soil. *Better Crops Plant Food* 66, 24–25.

Ron Vaz, M.D., Edwards, A.C., Shand, C.A. and Cresser, M.S. (1993) Phosphorus fractions in soil solution: influence of soil acidity and fertilizer additions. *Plant and Soil* 148, 175–183.

Ryan, J., Curtin, D. and Chemma, M.A. (1985) Significance of iron oxides and calcium carbonate particle size in phosphate sorption by calcareous soils. *Soil Science Society of America Journal* 49, 74–76.

Ryden, J.C., Syers, J.K. and Harris, R.F. (1973) Phosphorus in runoff and streams. *Advances in Agronomy* 25, 1–45.

Sample, E.C., Soper, R.J. and Racz, G.J. (1980) Reactions of phosphate fertilizers in soils. In: Khasawneh, F.E., Sample, E.C. and Kamprath, E.J. (eds) *The Role of Phosphorus in Agriculture*. American Society of Agronomy, Madison, Wisconsin, pp. 263–310.

Sawhney, B.L. (1977) Predicting phosphate movement through soil columns. *Journal of Environmental Quality* 6, 86–89.

Sharpley, A.N. (1981) The contribution of phosphorus leached from crop canopy to losses in surface runoff. *Journal of Environmental Quality* 10, 160–165.

Sharpley, A.N. and Menzel, R.G. (1987) The impact of soil and fertilizer phosphorus on the environment. *Advances in Agronomy* 41, 297–324.

Sharpley, A.N., Ahuja, L.R. and Menzel, R.G. (1981) The release of soil phosphorus to runoff in relation to the kinetics of desorption. *Journal of Environmental Quality* 10, 386–391.

Spratt, E.D. (1978) Residual fertilizer P benefits wheat and oilseed production. *Better Crops Plant Food* 62, 24–26.

Taylor, A.W., Lindsay, W.L., Huffman, E.O. and Gurney, E.L. (1963) Potassium and ammonium taranakites, amorphous aluminium phosphate, and variscite as sources of phosphate for plants. *Soil Science Society of America Proceedings* 27, 148–151.

Tisdale, S.L., Nelson, W.L. and Beaton, J.D. (1985) *Soil Fertility and Fertilizers*, 4th edn. Macmillan Publishing Company, New York, 754 pp.

Tunney, H. (1990) A note on a balance sheet approach to estimating the phosphorus fertilizer needs of agriculture. *Irish Journal of Agriculture Research* 29, 149–154.

White, E.M. (1973) Water-leachable nutrients from frozen or dried prairie vegetation. *Journal of Environmental Quality* 2, 104–107.

White, R.E. (1980) Retention and release of phosphate by soil and soil constituents. In: Tinker, P.B. (ed.) *Soils and Agriculture*. Blackwell Scientific Publications, Oxford, pp. 71–114.

Wild, A. (1950) The retention of phosphate by soil: a review. *Journal of Soil Science* 1, 221–238.

7 Setting and Justifying Upper Critical Limits for Phosphorus in Soils

E. Sibbesen[1] and A.N. Sharpley[2]
[1]Danish Institute of Plant and Soil Science, Department of Soil Science, Research Centre Foulum, DK-8830 Tjele, Denmark; [2]USDA-ARS, Pasture Systems and Watershed Management Research Laboratory, Curtin Road, University Park, PA 16802-3702, USA

INTRODUCTION

There are two main issues regarding the level of soil phosphorus (P): (i) a certain critical level of soil P is necessary for economic crop production; and (ii) part of the soil P is lost from agricultural land to the aquatic environment by wind erosion, surface runoff and water erosion, and by leaching. Upper critical limits for P in soil should be set relative to the two issues.

The mobility of P in soil is low compared with other plant nutrients because of the generally low solubility of phosphate compounds and strong P-binding capacity of soil material. Plant roots, therefore, have a 'contact problem' relative to both soil P and applied fertilizer P. A plant root gets most of its P from within 2 mm of the root surface during its period of active P uptake (Nye and Tinker, 1977; Barber, 1985). Agricultural crops generally take up only 5–10% of the applied fertilizer P the first year (Greenwood *et al.*, 1980) and gradually less the following years. Often 90% of the P uptake originates from residual P in soil. Freshly applied fertilizer P cannot compensate for a low soil-P status (Johnston *et al.*, 1986). Therefore, to ensure P nutrition of crops, soils have to have a sufficient pool of plant-available P, i.e. a sufficient P status. The 'contact problem', however, implies that this pool must be considerably larger than the P uptake of a single crop. The question is how large relative to optimum P nutrition of crops.

The widespread use of P fertilizers in this century has increased soil-P status of agricultural land from very low levels to medium and high levels in most industrialized countries. Numerous field experiments on types and rates of P fertilizers have been conducted and related to soil-P testing for the guidance of P fertilization. Even so, in many regions of Europe and the USA, soil-P status levels have increased above what is necessary for the crops (Schachtschabel and Köster, 1985; Gartley and Sims, 1994; Sibbesen and Runge-Metzger, 1995). Part of the reason may be that it is difficult to define the optimum soil-P status level or that some farmers have practised 'insurance fertilization' or have been wrongly advised (Olson et al., 1982). In many regions of intensive confined livestock production, amounts of P produced in manure often exceed local crop requirements, leading to disposal rather than utilization of manure (Sharpley et al., 1993; Kingery et al., 1994; Sibbesen and Runge-Metzger, 1995).

The loss from soil of bioavailable P (BAP), both as dissolved P (DP) and as particulate P (PP) in runoff and erosion is of increasing concern in Europe and the USA because of its impact on surface-water quality (Kristensen and Hansen, 1994; Sharpley et al., 1994b; USEPA, 1994). Also, the nonpoint-source load from agricultural land accounts for an increasing proportion of the total P load (Daniel et al., 1994), as P losses from point sources have been reduced by more effective identification and treatment than non-point sources.

The loss of PP is mainly a function of erosion. Therefore, erosion control measures can reduce PP losses. However, loss of dissolved phosphate (DP) is more difficult to reduce. Control measures are largely limited to preventing soil-P accumulation to environmentally sensitive levels.

Environmentally sensitive or threshold soil-P levels are those above which the potential for P loss in runoff exceeds any crop production concerns. Quantification of these levels is critical for development of P management guidelines for water quality, as well as crop-production goals.

There is no direct relation between soil-P content and P transfer to surface waters or eutrophic response of P-sensitive waters. In order to estimate defensible, upper, critical, soil-P limits that are environmentally sensitive, we must first develop analytical methods that measure soil-P availability relevant to the release of soil P to runoff; quantify the relationship between soil and runoff P; and identify transport potential for a site.

This chapter will deal with the following.

1. Levels of optimum soil-P status relative to crop production estimated in Europe and North America and based on common soil-P test methods.
2. Applicability of common soil-P test methods and more environmentally orientated methods for estimating potential release of DP and bioavailable PP to surface runoff and from eroded sediments.
3. How we may implement environmental soil-test programmes.
4. Attempts to identify upper, critical limits for P in soils.

OPTIMUM SOIL-PHOSPHORUS STATUS RELATIVE TO CROP PRODUCTION

A certain pool of plant-available P has to be present in the soil for optimum, economic crop production. Therefore, the obvious question is how large, or in other words what is the optimum soil-P status relative to crop production? The farmer has to decide each year how much fertilizer P should be applied to achieve optimum yield. At least in theory, there is a relation between the optimum P-application rate and soil-P status. At low soil-P status the optimum P-application rate is greater than the P amount removed with the harvested crops. The farmer, therefore, applies more P than is removed and the soil-P status gradually increases from year to year. As soil-P status increases, the optimum P-application rate decreases (Colwell, 1963; Munk, 1985). The optimum soil-P status is reached when the optimum P-application rate equals crop P removal. If the farmer continues to apply more P than is removed, particularly if the soil-P status is at or above optimum, he will lose profit and the soil-P status will continue to increase until there is no crop response to applied P.

In practice, the system is much more complicated. The soil-P status normally varies tremendously over a field and down through the soil profile. Measuring the P status of a soil also gives problems. There are many different soil-P test methods with varying ability to assess the P status of soils (Sibbesen, 1983) and P-fertilizer residues. Even if the P status is measured correctly, the next problem is to estimate the optimum fertilizer-P application rate. This requires expensive field experiments with several P-application rates, preferably running for several years. Further, for a given soil-P status, the optimum P-application rate varies with crop-yield potential (Munk, 1985), type of crop (Johnston *et al.*, 1986), climate or year (Holford *et al.*, 1985; Cox, 1992), P-fertilizer type, fertilizer application method, reaction between soil and fertilizer P and finally the price ratio between crop and P fertilizer. The difficulties in assessing the optimum P-application rate are further increased by the fact that P fertilizers have a long-term effect, which may provide profit for many years, even though the effect decreases with time.

Data from field trials on P are often complicated to interpret. The choice of yield model and method for calculating the so-called critical soil-P test level, which is an estimate of the optimum soil-P status, greatly influences the level (Dahnke and Olson, 1990). Another complicating factor for the interpretation of long-term field experiments is that the annually repeated tillage gradually moves soil from plot to plot (Sibbesen, 1986). Even for 5–10-year-old field experiments with 5–10 m wide plots this mixing may affect the soil-P status in the centre of the plots (Sibbesen *et al.*, 1995). This means that the long-term effect of the treatments on the soil-P status and crop behaviour of each plot is blurred by the influence from

Table 7.1. Critical levels of soil-test P calibrated in field experiments in Europe, USA and Canada.

Soil-P test method*	Critical level (mg kg^{-1})	Calibration basis	Crop	Number of trials	Duration of trials (years)	Region, soils	Ref.
DL + CAL	15 26 37	90% rel. yield 95% rel. yield 97.5% rel. yield	2/3 cereals + 1/3 sug. beets, pot., maize	150	1–3	Germany, various	A
CAL	74† 83 118†	Economic P applic., Mitcherlich	72% cereals 28% sug. beets, pot., maize	82	3–13	Germany, loess + loamy soils	B
CAL DL Olsen H$_2$O	30‡ 37‡ 13‡ 5	No yield response to P applic. above level	W. wheat + w. barley + sug. beets	105	1	Germany, loess	C
H$_2$O	18§	As ref. B		55	6–10	Germany	D
H$_2$O	10§	Max. yield	3/4 cereals + 1/4 sug. beets	6	15	Germany, loess	E
DL	50–60	95% rel. yield	Rotation ¶	1	42	Germany, phaeozem	F
H$_2$O	9 14 20	90% rel. yield 95% rel. yield 97.5% rel. yield	Potatoes	37	1	Netherlands	G
Olsen	25 20 33 20	Max. yield minus 1 SE = 97–98% rel. yield, Mitcherlich	Potatoes Sugar beets S. barley W. wheat	1	9	England, sandy clay loam	H
Olsen	21–35	Yield reduct. below level	Cereals + root crops	7	20	Denmark, various	I

Method	Value	Criterion	Crop			Location	Ref
Olsen	10	As ref. C	W. wheat	4	6	Canada, chernozems	J
Mehlich-1	14	95% rel. yield	Maize + soyb.	41	2–3	Alabama, Low CEC soils	K
Bray-1	15 22	As ref. C	Maize Wheat	4	12	Nebraska, mollisols	L
Mehlich-3 Bray-1	19§ 18§	Cate and Nelson (1971)	Maize	67	1	Pennsylv., various	M
Mehlich-1 Mehlich-3	16 27	95% rel. yield	Maize + soyb.	1	33	N. Carolina, fine sandy loam	N
Bray-1 Mehlich-3 Olsen	11 11 5	95% rel. yield	Maize	25	1	Iowa, various	O
Bray-1	16–20	Econ. P applic.	Maize + soyb.	1	15	Iowa, typic haplaquoll	P

* See Table 7.3.
† Critical levels refer to yield levels of 37–55, 55–70 and 70–105 grain equivalents ha^{-1} where one grain equivalent = 0.1 t grain, 0.4 t sugar beets or 0.5 t potatoes.
‡ Calculated from relations in Table 7.2.
§ Assuming a soil density of 1.1 g cm^{-3}.
¶ Lucerne, lucerne, potatoes, w. rye, sugar beets, s. barley.
A: Köster and Schachtschabel (1983); B: Munk (1985); C: Baumgärtel (1989); D: Munk and Rex (1990); E: Jungk et al. (1993); F: Stumpe et al. (1994); G: Van der Paauw (1977); H: Johnston et al. (1986); I: G.H. Rubaek and E. Sibbesen, Foulum, DK, personal communication (1995); J: Read et al. (1973); K: Whitney et al. (1985); L: McCallister et al. (1987); M: Beegle and Oravec (1990); N: McCollum (1991); O: Mallarino and Blackmer (1992); P: Webb et al. (1992).

Table 7.2. Ratios between average amount of soil-test P extracted by different soil-P test methods (see Table 7.3).

Region	Soils	Number of soils	DL/ Olsen	CAL/ Olsen	AL/ Olsen	H_2O/ Olsen	Bray-1/ Olsen	Mehl.-3/ Olsen	Bray-1/ Mehl.-3	Mehl.-1/ Mehl.-3	Ref.
Germany	Loess	191	2.36	1.98	3.34	0.415					1
	Sandy	147	1.60	1.38	2.91	0.477					2
Germany	Loess	39	2.81	2.27							3
Norway		189*			3.29	0.375					4
Italy	Calcareous	54						2.56			5
	Noncalcar.	66						1.63			5
USA, Oklahoma		310							1.04		5
USA, north-central		91					3.01	2.85	1.06	0.540	6
USA, Delaware		400								0.528	7
USA, Georgia		450							1.13	0.588	8
USA, Pennsylvania		67							0.948		9
USA, Iowa		25					2.05	2.37	0.864		10
Canada, Ontario		88					2.56	3.03	0.845		11
Weighted average			2.11	1.78	3.20	0.435	2.70	2.56	1.06	0.558	

* Excluding soils from levelled areas.
1: Schachtschabel (1973); 2: Baumgärtel (1989); 3: Semb (1986); 4: Buondonno et al. (1992); 5: Hanlon and Johnson (1984); 6: Wolf and Baker (1985); 7: Sims (1989); 8: Gascho et al. (1990); 9: Beegle and Oravec (1990); 10: Mallarino and Blackmer (1992); 11: Bates (1990).

Table 7.3. Description of soil-P test methods.

Soil-P test method		Extractant	pH	Soil solution ratio	Shaking time	Ref.
DL	Double lactate	0.02 M Ca-lactate + 0.02 M HCl	3.7	1:50*	1.5 h	1
CAL	Calcium lactate	0.05 M Ca-lactate + 0.05 M Ca-acetate + 0.03 M CH_3COOH	4.1	1:20*	2 h	2
AL	Ammonium lactate	0.1 M NH_4-lactate + 0.4 M CH_3COOH	3.1	1:20*	2 h	3
Olsen	Bicarbonate	0.5 M $NaHCO_3$	8.5	1:20*	30 min	4
H_2O or Pw	Water	Deionized water		1:60 †‡	1 h‡	5
Bray-1	Bray and Kurtz No. 1	0.03 M NH_4F + 0.025 M HCl	3.0	1:10*§	5 min§	6
Mehlich-1	Mehlich No. 1	0.05 M HCl + 0.0125 M H_2SO_4	1.2	1:5*	5 min	7
Mehlich-3	Mehlich No. 3	0.2 M CH_3COOH + 0.25 M NH_4NO_3 + 0.015 M NH_4F + 0.013 M HNO_3 + 0.001 M EDTA	2.5	1:10‡	5 min	8

* Weight basis.
† Volume basis.
‡ 1.2 ml soil + 2 ml H_2O, leave for 22 h, add 70 ml H_2O, shake for 1 h.
§ The original method had a soil solution ratio of 1:7 and a shaking time of 1 min.
1: Egnér and Riehm (Riehm, 1955); 2: Schüller (1969); 3: Egnér et al. (1960); 4: Olsen et al. (1954); 5: Sissingh (1971) and Van der Paauw (1971); 6: Bray and Kurtz (1945); 7: Nelson et al. (1953); 8: Mehlich (1984).
Pw, water-extractable phosphorus; EDTA, ethylenediamine tetra-acetic acid.

neighbouring plots. In such long-term experiments, available soil P seems to disappear with time in plots with high P-application rates. This is probably part of the reason for the common view that P is needed in surplus of the crop offtake to maintain a given soil-P test level, a practice observed by Olson *et al.* (1982). Conversely, in such experiments, unmanured soil does go on delivering P to the crops for many years. This experimental error has no doubt caused a lot of confusion in our understanding of the behaviour of P in the soil–plant system.

Nevertheless, to get an idea of the critical levels of soil-test P in Europe, the USA and Canada, we carried out a literature review (Table 7.1). It should be noted that the calibrations had been done in many different ways and that several different soil-P test methods had been used. To be able to compare critical levels obtained with different methods, some weighted mean ratios between P amounts extracted with the different methods were calculated (Table 7.2). The rationale for this, of course, can be questioned, as these ratios vary somewhat between soil types (Sharpley *et al.*, 1984) as is also the case in Table 7.2. The P-test methods included are described in Table 7.3. Finally, the critical levels of Table 7.1 were plotted in Fig. 7.1, which was constructed so that the same position on the different scales of the soil-P test methods should reflect a similar soil-P status (see also Chapter 8, this volume).

The critical soil-P test levels calibrated in Europe (A–I reference codes in Table 7.1 and Fig. 7.1) were generally higher than the ones calibrated in the USA and Canada (J–P), probably because the yield potential is larger in Europe. However, even within the same country, the critical levels obtained from different experiments can differ greatly. For example, in Germany the critical (CAL) levels found by Munk (1985) (B) were much greater than the ones found by Köster and Schachtschabel (1983) (A) and Baumgärtel (1989) (C). Part of the reason may be that all experiments of Munk (1985) had run for at least 3 years. The relation between P-test value and crop-yield response to P fertilization is often very scattered in 1-year experiments (Munk and Rex, 1990). The different soil-P status of subsoils, which are normally not sampled, may have contributed to the difference (Baumgärtel, 1989). Johnston *et al.* (1986) found that the critical soil-P test level was highest for spring barley, followed by potatoes and winter wheat. The great difference of the critical CAL levels (B) obtained with low, medium and high yield levels by Munk (1985) should be noted, and so should the different levels obtained by 90, 95 and 97.5% relative yield (A, G). There is no direct relation between critical levels based on relative yield and the optimum soil-P status.

From the experiments in Table 7.1, it may be concluded that the optimum soil-P status relative to crop growth varies with local conditions, governed primarily by the yield potential and secondly by the choice of crops.

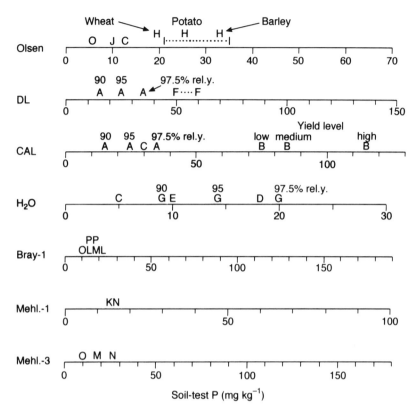

Fig. 7.1. Critical-crop response levels of different soil-P test methods (Table 7.3) calibrated from field experiments and indicated by reference codes A–P (Table 7.1). The scales were made so that the same position on the different scales should reflect a similar soil-P status (based on Table 7.2).

ROUTINE SOIL-PHOSPHORUS TEST METHODS FOR ESTIMATING POTENTIAL RELEASE OF PHOSPHORUS TO SURFACE RUNOFF

The literature and experience relating the level of P in soil to that in surface runoff are limited. Routine soil-P test methods have not been fully evaluated for their ability to predict P concentrations in surface runoff. A linear relationship between level of soil-test P and concentration of DP in surface runoff was documented by Romkens and Nelson (1974), using simulated rainfall and later confirmed by Vaithiyanathan and Correll (1992), Daniel *et al.* (1993) and others (Table 7.4). In fact, soil-test P accounted for 58–98% of the variation in DP concentration of surface runoff. Clearly, there is a greater potential for release of DP to surface

Table 7.4. Regression slope and coefficient of determination of the linear relationship between the dissolved P concentration of runoff (mg l^{-1}, y variable) and soil-test P (all studies used the Bray-1 method, mg kg^{-1}, x variable).

Land use	Soil type †	Location	Soil depth (cm)	Number obs.	Regression slope	r^2	Ref.
Grassed							
Fescue	Captina	Arkansas	0–5	15	0.0070	0.68**	1
Native grass	Renfrow	Oklahoma	0–5	4	0.0061	0.98*	2
Fescue	Tokamaru	New Zealand	0–2.5	24	0.0064	0.85***	3
Native grass	Kirkland	Oklahoma	0–5	11	0.0041	0.95***	4
Cultivated							
Maize	Griswold	Wisconsin	0–2.5	12	0.0014	0.58*	5
Alfalfa, cotton, wheat	McLain	Oklahoma	0–5	7	0.0125	0.89**	2
Fallow	Russell	Indiana	n/a	~150	0.0083	0.81***	6
Maize, cotton	Providence	Mississippi	0–2.5	13	0.0106	0.69**	7

† All soils were of silt loam texture.
*, **, and *** represent significance at $P = 0.05$, 0.01 and 0.001, respectively, as determined by analysis of variance for paired data.
1: Daniel *et al.* (1991); 2: Olness *et al.* (1975); 3: Sharpley *et al.* (1978); 4: Sharpley *et al.* (1986); 5: Andraski *et al.* (1985); 6: Romkens and Nelson (1974); 7: Schreiber (1988).
n/a, not available.

runoff as soil-test P increases. However, the relationship between soil-test P and DP concentration of surface runoff varied between studies as a function of cropping system and soil type (Table 7.4). Varying ratio and contact time between water and soil during runoff have probably also affected the relationships (Sharpley *et al.*, 1981a). It should be noted that, for the same level of soil-test P, generally less P was dissolved from grassland than from cultivated land. This difference may result from less interaction of surface runoff with surface soil for grass than for other crops, due to a greater vegetative cover and surface-soil protection by grass (Sharpley *et al.*, 1981b; Sharpley, 1985). However, recent broadcast applications of fertilizer or manure may reverse this trend, particularly if the application is unincorporated for grass but incorporated in the cultivated soil.

In a more detailed study at one site, Pote *et al.* (1996) found that the DP concentration of surface runoff from a Captina silt loam under fescue in Arkansas was related to Mehlich-3-extractable P content of surface soil (0–2 cm) (Fig. 7.2). This relationship was linear over the range of Mehlich-3 P values found (0–500 mg kg^{-1}). At higher soil P contents, this relationship may be curvilinear (Fig. 7.3). A wide range of soil-test P was obtained by previous application of poultry litter at varying rates. According to Fig. 7.2, approximately 200 mg kg^{-1} Mehlich-3 P in the surface soil would produce a DP concentration of 1 mg l^{-1} in surface runoff.

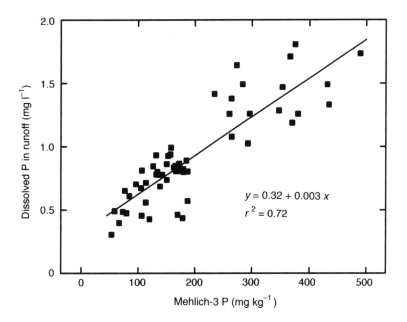

Fig. 7.2. Relationship between soil-test P content of surface soil (0–2 cm) and dissolved P concentration of surface runoff from fescue in Arkansas (adapted from Pote *et al.*, 1996).

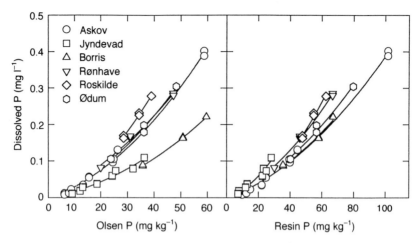

Fig. 7.3. Relationship between dissolved-P concentration and Olsen and resin P. Danish soils of varying P status obtained from 20–100-year-old field experiments on soils ranging in texture from coarse sandy soils to fine sandy loams. Laboratory studies made on moist, incubated soils using a water–soil ratio of 50 ml g^{-1} and a shaking time of 3 h.

For a given cropping system, the soil type, as well as other edaphic and agronomic factors, will influence the relationship between soil-test P and DP concentration of surface runoff. In other words, for a given level of soil-test P, the concentration of P maintained in surface runoff will be influenced by soil type, because of differences in P-buffering capacity between soils, caused by varying levels of iron (Fe) and aluminium (Al) hydroxyoxides, clay, carbonates, and organic matter.

A study of soil-P desorption to water (water–soil ratio = 50 ml g^{-1}, reaction time = 3 h) included seven Danish soils originating from 20–100-year-old field experiments on P. The soils, ranging in texture from coarse sands to fine sandy loams, had a varying P status within each soil. The soils were incubated moist for 5 months before the study. Concave relationships were obtained for the DP relative to P test levels of Olsen P and resin P (Fig. 7.3). The results indicate that the potential for release of DP from soil increases disproportionately with increasing soil-P status. The relationships of the seven soils were closer to each other when related to resin P than to Olsen P.

While these results showed promise, they are limited, for various reasons. Firstly, they do not account for PP, part of which may become available to the aquatic plant growth through desorption from the eroded sediments. Secondly, routine soil-P tests were developed for giving advice for crop production. Thus, their influence on the extraction of soil P is very different from the P-desorption processes which occur under natural surface runoff and erosion. Non-destructive soil-P tests, which use water (Sissingh, 1971; Van der Paauw, 1971) or various P sinks as extractants, such as anion-exchange resin (Amer *et al.*, 1955; Sibbesen, 1978) or Fe hydroxide-impregnated filter-paper (Menon *et al.*, 1989), have proved to be superior to common routine soil-P test methods as regards assessing the soil-P status relative to P uptake of plants (Sibbesen, 1983). Whereas routine soil-P test methods use acids, bases and complexing agents, some non-destructive methods, which maintain the soil approximately chemically intact, are probably superior to the routine soil-P tests methods for predicting potential release of soil P to surface-runoff water and of sediment P to surface waters.

Although the concentration of DP in surface runoff was related to the soil-P test level (Table 7.4 and Figs 7.2 and 7.3), the amount of DP lost (kg ha^{-1}) was not significantly related in any of the studies; regression coefficients ranged from only 0.04 (Andraski *et al.*, 1985) to 0.26 (Sharpley *et al.*, 1978). This suggests that the variability in volume of surface runoff as a result of climatic, edaphic and agronomic factors plays a larger role in determining P loss than the soil-P status of surface soil. Although soil-test P is related to the DP concentration of surface runoff, it is not a reliable indicator of DP loss. Thus, care should be taken in using soil-test P as the sole criterion for determining the potential for P enrichment of surface runoff from subsequent fertilizer or manure application.

SOIL PHOSPHORUS RELATIVE TO BIOAVAILABLE PHOSPHORUS

Less information is available on the relationship between soil P and BAP in surface runoff, including eroded sediments. Bioavailable P consists of DP and a fraction of the PP available for uptake by algae. In the above-mentioned study of surface runoff and erosion from a Captina silt loam under fescue in Arkansas, Pote *et al.*, (1996) found that the content of Fe-oxide strip P (strip P) of surface soil (0–2 cm) was related ($P < 0.001$) to the BAP concentration of surface runoff (Fig. 7.4.). It appears that above about 120 mg kg^{-1} strip P, the relationship becomes curvilinear and BAP increases more rapidly. In a similar study in Oklahoma, the concentration of BAP in surface runoff, including sediments from a Kirkland silt loam soil under native grass, was also related to the strip-P content of surface soil (Fig. 7.5). If the strip-P content of grassed soils had exceeded 120 mg kg^{-1}, it is evident that BAP would have increased curvilinearly (Figs 7.4 and 7.5). The greater curvilinearity for wheat than for grass was probably due to the fact that overall, bioavailable PP comprised a greater portion of BAP in the surface runoff and sediments from wheat (56%) than from native grass (19%). The transport of bioavailable PP in surface runoff and sediments is dependent on both erosion potential and surface soil-P content.

Clearly, surface-soil strip-P content by itself is not a reliable estimate of

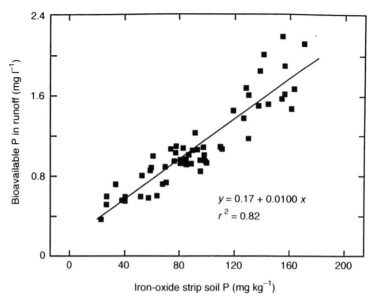

Fig. 7.4. Relationship between iron-oxide strip-P content of surface soil (0–3 cm) and bioavailable-P concentration of surface runoff including sediments from fescue in Arkansas (adapted from Pote *et al.*, 1996).

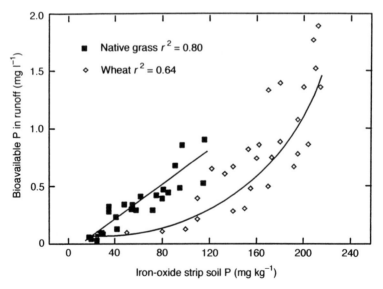

Fig. 7.5. Relationship between iron-oxide strip-P content of surface soil (0–5 cm) and bioavailable-P concentration of surface runoff including sediments from native grass and wheat watersheds at El Reno, Oklahoma.

the potential for transport of BAP in surface runoff and erosion from agricultural land. An estimate of erosion is certainly more important.

PHOSPHORUS SORPTION SATURATION

In the Netherlands, the national strategy is to limit P entry into both surface water and groundwater. One of the approaches for accomplishing this goal is the identification of a soil-P saturation level, above which P-application rates should not exceed crop-removal rates (Van der Zee *et al.*, 1987; Breeuwsma and Silva, 1992). The P-saturation approach is based on the fact that the potential for soil-P desorption increases as sorbed P accumulates in soil. To determine the critical level of soil-P accumulation, Dutch regulations have set a critical limit of 0.1 mg l^{-1} as DP in groundwater at the depth equal to the mean highest water level (Breeuwsma and Silva, 1992). The degree of accumulation is expressed in terms of the phosphate saturation, given by:

$$\text{P-sorption saturation} = \frac{\text{Extractable soil P}}{\text{P-sorption capacity}} \times 100 \qquad \text{equation 1}$$

where the units of extractable soil P and P-sorption capacity are unit mass of DP for a given soil (mg kg^{-1}). In the Netherlands, extractable soil P and P-sorption capacity are determined from the content of oxalate-extractable P, Al and Fe of non-calcareous soils (Breeuwsma and Silva, 1992). A P sorption of 25% has been established as the critical value above which the potential for P leaching becomes unacceptable (Van der Zee *et al.*, 1990).

Using simulated rainfall (2.54 cm h^{-1} for 30 min) on ten soils, ranging from sandy loam to clay in texture, Sharpley (1995) found different relationships between DP in surface runoff and Mehlich-3 P content of the surface-soil layer (0–1 cm). For example, a Mehlich-3 P content of 200 mg kg^{-1} would support a DP concentration of 0.28 mg l^{-1} in the surface runoff from one soil but of 1.36 mg l^{-1} in the surface runoff from another. However, when the P-sorption saturation of the surface soil was calculated from equation (1), using Mehlich-3 P as extractable soil P and the Langmuir P-sorption maximum as P-sorption capacity, a single relationship (r^2 of 0.86) described the concentration of DP as a function of P-sorption saturation for all soils (Fig. 7.6). Therefore, P saturation better describes the effect of soil type on the differential release of soil P to surface runoff than common soil-P test measures. Using the relation between surface-runoff P and soil-P saturation, we can determine either a P saturation that will support an 'acceptable' P concentration in surface runoff or vice versa, a P concentration that could be expected from a soil of given P saturation. For example, a P saturation of 25%, the critical value used in the

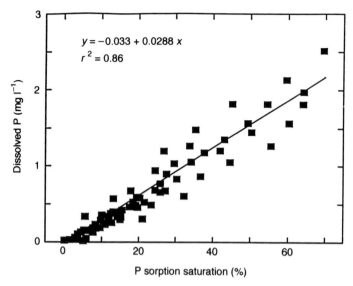

Fig. 7.6. Relationship between dissolved-P concentration of surface runoff obtained from simulated rainfall (2.54 cm h^{-1}, 30 min) and soil-P sorption saturation of surface soil (0–1 cm) 7 days after poultry-litter application. P-sorption saturation was calculated from Mehlich-3 P and Langmuir P-sorption maximum. Ten soils ranging from sandy loam to clay. (From Sharpley, 1995.)

Netherlands, would support a DP concentration in surface runoff of 0.69 mg l^{-1} using Mehlich-3 P and 1.58 mg l^{-1} using strip P. Clearly, the actual value of P-sorption saturation for a given soil will vary with the method used to estimate extractable P in equation (1). However, once calibrated for specific methodology, the P sorption saturation approach has the potential to describe the release of P from a wide range of soils to surface runoff.

An added advantage of the P-saturation approach is that it not only describes the potential for P release from soil but also indicates how close the P-sorption sites of a soil are to being saturated. In other words, measuring P saturation both describes the potential of a soil to enrich surface runoff with DP (high degree of P saturation) and also helps to predict how much of the P added in fertilizers and manures will be retained by the soil in a form that is relatively resistant to loss in surface runoff (low degree of P saturation).

The added complexity of this approach in terms of obtaining a reliable estimate of soil-P sorption capacity, compared with standard soil-test methods, may limit its acceptability at the present time. Also, greater variation from the linear relationship at high P-sorption saturations (> 50%) may limit use of this approach to lower saturation values (< 40%). However, since it accounts for soil properties affecting P sorption and desorption, the P-saturation approach provides a greater degree of flexibility across soil types than soil-test P alone in estimating the potential for loss in surface runoff from a given site. This approach could then be used on soils that have already been identified as being vulnerable to P loss, as a result of high soil P and potential for surface runoff and erosion.

IMPLEMENTING ENVIRONMENTAL SOIL-TEST PROGRAMMES RELATIVE TO SURFACE RUNOFF AND EROSION

Several states in the USA have identified critical or threshold soil-test P levels above which the potential for unacceptable P losses in runoff exceeds any crop-response concerns (Table 7.5). Attempts have been made to base manure management recommendations on these soil P levels.

Because of the economic implications of reduced manure applications when based on P rather than crop nitrogen (N) requirements and lack of supporting documentation relating runoff and soil P, many farmers are contesting the use of threshold soil-P levels as a basis for determining manure-application rates. Clearly, more information is needed on the relationship between runoff and soil P as a function of site, P source and management characteristics.

However, it is extremely time-consuming and costly to obtain data on DP concentrations in surface runoff and unrealistic to expect that a large database relating soil-test P to DP will be available in the near future. Because of this, one role soil-testing laboratories could play in improved soil-P management would be to provide predictions of 'readily desorbed P', based on soil-test P and other measured soil properties (e.g. pH, texture and organic matter content). These data could then be integrated with other information specifically related to surface-runoff volume, in order to rate the potential of individual fields to be significant sources of P to nearby surface waters.

Soil-testing laboratories could further contribute to more environmentally efficient P management by offering special tests for P that could be conducted on samples from areas with a high potential for P losses in surface runoff and erosion. Fields for more intensive sampling and testing could be identified based on data available in routine soil tests and supplemental information related to the potential for surface-runoff volume and erosion. Once high-risk areas are known, advisory agencies could conduct more intensive sampling of the upper 0–5 cm, focusing on the most erosion- and surface-runoff-prone areas. Together, these data would not only identify fields where additional P should not be applied, but also specific sites where more intensive soil-conservation practices would be needed because of topographical and hydrological considerations.

Examples of special tests include direct measurements of readily

Table 7.5. Soil test P interpretations and management guidelines in USA (from Sharpley et al., 1994a, and Gartley and Sims, 1994).

State	Critical value	Management recommendation	Rationale
Arkansas	Mehlich-3 P: 150 mg kg^{-1} SR: No	At or above 150 mg kg^{-1} soil P: 1. Apply no P from any source 2. Provide buffers next to streams 3. Overseed pastures with legumes to aid P removal 4. Provide constant soil cover to minimize erosion	CV: Ohio sewage sludge data MR: reduce soil P and minimize movement of P from field
Delaware	Mehlich-1 P: 120 mg kg^{-1} SR: No	Above 120 mg kg^{-1} soil P: Apply no P from any source until soil P is significantly reduced	CV: greater P-loss potential from high-P soils MR: protect water quality by minimizing further soil-P accumulations
Ohio	Bray-1 P: 150 mg kg^{-1} SR: Yes	Above 150 mg kg^{-1} soil P: 1. Institute practices to reduce erosion 2. Reduce or eliminate P additions	CV: greater P-loss potential from high-P soils as well as role of high soil P in zinc deficiency MR: protect water quality by minimizing further soil-P accumulations
Oklahoma	Mehlich-3 P: 130 mg kg^{-1} SR: No	30 to 130 mg kg^{-1} soil P: Half P rate on > 8% slopes 130 to 200 mg kg^{-1} soil P: Half P rate on all soils and institute practices to reduce runoff and erosion Above 200 mg kg^{-1} soil P: P rate not to exceed crop removal	CV: greater P-loss potential from high-P soils MR: protect water quality, minimizing further soil-P accumulations, and maintain economic viability

State	Soil test	Recommendation	Rationale
Michigan	Bray-1 P: 75 mg kg^{-1}	Above 75 mg kg^{-1} soil P: P application must not exceed crop removal Above 150 mg kg^{-1} soil P: Apply no P from any source	CV: minimize P loss by erosion or leaching in sandy soils MR: protect water quality and encourage wider distribution of manures
	SR: No		
Texas	Bray-1 P or Texas A&M P: 200 mg kg^{-1}	Above 200 mg kg^{-1} soil P: P addition not to exceed crop removal	CV: greater P loss potential from high-P soils MR: protect water quality by minimizing further soil-P accumulations
	SR: No		
Wisconsin	Bray-1 P: 75 mg kg^{-1}	Above 75 mg kg^{-1} soil P: 1. Rotate to P-demanding crops 2. Reduce manure-application rates Above 150 mg kg^{-1} soil P: Discontinue manure applications	CV: soils will remain non-responsive to applied P for 2–3 years MR: minimize further soil-P accumulations
	SR: For cost-share programmes		

SR represents whether state regulations govern P application rates; CV represents critical-value rationale and MR management-recommendation rationale.

desorbable P, Fe-oxide strip P, biologically available P and P-sorption saturation. Incorporation of them in a soil-testing programme is primarily a matter of realizing the important role that these tests can play in improving the environmental efficiency of modern agriculture.

Finally, soil-testing laboratories must provide interpretative and educational information to those involved in assessing the impact of P losses in surface runoff, including programmes to explain the meaning and use of current and newly developed soil-P tests. Since it is highly unlikely that dozens of multiyear, multisite studies relating surface-runoff P to soil-test P and other factors controlling runoff volume (climate, topography, etc.) will be conducted in the next 5–10 years, interim steps are needed to identify soils and areas that now contribute significant amounts of runoff P to surface waters. There is a definite need to continue efforts to identify soils that are excessive in P and develop a logical approach to establish environmental upper limits for soil-test P. Just as important is the need to prevent soils that are near this state from getting worse, by developing fertilizer and manure recommendations that maintain agronomic profitability and minimize environmental risk from high P-soils.

CONCLUSION

It is proposed that upper critical limits for P in agricultural soils should be set relative to both optimizing P nutrition of crops and minimizing P losses to the aquatic environment.

In theory, the optimum soil-P status relative to crop production is the soil-P status when the optimum P-application rate equals the amount of P removed with the harvested crop. In practice, determination of the optimum soil-P status is extremely difficult, due to problems with determining both the soil-P status (methodology and spatial variation) and the optimum P-application rates. The latter vary with crop and crop yield potential and with P-fertilizer type, application method and effects on crops and soil-P status in the short and long term, as well as with price relations between crops and P fertilizers. Determination of the optimum P-application rate requires costly field experiments, which are difficult to interpret and therefore have been interpreted in several different ways.

So-called critical levels of soil-test P determined from field experiments are generally higher in Europe than in the USA and Canada, but also vary within countries, probably mainly due to differences in crop-yield potential.

Concentration of DP in surface runoff is related to soil-test P, but the relation varies with soil type, cropping and runoff episode. Dividing the P-test value of a soil by its P-sorption capacity seems to account for the variation due to soil type and to give an indication of how close the P-sorption sites are to being saturated. Concentration of BAP in surface runoff containing eroded sediments is related to P extracted by Fe-oxide strip.

Critical soil-test P values suggested by some USA states relative to P-loss potential are four to eight times above critical levels related to crop production.

Variability in runoff volume and erosion as a result of climatic, topographic and agronomic factors plays a larger role than soil-test P in determining the amount of P losses from agricultural land. A comprehensive approach is needed for reliable and yet flexible recommendations of fertilizer- and manure-P management integrating soil-test P with estimates of potential runoff and erosion losses.

ABSTRACT

There are two main issues regarding the level of soil phosphorus (P): (i) a certain critical level is necessary for economic crop production, and (ii) part of the soil P is lost from agricultural land to the aquatic environment by wind erosion, surface runoff, water erosion and leaching. The chapter deals with attempts to set levels of soil P relative to the first issue and relative to the potential for P losses by mainly surface runoff and water erosion. Implementation of environmental soil-test programmes relative to surface runoff and erosion is also discussed.

Calculations were made of ratios between amounts of soil P extracted with different P-test methods used in Europe, the USA and Canada. Then so-called critical levels of soil-test P, calibrated from P-fertilization field trials, could be compared, even if different P-test methods had been used. The critical levels of soil-test P calibrated in Europe were generally higher than the ones calibrated in the USA and Canada, but large differences in estimated levels were also observed within countries and the levels also varied with type of crop and yield potential.

Soil-test P level was related to concentration of dissolved phosphate (DP) in surface runoff, but the relation was found to differ between soil types, crops and runoff episodes. Inclusion of P-sorption capacity levelled out the variation due to soil type. P extracted with Fe-oxide strips from a soil suspension was related to bioavailable P (BAP). The variability in runoff volume and erosion as a result of climatic, topographic and agronomic factors, however, plays a larger role than soil P in determining the losses of P in plot experiments.

Critical soil-test P values, suggested by some USA states relative to P-loss potential, were four to eight times above the critical levels of soil-test P relative to crop production. Thus, a comprehensive approach is needed for reliable and yet flexible recommendations of fertilizer- and manure P-management, integrating soil-test P with estimates of potential runoff and erosion losses.

REFERENCES

Amer, F., Bouldin, D.R., Black, C.A. and Duke, F.R. (1955) Characterization of soil phosphorus by anion exchange resin adsorption and P^{32} equilibrium. *Plant and Soil* 6, 391–408.

Andraski, B.J., Mueller, D.H. and Daniel, T.C. (1985) Phosphorus losses in runoff as affected by tillage. *Soil Science Society of America Journal* 49, 1523–1527.

Barber, S. (1985) *Soil Nutrient Bioavailability. A Mechanistic Approach.* John Wiley and Sons, New York, 398 pp.

Bates, T.E (1990) Prediction of phosphorus availability from 88 Ontario soils using five phosphorus soil tests. *Communications in Soil Science and Plant Analysis* 21, 1009–1023.

Baumgärtel, G. (1989) Phosphat-Düngerbedarf von Getreide und Zuckerrüben im Südniedersächsischen Lössgebiet. *Zeitschrift für Pflanzenernaehrung und Bodenkunde* 152, 447–452.

Beegle, D.B and Oravec, T.C. (1990) Comparison of field calibrations for Mehlich 3 P and K with Bray–Kurtz P1 and ammonium acetate K for corn. *Communications in Soil Science and Plant Analysis* 21, 1025–1036.

Bray, R.H. and Kurtz, L.T. (1945) Determination of total, organic and available forms of phosphorus in soils. *Soil Science* 59, 39–45.

Breeuwsma, A. and Silva, S. (1992) *Phosphorus Fertilisation and Environmental Effects in the Netherlands and the Po Region (Italy).* Agricultural Research Department Report 57, The Winand Staring Centre for Integrated Land, Soil and Water Research, Wageningen, the Netherlands, 39 pp.

Buondonno, A., Coppola, E., Felleca, D. and Violante, P. (1992) Comparing tests for soil fertility: 1. Conversion equations between Olsen and Mehlich 3 as phosphorus extractants for 120 soils of south Italy. *Communications in Soil Science and Plant Analysis* 23, 699–716.

Cate, Jr, R.B. and Nelson, L.A. (1971) A simple statistical procedure for partitioning soil test correlation data into two classes. *Soil Science Society of America Proceedings* 35, 658–660.

Colwell, J.D. (1963) The estimation of the phosphorus fertilizer requirements of wheat in southern New South Wales by soil analysis. *Australian Journal of Experimental Agriculture and Animal Husbandry* 3, 190–197.

Cox, F.R. (1992) Range in soil phosphorus critical levels with time. *Soil Science Society of America Journal* 56, 1504–1509.

Dahnke, W.C. and Olson, R.A. (1990) Soil test correlation, calibration, and recommendation. In: Westerman, R.L. (ed.) *Soil Testing and Plant Analysis*, 3rd edn. Soil Science Society of America, Madison, Wisconsin, USA, pp. 45–71.

Daniel, T.C., Sharpley, A.N. and Logan, T.J. (1991) Effect of soil test phosphorus on the quality of runoff water: research needs. In: Blake, K.P. (ed.) *Proceedings of the National Livestock, Poultry and Aquaculture Waste Management Workshop*, April 1991, Kansas City, Missouri. American Society of Agricultural Engineering, St Joseph, Michigan, 13 pp.

Daniel, T.C., Edwards, D.R. and Sharpley, A.N. (1993) Effect of extractable soil surface phosphorus on runoff water quality. *Transactions of the American Society of Agricultural Engineering* 36, 1079–1085.

Daniel, T.C., Sharpley, A.N., Edwards, D.R., Wedepohl, R. and Lemunyon, J.L.

(1994) Minimizing surface water eutrophication from agriculture by phosphorus management. *Journal of Soil and Water Conservation* 49, 30–38.

Egnér, H., Riehm, H. and Domingo, W.R. (1960) Untersuchungen über die chemische Bodenanalyse als Grundlage für die Beurteilung des Nährstoffzustandes der Böden. II. Chemische Extraktionsmethoden zur Phospor- und kaliumbestimmung. *Kungliga Landbrukshögskolans Annaler* 26, 199–215.

Gartley, K.L. and Sims, J.T. (1994) Phosphorus soil testing: environmental uses and implications. *Communications in Soil Science and Plant Analysis* 25, 1565–1582.

Gascho, G.J., Gaines, T.P. and Plank, C.O. (1990) Comparison of extractants for testing coastal plain soils. *Communications in Soil Science and Plant Analysis* 21(13–16), 1051–1077.

Greenwood, D.J., Cleaver, T.J., Turner, M.K., Hunt, J., Niendorf, K.B. and Loquens, S.M.H. (1980) Comparison of the effects of phosphate fertilizer on the yield: phosphate content and quality of 22 different vegetable and agricultural crops. *Journal of Agricultural Science* 95, 457–469.

Hanlon, E.A. and Johnson, G.V. (1984) Bray/Kurtz, Mehlich III, AB/D and ammonium acetate extractions of P, K and Mg in four Oklahoma soils. *Communications in Soil Science and Plant Analysis* 15, 277–294.

Holford, I.C.R., Morgan, J.M., Bradley, J. and Cullis, B.R. (1985) Yield responsiveness and response for the evaluation and calibration of soil phosphate tests for wheat. *Australian Journal of Soil Research* 23, 167–180.

Johnston, A.E., Lane, P.W., Mattingly, G.E.G. and Poulton, P.R. (1986) Effects of soil and fertilizer P on yields of potatoes, sugar beet, barley and winter wheat on a sandy clay loam soil at Saxmundham, Suffolk. *Journal of Agricultural Science, Cambridge* 106, 155–167.

Jungk, A., Claassen, N., Schulz, V. and Wendt, J. (1993) Pflanzenverfügbarkeit der Phosphatvorräte ackerbaulich genutzter Böden. *Zeitschrift für Pflanzenernaehrung und Bodenkunde* 156, 397–406.

Kingery, W.L., Wood, C.W., Delaney, D.P., Williams, J.C. and Mullins, G.L. (1994) Impact of long-term land application of broiler litter on environmentally related soil properties. *Journal of Environmental Quality* 23, 139–147.

Köster, W. and Schachtschabel, P. (1983) Kurzmitteilung-Beziehung zwischen dem durch Phosphatdüngung erzielbaren Mehrertrag und dem Phosphatgehalt im Boden. *Zeitschrift für Pflanzenernaehrung und Bodenkunde* 146, 539–542.

Kristensen, P. and Hansen, H.O. (1994) *European Rivers and Lakes – Assessment of Their Environmental State*. EEA Environmental Monograph No. 1, European Environmental Agency, Copenhagen, 122 pp.

McCallister, D.L., Shapiro, C.A., Raun, W.R., Anderson, F.N., Rehm, G.W., Engelstad, O.P., Russelle, M.P. and Olson, R.A. (1987) Division S-8-fertilizer technology and use. *Soil Science Society of America Journal* 51, 1646–1652.

McCollum, R.E. (1991) Buildup and decline in soil phosphorus: 30-year trends on a typic umprabuult. *Agronomy Journal* 83, 77–85.

Mallarino, A.P. and Blackmer, A.M. (1992) Comparison of methods for determining critical concentrations of soil test phosphorus for corn. *Agronomy Journal* 84, 850–856.

Mehlich, A. (1984) Mehlich 3 soil test extractant: a modification of Mehlich 2 extractant. *Communications in Soil Science and Plant Analysis* 15(12), 1409–1416.

Menon, R.G., Hammon, L.L. and Sissingh, H.A. (1989) Determination of plant-available phosphorus by the iron hydroxide-impregnated filter paper (P_i) soil test. *Soil Science Society of America Journal* 53, 110–115.

Munk, H. (1985) Ermittlung wirtschaftlich optimaler Phosphatgaben auf Löss- und Geschiebelehmböden auf Basis der CAL-Methode. *Zeitschrift für Pflanzenernaehrung und Bodenkunde* 148, 193–213.

Munk, H. and Rex, M. (1990) Zur Eichung von Bodenuntersuchungsmethoden auf Phosphat. *Agribiological Research* 43, 164–174.

Nelson, W.L., Mehlich, A. and Winters, E. (1953) The development, evaluation and use of soil tests for phosphorus availability. In: Pierre, W.H. and Norman, A.G. (eds) *Soil and Fertilizer Phosphorus*. Agronomy No. 4, American Society of Agronomy, Madison, Wisconsin, pp. 153–188.

Nye, P. and Tinker, P.B. (1977) *Soil Movement in the Soil-Root System*. Studies in Ecology Vol. 4, Blackwell Scientific Publications, Oxford, 342 pp.

Olness, A.E., Smith, S.J., Rhoades, E.D. and Menzel, R.G. (1975) Nutrient and sediment discharge from agricultural watersheds in Oklahoma. *Journal of Environmental Quality* 4, 331–336.

Olsen, S.R., Cole, C.V., Watanabe, F.S. and Dean, L.A. (1954) *Estimation of Available Phosphorus in Soils by Extraction with Sodium Bicarbonate*. US Department of Agriculture Circular No. 939, Washington DC, 19 pp.

Olson, R.A., Frank, K.D., Grabouski, P.H. and Rehm, G.W. (1982) Economic and agronomic impacts of varied philosophies of soil testing. *Agronomy Journal* 74, 492–499.

Pote, D.H., Daniel, T.C., Sharpley, A.N., Moore, Jr, P.A., Edwards, D.R. and Nichols, D.J. (1996) Relating extractable soil phosphorus to phosphorus losses in runoff. *Soil Science Society of America Journal* 60, 855–859.

Read, D.W.L., Spratt, E.D., Bailey, L.D., Warder, F.G. and Ferguson, W.S. (1973) Residual value of phosphatic fertilizer on chernozemic soils. *Canadian Journal of Soil Science* 53, 389–398.

Riehm, H. (1955) Die Untersuchung von Böden. In: Thun, R., Herrmann, R. and Knickmann, E. (eds) *Methodenbuch Band I*. Neumann Verlag, Berlin.

Romkens, M.J.M and Nelson, D.W. (1974) Phosphorus relationships in runoff from fertilized soil. *Journal of Environmental Quality* 3, 10–13.

Schachtschabel, P. (1973) Beziehungen zwischen dem Phosphorgehalt in Böden und jungen Haferpflanzen. *Zeitschrift für Pflanzenernährung und Bodenkunde* 135, 31–43.

Schachtschabel, P and Köster, W. (1985) Kurzmitteilung-Beziehung zwischen dem Phosphatgehalt im Boden und der optimalen Phosphatdüngung in langjährigen Feldversuchen. *Zeitschrift für Pflanzenernährung und Bodenkunde* 148, 459–464.

Schreiber, J.D. (1988) Estimating soluble phosphorus (PO_4-P) in agricultural runoff. *Journal of Mississippi Academy of Science* 33, 1–15.

Schüller, H. (1969) Die CAL-Methode, eine neue Methode zur Bestimmung des pflanzenverfügbaren Phoshates in Böden. *Zeitschrift für Pflanzenernaehrung und Bodenkunde* 123, 48–63.

Semb, G. (1986) Sammenligning av AL-og natriumbikarbonatløselig fosfor i jord med pH over 6,6. *Jord og Myr* 10, 185–193.

Sharpley, A.N. (1985) Depth of surface soil–runoff interaction as affected by rainfall, soil slope and management. *Soil Science Society of America Journal* 49, 1010–1015.

Sharpley, A.N. (1995) Dependence of runoff phosphorus on soil phosphorus content. *Journal of Environmental Quality* 24, 920–926.

Sharpley, A.N., Syers, J.K. and Tillman, R.W. (1978) An improved soil-sampling procedure for the prediction of dissolved inorganic phosphate concentrations in surface runoff from pasture. *Journal of Environmental Quality* 7, 455–456.

Sharpley, A.N., Ahuja, L.R., Yamamoto, M. and Menzel, R.G. (1981a) The kinetics of phosphorus desorption from soil. *Soil Science Society of America Journal* 45, 493–496.

Sharpley, A.N., Ahuja, L.R. and Menzel, R.G. (1981b) The release of soil phosphorus to runoff in relation to the kinetics of desorption. *Journal of Environmental Quality* 10, 386–391.

Sharpley, A.N., Jones, C.A. and Gray, C. (1984) Relationships among soil P test values for soils of differing pedogenesis. *Communications in Soil Science and Plant Analysis* 15, 985–995.

Sharpley, A.N., Smith, S.J. and Menzel, R.G. (1986) Phosphorus criteria and water quality management for agricultural watersheds. *Lake Reservoir Management* 2, 177–182.

Sharpley, A.N., Smith, S.J. and Bain, W.R. (1993) Nitrogen and phosphorus fate from long-term poultry litter applications to Oklahoma soils. *Soil Science Society of America Journal* 57, 1131–1137.

Sharpley, A.N., Sims, J.T. and Pierzunski, G.M. (1994a) Innovative soil P availability indices: assessing inorganic phosphorus. In: Havlin, J.L., Jacobsen, J.S., Leikam, D.F., Fixen, P.E. and Hergert, G. (eds) *Soil Testing: Prospects for Improving Nutrient Recommendations*. Soil Science Society of America Special Publication No. 40, American Society of Agronomy, Madison, Wisconsin, pp. 115–142.

Sharpley, A.N., Chapra, S.C., Wedepohl, R., Sims, J.T., Daniel, T.C. and Reddy, K.R. (1994b) Managing agricultural phosphorus for protection of surface waters: issues and options. *Journal of Environmental Quality* 23, 437–451.

Sibbesen, E. (1978) An investigation of anion exchange resin method for soil phosphorus extraction. *Plant and Soil* 50, 305–321.

Sibbesen, E. (1983) Phosphate soil tests and their suitability to assess the phosphate status of soil. *Journal of the Science of Food and Agriculture* 34, 1368–1374.

Sibbesen, E. (1986) Soil movement in long-term field experiments. *Plant and Soil* 91, 73–85.

Sibbesen, E. and Runge-Metzger, A. (1995) Phosphorus balance in European agriculture – status and policy options. In: Tiessen, H. (ed.) *Phosphorus in the Global Environment: Transfers, Cycles and Management*. SCOPE 54, John Wiley & Sons, Chichester, pp. 43–57.

Sibbesen, E., Skjøth, F. and Christensen, B.T. (1995) Soil and substance movement between plots in long-term field experiments. In: Christensen, B.T. and Trentemøller, U. (eds) *The Askov Long-term Experiments on Animal Manure and Mineral Fertilizers: 100th Anniversary Workshop*. SP Report No. 29, Danish Institute of Plant and Soil Science, Tjele, pp. 136–153.

Sims, J.T. (1989) Comparison of Mehlich 1 and Mehlich 3 extractants for P, K, Ca, Mg, Cu, and Zn in Atlantic coastal plain soils. *Communications in Soil Science and Plant Analysis* 20, 1707–1726.

Sissingh, H.A. (1971) Analytical technique on the Pw method, used for assessment of the phosphate status of arable soils in the Netherlands. *Plant and Soil* 34, 483–486.

Stumpe, H., Garz, J. and Scharf, H. (1994) Wirkung der Phosphatdüngung in einem 40 jährigen Dauerversuch auf einer Sandlöss-Braunschwarzerde in Halle. *Zeitschrift für Pflanzenernährung und Bodenkunde* 157, 105–110.

US Environmental Protection Agency (USEPA) (1994) *National Water Quality Inventory: 1992 Report to Congress*. USEPA 841-R-94-001, Office of Water, US Government Printing Office, Washington, DC.

Vaithiyanathan, P. and Correll, D.L. (1992) The Rhode River watershed: phosphorus distribution and export in forest and agricultural soils. *Journal of Environmental Quality* 21, 280–288.

Van der Paauw, F. (1971) An effective water extraction method for determination of plant-available soil phosphorus. *Plant and Soil* 34, 467–481.

Van der Paauw, F. (1977) Die stellung der P-Wassermethode zur Erfassung des P-Angebotes des Bodens. *Landwirtschaftliche Forschung* 34(2), 109–120.

Van der Zee, S.E.A.T.M., Fokkink, L.G.J. and van Riemsdijk, W.H. (1987) A new technique for assessment of reversibly adsorbed phosphate. *Soil Science Society of America Journal* 51, 599–604.

Van der Zee, S.E.A.T.M., van Riemsdijk, W.H. and de Haan, F.A.M. (1990) *Het protocol fosfaatverzadigde gronden*. Vakgroep Bodemkunde en Plantevoeding, Wageningen Agricultural University, Wageningen, the Netherlands.

Webb, J.R., Mallarino, A.P. and Blackmer, A.M. (1992) Effects of residual and annually applied phosphorus on soil test values and yields of corn and soybean. *Journal of Production Agriculture* 5, 148–152.

Whitney, D.A., Cope, J.T. and Welch, L.F. (1985) Prescribing soil and crop nutrient needs. In: Engelstad, O.P. (ed.) *Fertilizer Technology and Use*, 3rd edn. Soil Science Society of America, Madison, Wisconsin, USA, pp. 25–52.

Wolf, A.M. and Baker, D.E. (1985) Comparisons of soil test phosphorus by Olsen, Bray PI, Mehlich I and Mehlich III methods. *Communications in Soil Science and Plant Analysis* 16, 467–484.

8 Phosphorus Fertilizer Strategies: Present and Future

H. Tunney,[1] A. Breeuwsma,[2] P.J.A. Withers[3] and P.A.I. Ehlert[4]

[1] Teagasc, Johnstown Castle Research Centre, Wexford, Ireland; [2] The Winand Staring Centre (SC-DLO), PO Box 125, NL-6700 AC Wageningen, the Netherlands; [3] ADAS Bridgets Research Station, Martyr Worthy, Winchester, Hants SO21 1AP, UK; [4] Research Institute for Agrobiology and Soil Fertility (AB-DLO), PO Box 129, NL 9750 AC Haren (Gr), the Netherlands

INTRODUCTION

The importance of recycling organic manures to supply phosphorus (P) and other nutrients for crop production has been recognized by farmers for thousands of years. Two thousand years ago, Virgil in his book of verse, the *Georgics*, advised farmers to recycle organic manure (Rhoades, 1952). Phosphorus was the first element to be recognized as an essential nutrient for plants. It was only in the middle of the last century that a secure supply of relatively inexpensive P fertilizer began to emerge (Johnston and Poulton, 1992). In 1842, J.B. Lawes of Rothamsted took out a patent for the manufacture of superphosphate and started commercial production in London in 1843. From about that time agriculturalists began to address the question of P recommendations and the correct quantity to apply to meet the needs of crops.

We now have a century and a half of a tradition of P fertilizer recommendations, which has involved scientists in research institutes and universities, the chemical and fertilizer industries, the fertilizer trade and farmers and their advisers. The success of this development is evident in all developed countries, where serious plant and animal health problems due to P deficiency have been eliminated. Most of the work on improving P recommendations has developed over the past 50 years. Many experiments have demonstrated the response of crops to chemical P fertilizers in most countries of the world and the agronomic benefit associated with building up a satisfactory level of P fertility. Figure 8.1 shows an example of the dramatic increase in crop yield that can be obtained by applying P where soil P is low. The initial soil-test P value was 2 mg l^{-1} (Morgan, 1941) in this

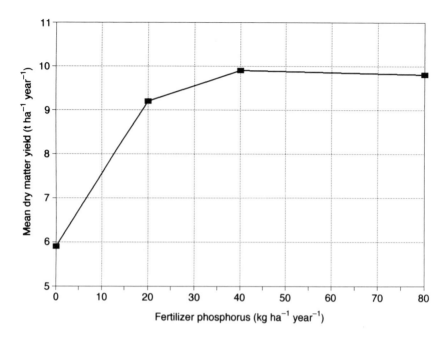

Fig. 8.1. Effect of phosphorus fertilizer on mean grass yield, on three sites over 4 years (1967–1970) (from Ryan and Finn, 1976).

experiment. For definitions and references for soil-P tests, see Table 8.1. High fertilizer P dressings to low-P soils may give yields equal to those with a residual P fertility (Johnston and Poulton, 1992).

Soil testing is another development that also started in the last century (Liebig, 1872; Dyer, 1894, 1901). It is now widely accepted and used, particularly over the past 50 years. These tests were used initially to identify soils low in nutrients and are now widely used as a basis for P fertilizer recommendations and monitoring P fertility.

As P fertilizer strategies have developed, a number of gradual changes have been observed on intensive farms in many developed countries.

- Fertilizer P and animal manure use increased and is now stabilizing.
- A surplus of P inputs over outputs developed.
- Soil-test P levels increased.
- A visual response in the field to P fertilizer disappeared.
- Eutrophication of waters started to be a problem in predominantly agricultural areas.

Table 8.1. Examples of methods of soil analyses for P used in selected European countries.

P test	Method (soil : solution ratio)	Country	Reference
Pw	1 : 60 (v/v), extraction with water at 20°C, 22 h incubation, 1 h shaking	Belgium, the Netherlands, Switzerland	Sissingh, 1971
Pw modified	1 : 50 (v/v), extraction with water at 20°C, 2 h shaking	Germany (Hanover)	Schachtschabel and Koster, 1985 (after Sissingh, 1971)
P-AL	1 : 20 (w/v), 0.1 M ammonium lactate + 0.4 N acetic acid, pH 3.75, 2 h shaking	Belgium, the Netherlands	Egnér et al., 1960
P-DL	1 : 50 (w/v), 0.02 M calcium lactate + 0.02 M hydrochloric acid, pH 3.7, 1.5 h shaking	Belgium, Germany	Egnér and Riehm, 1955
P-CAL	1 : 20 (w/v), 0.05 M calcium lactate + 0.05 M calcium acetate + 0.3 M acetic acid, pH 4.1, 2 h shaking	Austria, Belgium, Germany	Schüller, 1969
P-NH$_4$Ac + EDTA	1 : 5 (w/v), ammonium acetate + EDTA, pH 4.65	Belgium, Switzerland, Finland	van den Hende and Cothenie, 1960
P-EUF	Electroultrafiltration	Austria, Germany	Nemeth, 1979
P Dyer	1 : 5, citric acid 2%, 4 h shaking	France	Dyer, 1894
P Joret-Hebert	1 : 25, ammonium oxalate 0.2 M, 2 h shaking	France	Joret and Hebert, 1955
P Olsen	20 : 1 (w/v), 0.5 M sodium bicarbonate, pH 8.5, 1 h shaking	Denmark, France, England, Wales, Northern Ireland, Italy	Olsen et al., 1954
P Morgan	6 : 30 (v/v), 10% sodium acetate, pH 4.8, 0.5 h shaking	Ireland	Morgan, 1941

EDTA, ethylenediamine tetra-acetic acid.

Countries with water-quality problems are now focusing on P use (both fertilizers and manures) and recommendations, just as happened with nitrogen (N) about a decade ago.

Rapid soil-P tests were developed for estimating the P requirements of crops and do not have a clear relation with a mass balance for P in soil; nor do they provide information on mineralization, fixation, release, leaching, surface runoff or erosion. The identification of the quantitative contribution of these processes in the P cycle would help with more accurate estimates of sustainable P use.

This chapter addresses some of the approaches to P recommendations used at present in a number of European countries and goes on to consider proposals for P recommendations for sustainable agriculture. In this context, sustainable agriculture means optimum agricultural production that would not lead to problems of water pollution resulting from overenrichment with P.

CHEMICAL SOIL TESTS FOR PHOSPHORUS

Methods

Soil-P tests are used in many countries as a basis for fertilizer recommendations. There have been many recent publications and reviews relating to the different soil-P tests being used (e.g. Sibbesen, 1983; Sharpley *et al.*, 1984, 1994a; Buondonno *et al.*, 1992; Houba *et al.*, 1992; Sims, 1993; Campbell, 1994; van Raij, 1994).

The small size of the sample relative to the area sampled presents problems for accuracy of sampling. Generally, about 20 cores (can vary from 15 to 40) from an area of between 1 and 4 ha collects 0.5–1 kg of soil; the plough layer can contain about 2×10^6 kg soil ha^{-1}. Sometimes much larger areas are sampled and sampling can vary, but it is usually between 0–5 and 0–25 cm deep. This means that the soil sample is only about one part in 10 million of the bulk sample it is considered to represent and as little as 1–5 g of the sample is analysed. In addition, there may be wide variation in soil-test P values within different parts of a single field, due to spatial variability (Froment *et al.*, 1995).

Different extractants are used to extract what is often referred to as plant-available P from soil. However, these extractants, with the exception of water, are destructive in their action and recover variable amounts of P from different soil pools. Such extractants have been selected because, in field experiments, they have been shown to give a reasonable correlation between extracted P and uptake, dry-matter yield and crop response to added fertilizer P where soil P is relatively low.

Environmental aspects associated with the application of P fertilizers have recently been reviewed (Withers and Sharpley, 1995). In particular, the excessive accumulation and subsequent risk of desorption of soil P from previous P applications, as either fertilizers or manures, represents a potential eutrophication risk. In order to minimize the unnecessary build-up of excessive soil P, recommendations on fertilizers and manures need to be such that only agronomically justified amounts are applied. Since eutrophication problems extend across national boundaries, this chapter compares fertilizer P recommendations in different countries in order to provide a focus for the development of environmentally sustainable P fertilizer strategies.

More than ten different soil extractants are used to test for soil P in Europe (Hanotiaux and Vanoverstraeten, 1990) and almost double that number worldwide; this number is larger than for any other plant nutrient.

No one test is superior to all others and there has been an ongoing search for an improved test. Most countries or regions have decided, over the past 50 years, on one test (or sometimes two) and have continued with it because they consider that recommendations derived from the accumulated database and experimental results are sufficiently accurate for their needs, which are usually related to crop production. In countries where two tests are used, one may be for calcareous soils and the other for non-calcareous soils (e.g. France) or one for grassland and one for cultivated soil (e.g. the Netherlands). A change in the soil-test method may raise questions about the reliability of past recommendations. In Ireland, for example, the Morgan P test has been used for almost 50 years, even though it is now the only country in Europe to use this test. Similarly in the USA, only two states – New York and Maryland – use the Morgan P test. In England, Wales and Northern Ireland a change was made from Morgan's and other reagents to Olsen's in 1971. There is interest in a single extractant that can be used for all or most nutrients. In the Netherlands, for example, recalibration of current fertilizer recommendations based on 0.01 M calcium chloride ($CaCl_2$) is being investigated.

One important consequence of the large number of different tests is that soil-test results and consequently P recommendations cannot be compared readily between countries or states. In Ireland, for example, the Olsen test is used in the six counties of Northern Ireland (UK) and Morgan's test is used in the 26 counties of the Republic of Ireland. It is therefore difficult to compare soil-test results and P recommendations, even though soil, farming and climatic conditions are similar. Likewise, in many parts of Belgium, Denmark, the Netherlands and northern Germany, the soil and farming systems are broadly similar, and yet these four countries have different soil-P tests. Some soil-P tests may be more suited to certain soil types; for example, the Olsen test, based on sodium bicarbonate, is considered more suitable for neutral and alkaline soils than an acid-based extractant such as Mehlich 3 or Morgan.

The relative merits of different soil-P tests have been extensively discussed. Harmonizing soil-P test methods between different countries may be difficult because of technical and historical considerations. Perhaps the best approach, in the short term, may be to standardize methods against a common reference method or methods, which should be relatively independent of soil factors like pH. Such methods would include resin-, water- or 0.01 M CaCl$_2$-extractable P. Soil-P tests used routinely on samples taken from farms in a number of European countries are summarized in Table 8.1. The soil samples are generally taken to a depth of up to 10 cm for grassland and up to 30 cm for cultivated land. Most of the soil sampling is carried out in autumn and spring. From Table 8.1 it can be seen that the soil-to-solution ratio varies from 1:5 to 1:60 and that shaking time varies from 0.5 h up to 4 h.

Table 8.2 summarizes P tests used in selected European countries and the classification of low, medium and high P levels. It shows general trends and is for comparison purposes only; it does not assume that each P fertilizer recommendation scheme uses the same approach. Although there are differences between countries, there are also similarities; for example, the classification of soil fertility by the Chamber of Agriculture of Hanover, Germany, is similar to that of the Netherlands.

Although Table 8.1 deals only with European countries, the Olsen P test, often slightly modified, is used widely in Australia and New Zealand. Sharpley *et al.* (1994a) indicate that the most widely used tests in the USA are Mehlich 1 (Nelson *et al.*, 1953), Mehlich 3 (Mehlich, 1984), Bray and Kurtz No. 1 (Bray and Kurtz, 1945), Olsen (Olsen *et al.*, 1954), Morgan (Morgan, 1941) and modified Morgan (McIntosh, 1969). More information on comparisons between different soil-P tests is given by Sibbesen and Sharpley (Chapter 7, this volume).

Even using the same test there are differences in classification. For example, in Italy (e.g. Emilia–Romagna region), Olsen P concentrations used to classify soils as low, medium and high are much lower than in Denmark or the UK; the difference can be striking. For example, in Denmark, soils are low in P if they contain less than 20 mg kg^{-1} but in Italy a level of 20 mg kg^{-1} is considered very high. In Italy, target soil-P concentrations have been defined for both different soil types and different crops. Hence a soil-test P value of 15 mg kg^{-1} is considered very high for a sandy soil but normal (medium) for a clay soil. Similarly, satisfactory soil-P levels for responsive crops (potatoes, sugar beet) are a little higher than for less responsive crops (e.g. cereals). For permanent grassland, Olsen soil-test P levels, in mg P kg^{-1} soil, of < 12, 12–15, 16–20 and > 20 are classified as low, medium, high and very high, respectively, for a medium (between sandy and clayey) soil. The identification of more precise critical thresholds of Olsen P for different crops has also been undertaken for selected sites in the UK (Johnston and Poulton, 1992), but this approach has not been extended to national recommendations. In New Zealand, both individual and range values for the biologically optimum Olsen P level (or target P levels) for grassland on different soil types have been identified (Roberts *et*

Table 8.2. Comparison of soil fertility classification based on soil-test P levels used in some European countries (mg P kg^{-1} or l^{-1} of soil).

Method (see Table 8.1)	Country	A very low	B low	C medium optimal	D high	E very high	Reference
P-AL	Belgium (south)	0–85	86–115	116–185	186–305	306+	Hanotiaux and Vanoverstraten, 1990
	Belgium (north)	0–70		70–180		180+	Hanotiaux and Vanoverstraten, 1990
	Netherlands (grassland)	0–80	80–140	140–180	180–240	240+	Anon., 1994
P-CAL	Belgium	0–80	80–150	150–310		310+	Hanotiaux and Vanoverstraten, 1990
	Germany (Western) (P-CAL/P-DL)	0–44	45–83	84–144	145–201	201+	Finck, 1982
P-DL	Germany (former East)	0–20	21–39	40–55	56–79	80+	Egnér and Riehm, 1955
P-NH$_4$Ac + EDTA	Belgium, Finland, Switzerland	0–25	25–50	50–90	90–180	180+	Hanotiaux and Vanoverstraten 1990
P Morgan	Ireland (Republic)	0–3	4–6	7–10	11–15	15+	Teagasc, 1994
P Olsen	Italy	0–10		11–20	20–30	30+	A. Giapponesi, 1995 (personal communication)
	United Kingdom	0–9	10–15	16–25	26–45	46+	MAFF, 1994
P Dyer	France	0–44	45–83	84–131	132–175	175+	Dyer, 1894
P Joret–Hebert	France	0–31	32–62	63–96	97–122	122+	Quemener, 1985
Pw	Netherlands (arable)	0–5	5–9	9–14	14–28	28+	Anon, 1992
Pw (modified)	Germany (Hanover)	0–4	5–10	11–18	19–30	31+	Werner, 1987

EDTA, ethylenediamine tetra-acetic acid.

al., 1994). Such approaches help to quantify the ability of different soils to maintain an adequate level of P in the soil solution for optimum uptake by the crop, i.e. a measure of soil-P buffer capacity.

The intensity of soil sampling varies widely between countries. In Europe, the Netherlands has the highest soil-sampling intensity, with an average of more than one soil sample per 10 ha per year, and Ireland would be intermediate, with an average of about one soil sample per 100 ha per year, while southern European countries have less. The use of soil testing is relatively low in most countries and almost non-existent in some developing countries. The more intensive farming systems use soil testing most. It is interesting that there appears to be a trend for countries with the highest soil testing intensity to have the highest P use, highest P surplus and highest soil-test P values. However, there are still areas of P-deficient soil in most developed countries.

Soil-P tests are increasingly being considered as tests for estimating excess P in the soil and as an indicator of P loss to water (Sims, 1993). There is little information on the suitability of soil-P tests for this purpose. The system of P loss from soil to water during rainfall is likely to be a simpler system than the uptake of P from the soil through the semipermeable membranes in plant roots. There may be a better relationship between soil-test P and P loss to water than with crop yield or uptake by plants. This may be particularly true where soil-P levels are high, as luxury P uptake by plants does not occur to any great extent, whereas loss to water is likely to be linearly related to the soil-test P (Sharpley *et al.*, 1994b).

At present, it is difficult to compare soil-P test results between countries; therefore, it would be helpful for European Union (EU) member states to work towards agreeing a common methodology for assessing soil-P levels required for optimum agricultural production and minimum risk of P loss from soil. Such studies are being carried out in Australia, New Zealand and the USA. One approach would be to agree on two or three of the most suitable tests (for example, water-soluble, Olsen and Mehlich-3 P) and correlate results from other methods with them. This approach should help facilitate the exchange of information and comparison of results on the minimum soil-P status for optimum agricultural production and water quality.

Soil-P tests are useful in identifying soils which have P levels far in excess of those required for optimum crop yields and which are therefore an environmental hazard. This information is already, or can be, used to reduce fertilizer costs by omitting P fertilizer or manures. However, for the majority of countries, maximum environmental thresholds for soil-P accumulation have not been defined, and more sensitive environmental soil tests for assessing the eutrophication risk are required (Withers and Sharpley, 1995).

Estimates of Long-term Availability of Soil Phosphorus Reserves

Very large reserves of soil P have been built up over the years on many intensively farmed soils. The soil reserves can be estimated as the product of the P removal in the agricultural production and the number of years that the optimum yield can be produced without the addition of fertilizer P. In Ireland, for example, it has been shown that optimum yields of grass can be obtained without the addition of P in fertilizer or manure for several years (Power, 1992). In this work, experiments were carried out on small plots on three farms in south-east Ireland that had not received large dressings of animal manures. Three grass cuts were taken per year, with yields averaging 12 t dry matter ha^{-1} $year^{-1}$ and an average annual P removal of about 40 kg P ha^{-1} $year^{-1}$. After 9 years of the experiment, more than 350 kg P ha^{-1} was removed on plots that had received no P and the yields were not significantly different between the zero P treatment and the treatments that were receiving up to 50 kg P ha^{-1} $year^{-1}$ (Tunney, 1995). This suggests a surplus P input in excess of 350 kg P during the 40 years before this experiment started. In addition, it indicates that most of this surplus input was available for plant uptake later. Power (1992) also showed that most of the surplus P had accumulated as calcium phosphate and that there has been little change in the organic P level in these experiments.

The average net removal of P from grassland in Ireland in milk, meat and water is less than 10 kg P ha^{-1} $year^{-1}$. This indicates that the soils in this experiment have adequate P for over 30 years' production if the animal manures are recycled. On cultivated soil where all the crop is sold off the farm, up to 30 kg P ha^{-1} $year^{-1}$ (with cereals, for example) may be removed. Similarly, Johnston and Poulton (1992) showed that, when a sandy clay loam was cropped without P applications for 9 years, the Olsen P values halved. More rapid estimates of soil P reserves can be made using small quantities of soil in pot experiments, as proposed by Stanford and deMent (1957), but in such experiments the very slow release of P from soil-P pools that are not readily available may not be measured. Residues of P accumulated between 1856 and 1901 in the 'exhaustion land' experiment at Rothamsted are still being released and giving larger yields of spring barley than on soils without such residues (Johnston and Poulton, 1977). For some soils, there may be a small but consistent yield loss as a result of withholding fertilizer P for periods longer than about 3 years (Withers, 1995). In New Zealand, farmers were not willing to reduce P inputs because of concern that it would affect production (A.H.C. Roberts, 1995, Ruakura, New Zealand, personal communication). Further information is needed on the ability of different P-rich soils to sustain maximum crop yields without fresh P inputs, because this could be important in terms of sustainable P use.

Trends in Soil-Test Phosphorus Levels and Fertilizer Use

Increasing trends in soil-test P levels have been observed in many countries (Tunney, 1990; Sims, 1993; Sharpley *et al.*, 1994a; Sibbesen and Runge-Metzer, 1995). In the UK, soil-P levels appear to have remained relatively stable over the past 25 years; however, a high proportion of arable soils (56%) and grassland soils (30%) have more than 26 mg l^{-1} of Olsen P (Skinner *et al.*, 1992). It appears that this is because in England and Wales P applied and P off-take are almost in balance (Salter *et al.*, in Chapter 21, this volume).

An example of increasing trends in soil-P levels is well illustrated in Fig. 8.2, where it can be seen that the average soil-P test level in Ireland increased steadily from less than 1 mg P l^{-1} soil in 1950 to about 9 mg P l^{-1} at present. The information summarized in Fig. 8.2 is based on soil samples from Irish farms, which were analysed at Johnstown Castle Laboratory. These results represent approximately 3 million soil samples from farms analysed over 45 years and are perhaps the best available example of the change in soil-P test levels with time. Figure 8.2 also shows that fertilizer P use on Irish farms has stabilized at about 62,000 t year^{-1} over the past 15 years, but the average soil-test P level continues to increase. This is

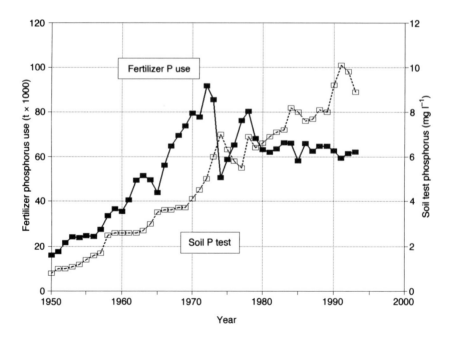

Fig. 8.2. Trends in phosphorus fertilizer use and in average soil-test phosphorus (Morgan) for samples from Irish farms over the past 45 years (from Tunney, 1990).

explained by a P balance-sheet study, which shows that inputs, at 77,000 t year^{-1}, are more than double the outputs of 31,000 t, giving an annual surplus of 46,000 t P (Tunney, 1990).

A study of a minicatchment (with 228 ha and eight farms) of the Dripsey River (which is a tributary on the north side of the Lee Valley in Co. Cork, Ireland) showed that many of the soils have high soil-P levels on the five intensive dairy farms in the catchment and lower levels on two beef farms and one sheep farm. Each field in the minicatchment was soil-sampled in 1993 and a scatter diagram (Fig. 8.3) summarizes the P and potassium (K) soil-test results (Tunney and Power, 1995), which show a trend of high soil K levels where soil P is high. A P balance sheet for the minicatchment for the 12 months of the study indicated a surplus of 18 kg P ha^{-1} and that some intensive dairy farms are building up excessive levels of both P and K in their soils. This does not make economic sense and is contributing to P loss to water. Water monitoring indicated that the loss of molybdenum reactive P (MRP) was 1.9 kg P ha^{-1} (and 3.9 kg total P ha^{-1}) for the 12 months from April 1993 to March 1994. The measurements suggest that half of this total may have come as a direct loss from farmyards. Similar trends have been noted on other dairy farms (Withers and Sharpley, 1995).

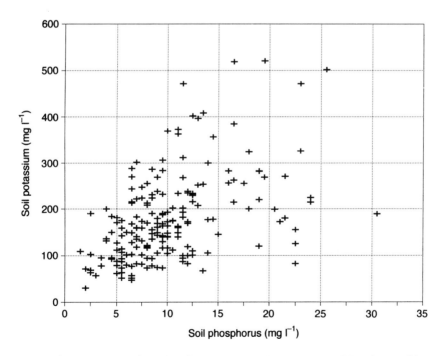

Fig. 8.3. Scatter diagram of soil phosphorus and potassium levels in a minicatchment of the River Lee in Ireland, showing high fertility, mostly on intensive dairy farms.

DIFFERENT PHOSPHORUS FERTILIZER RECOMMENDATIONS IN THE EUROPEAN UNION

Phosphorus Recommendations in Selected Countries

Many countries publish fertilizer recommendations for different crops at different soil-P levels (Anon., 1994; MAFF, 1994; Roberts *et al.*, 1994; Teagasc, 1994; COMIFER, 1995). The P recommendations, normally shown as kg P ha^{-1} year^{-1}, are usually based on soil-P values and there are three broad fertilizer strategies.

- No P fertilizer required for optimum production for a number of years when the test is high.
- Maintenance P required when the value is moderate.
- Build-up of P necessary when the value is low.

The approach used in Germany is illustrated in Fig. 8.4 (Vetter and Fruchtenicht, 1974) and shows, for example, that, at fertility class B, 1.5 times the maintenance fertilizer P recommendations should be applied, whereas, at fertility class D, only 0.5 the maintenance dressing should be applied.

Recommendations often include several application rates between the zero and build-up rates depending on soil-P. The continuing build-up of soil P levels, in Ireland for example, indicate that excess P is used in order

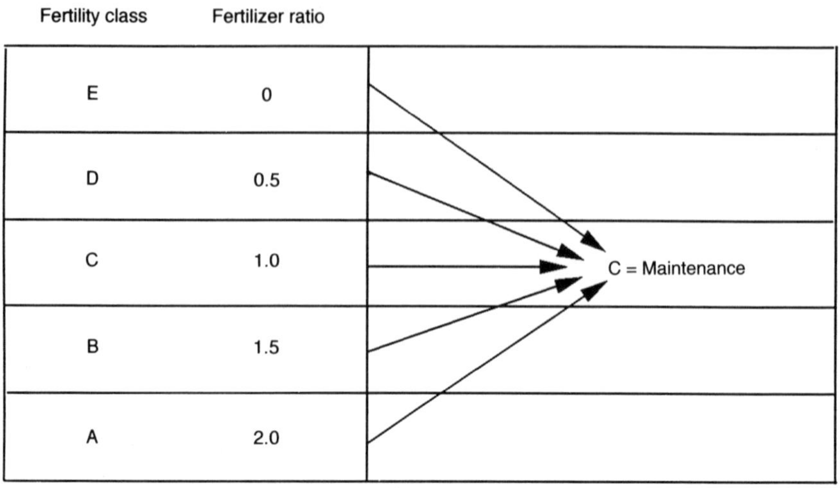

Fertility class	Fertilizer ratio
E	0
D	0.5
C	1.0
B	1.5
A	2.0

C = Maintenance

Fig. 8.4. Fertilizer recommendations for phosphorus based on the classes of soil fertility used in Germany (from Vetter and Fruchtenicht, 1974).

to be on the safe side or as an insurance, or because its use has not been questioned before as it is being now.

Actual recommended rates for a given soil-P level depend on the crop requirement (usually taken as crop off-take and referred to as maintenance), crop responsiveness and soil type. In some countries (e.g. UK, Denmark), soil-type effects are ignored. A comparison of fertilizer recommendations for countries using the Olsen method is given in Table 8.3. The amounts recommended show a considerable range, especially for grazed permanent pasture and potatoes. For other crops, there is as much as a twofold difference in recommendations at low soil-P levels. Most of these countries omit P fertilizer at soil-P levels over $40\,\text{mg kg}^{-1}$, with the exception of Denmark. Also in Denmark, the range in fertilizer P application rates is small ($20\,\text{kg P ha}^{-1}$) compared with the other countries.

Comparison of P recommendations in other studies shows an equally wide variation in recommendations where different tests are used (Hanotiaux and Vanoverstraten, 1990; Houba *et al.*, 1992; Baker and Tucker, 1994; MAFF, 1994; Teagasc, 1994).

Although a kilogram of P fertilizer is more than double the cost of a kilogram of N or K fertilizer, it is still relatively inexpensive, normally only about 5%, in terms of input costs. It is therefore not high on the list of priorities for reducing costs. However, on a large scale, it can represent an important cost. For example, in the EU, P fertilizer imports of over 2 million tonnes per year cost farmers more than 2 billion European currency units (ECU). If surplus P application were eliminated, this would lead to significant saving for farmers.

Fertilizer recommendation systems also vary in their level of sophistication. This probably reflects differences in the amount of soil-based agricultural research undertaken for different crops. For example, in New Zealand, where permanent pasture grazed all year round is the major crop, biological optimum 'target' levels of Olsen-extractable P have been identified for each of the major soil types, and amounts of fertilizer P required either to raise soil-P levels to this optimum or maintain soil-P levels at or above the optimum have been identified for different stocking rates. The biological optimum is defined as the soil-P level corresponding to 97% of maximum pasture yield (Roberts *et al.*, 1994). In the Emilia–Romagna region of Italy, which has a high proportion of clay-textured soils rich in limestone, P fertilizer recommendations for building up soil-P levels vary according to soil type, bulk density and working depth. In Belgium, Germany and the Netherlands, soil type is also taken into account when making fertilizer P recommendations. In contrast, in the UK and Denmark, all soil types are treated similarly, and recommendations are based only on crop responsiveness and soil-P level. There is clearly scope for unifying the basic principles upon which P fertilizer recommendations are based.

Table 8.3. Fertilizer recommendation systems (in kg P ha^{-1} year^{-1}) compared for four countries using Olsen P test.

	Olsen P (mg kg^{-1} or l^{-1})				
	7	14	21	28	>40
Maize					
New Zealand	55	25	15	15	15
England and Wales	35	26	17	0	0
Denmark	50	50	35	35	30
Italy	54	31	0	0	0
Grazed grass					
New Zealand	84	54	24	24	0
England and Wales	26	17	9	0	0
Denmark	35	35	25	25	20
Italy	72	31	0	0	0
Cereals					
New Zealand	22	10	0	0	0
England and Wales	42	17	17	17	0
Denmark	30	30	25	25	20
Italy	46	21	0	0	0
Sugar beet					
England and Wales	44	33	22	22	0
Denmark	50	50	35	35	30
Italy	84	24	22	0	0
Potatoes					
New Zealand	44	36	24	10	0
England and Wales	153	131	109	87	44
Denmark	35	35	30	30	25
Italy	90	29	22	0	0

Comparison of Fertilizer Recommendations in the United Kingdom and Ireland

The Olsen soil-P test is used in the UK (excluding Scotland) and the Morgan test is used in the Republic of Ireland. In 1994, an attempt was made to compare the P recommendations in the two countries. One hundred soil samples from mineral grassland soils that were received from farmers in the Republic of Ireland for routine soil analyses in March 1994, including Morgan P, were selected, subsampled and sent to Rothamsted Experimental Station, for the Olsen P test. The soils were selected to have a range of Morgan test P levels from 1 to 15 mg P l^{-1} of soil, which is considered the critical range for P recommendations. The results of this comparison showed a reasonably good relationship between the two tests: Olsen P = 5.8 + 2.91(Morgan P), with r^2 = 0.67. Based on this relationship, the P recommendations in the UK (MAFF, 1994) and in Ireland (Teagasc, 1994) are being compared (Poulton *et al.*, Chapter 21, this volume).

In general, P recommendations are higher in Ireland than in the UK. The largest difference is for grazed grassland, where the recommendations in Ireland are approximately double the UK recommendations; this is summarized in Fig. 8.5. The reason for this large difference is not clear, but it does not appear to be explained by the difference in P removal or difference in soil or climate. It may be explained by a greater built-in safety margin or insurance in the Irish recommendations, which is reflected in the continuing build-up in soil-P levels in Ireland (see Fig. 8.2).

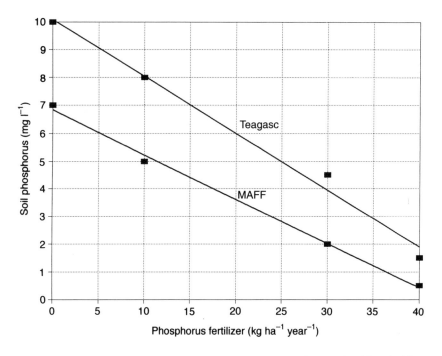

Fig. 8.5. Comparison of phosphorus fertilizer recommendations for grassland (grazing) by Teagasc (1994) for Ireland and MAFF (1994) for Northern Ireland (UK).

DIFFERENCE IN PHOSPHORUS RECOMMENDATIONS AND USE

Average Phosphorus Fertilizer Use

There is a wide variation in the average fertilizer use between countries; the situation in 12 EU countries is summarized in Table 8.4. When the P in animal manures is taken into account, the annual application of P is much higher; for example, in Ireland it is double and in the Netherlands four times higher than the figures shown in Table 8.4.

Table 8.4 indicates that the average use in some countries is more than double the use in others. For example, Germany and France use twice as much fertilizer P on average per hectare as the UK, where it has been shown that soil-P levels have been relatively stable over the past 25 years (Skinner et al., 1992). The European average P fertilizer use is much higher than in the USA.

Reasons for Differences in Phosphorus Use

Differences in P fertilizer recommendations between different countries are large, even for those using the same soil test, which may be the result of many factors. In general, P recommendations are based on optimum production with near-maximum yields, but in practice the average yield for a country may be much lower than the optimum on which the recommendation is based. In other words, many countries recommend the same P for all farmers, irrespective of yield potential, while some countries, such as France, base recommendations on expected yield.

Many countries (e.g. Ireland and the UK) give the same recommendations with a particular soil test irrespective of soil type, whereas other countries take soil type into account when making recommendations (e.g. Italy and New Zealand).

Table 8.4. Average chemical P fertilizer consumption in EU countries for 1989 in kg P ha^{-1} utilized agricultural area (from Tunney, 1992).

Belgium/Luxembourg	25	Greece	15
Germany	22	Ireland	12
France	21	United Kingdom	10
Italy	19	Portugal	9
Netherlands	17	Spain	9
Denmark	16		
		EU-12	15

An important difference is the extent to which the value of P in animal manures is taken into account. Farmers sometimes ignore the part of the P that is being recycled in animal manures when calculating fertilizer P requirements.. Some countries take the P in animal manures as being as effective as P in chemical fertilizer, whereas others assume that it is only 50% as effective. Also, P fertilizer such as superphosphate also supplies sulphur (S) and it is likely that in experiments on S-deficient soils there will be a response to S, which may be interpreted as a response to P. So, in areas low in S, the P recommendations may be higher than necessary.

For countries using Olsen, differences appear to be largely due to the interpretation of what soil-test P levels are low or high. Hence definitions of low and high relate not only to potential crop response but also to the general level of P accumulation in soils. For countries which have accumulated large amounts of P (e.g. Denmark and the UK), soil-P levels classified as low will be larger than in countries where the national average soil-P levels appear comparatively low. In Denmark, the average soil-P level is 46 mg kg^{-1} (Sibbesen and Runge-Metzger, 1995), whereas, in New Zealand, about 75% of sedimentary soils in sheep production and most arable soils have less than 20 mg l^{-1} (A.H.C. Roberts, 1995, Ruakura, New Zealand, personal communication). Differences in climate may also be important; for example, differences in P mineralization from soil organic matter and in plant uptake can be expected between Italy and Denmark. The more intensive farms have the highest soil-P levels and these farms are often used as examples to be followed by less intensive farms wishing to increase their productivity.

Differences also exist in the degree of complexity of fertilizer recommendations. Italy (Emilia–Romagna region) and New Zealand take account of soil type, which is not included the UK and Danish systems. This may reflect differences in the extent of national soil-P accumulation, since such factors would become less important as residual soil-P fertilizer increases. However, in Switzerland, where soil-P levels are relatively high, soil factors are still taken into account. Differences between and within countries must also relate to regional differences in climate, cropping intensity and soil type, which may partly explain the data shown in Table 8.3.

REQUIREMENTS FOR SUSTAINABLE PHOSPHORUS USE IN AGRICULTURE

Sustainable agriculture requires fertilization practices that give profitable production and prevent or minimize adverse environmental effects, for example, by losses of dissolved P to groundwater and surface waters. Negative environmental effects are caused by surface runoff from soils with high soil-P levels (Sharpley *et al.*, 1994a) and by leaching from P-saturated soils (Breeuwsma *et al.*, 1995). Obviously, the first step towards sustainable fertilizer use is to avoid overfertilization (with chemical fertilizer, animal manures or other sources) by following the minimum recommendations for optimum production. Recommendations need to be adapted if they are not compatible with environmental objectives within specific catchments at risk from eutrophication. This is necessary because environmental effects have not been (explicitly) taken into account in fertilizer recommendation in the past.

Where P recommendations exceed crop requirements, they should be examined for their effects on the environment. The high rates recommended at low soil-P levels are intended to raise the level to more optimal values. The P surplus recommended for this purpose can lead to P saturation in soils with shallow water-tables and therefore to leaching of P (Breeuwsma *et al.*, 1995). Also, at optimum soil-P levels, recommended applications may exceed removals sold off the farm or uptake by the crops, as illustrated in Fig. 8.6 for a crop rotation on cultivated soils in the Netherlands. In this case, an annual excess above crop requirements of 7 kg P ha^{-1} is estimated from modelling studies to be needed to maintain the P status of the soil at its optimum value, which is 13.3 mg water-extractable P (Pw) l^{-1} of soil (or 30 mg phosphate (P$_2$O$_5$) l^{-1} soil). More recent studies based on statistical analyses of long-term field experiments indicate that 17.5 kg P ha^{-1} is needed, on arable land, to maintain the Pw value of 13.3 (Ehlert *et al.*, 1995). The quantity required increases as soil-P status increases.

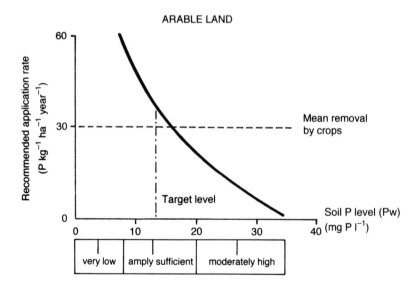

Fig. 8.6. Mean recommended phosphorus application rate (kg P ha^{-1} year^{-1}) as a function of soil-phosphorus level for arable land (potatoes, wheat, sugar beet, wheat, × 2) in the Netherlands.

On calcareous soils in south-west England, P fertilizer to replace crop removal did not maintain Olsen-extractable soil P levels over a number of years (Withers, 1995). These results may indicate that the apparent fixation losses are artefacts of the soil-P test method; most of the P that does not show up on the test may still be available to plants.

In the Netherlands, it appears that on arable soil some of the applied P may not be available for plant growth and does not contribute to a rise in soil-test P levels. This is in contrast to results on grassland in Ireland, where most of the P surplus applied in the past appears to become available later.

Sustainable fertilization strategies require a quantification of the different P losses for various soil-P levels, soil types, cropping systems, application methods and climate. Some tentative data have been derived from recent studies in the Netherlands. The studies were made to support the Dutch manure-management policies. One of the objectives is 'equilibrium' fertilization, defined as fertilization rates equal to crop uptake plus environmentally acceptable inevitable agricultural 'losses' of N or P.

Agriculturally inevitable 'losses' of P have been defined as the P surplus that is needed to maintain soil-test P levels at agriculturally optimum values (Oenema and van Dijk, 1994). In this study, data were obtained from experimental farms and field trials, and changes in soil-P levels were related to P surpluses by various methods. Differences in data sets and methods affect the assessment of inevitable 'losses' and standard errors are high. However, the results clearly indicate significant mean inevitable 'losses' of the order of 15–35 kg P ha^{-1}. For pastures, these losses are attributed to subsurface placement (injection) of manures (Ehlert et al., 1995). Broadcast chemical fertilizer did not show 'losses'.

These 'losses' can partially be attributed to leaching of P from the topsoil to the subsoil, as shown by the model calculations of van der Salm and Breeuwsma (1995) (Table 8.5). These authors used the ANIMO model with the new sorption module (Schoumans and Breeuwsma, Chapter 19, this volume). Mean leaching losses from the topsoil were within the range of 3–4.5 kg P ha^{-1} year^{-1} for a soil with an optimum soil-P level (P-AL 175 mg kg^{-1} P), a median degree of phosphate saturation (40%) and a surplus of 13–22 kg P ha^{-1} year^{-1}. Fixation losses were negligible in this case.

In situations with higher soil-P levels and degree of P saturation and/or surpluses, losses by leaching can also be significantly higher. For example, leaching losses from the topsoil in P-saturated soil can be as high as 18 kg ha^{-1} at the upper limit of the median soil-P levels, as shown by Schoumans and Breeuwsma (Chapter 19, this volume). Fixation can also play a significant role in soils with a relatively high phosphate-sorption capacity (S.E.A.T.M. van der Zee, personal communication) (Table 8.5). Immobilization may be important in soil where organic matter is increasing due to change in cropping system or management. Surface runoff is often an

Table 8.5. P losses (kg ha^{-1} year^{-1}) from the topsoil at optimum soil-P level.

Soil process	P loss	Reference
Runoff	0.5–2	Sherwood (1990) and Sharpley et al. (1995)
Leaching	3–4.5	van der Salm and Breeuwsma (1995)*
Fixation	0–11	S.E.A.T.M. van der Zee (personal communication)†
Immobilization	0–9	van der Salm and Breeuwsma (1995)‡

* Much higher values occur in P-saturated soils.
† Model calculations.
‡ Higher values occur after changing land use from cultivated soils into pastures.

important source of the loadings of surface waters, but the contribution to the total loss of P is most likely not important on the flat soils of the Netherlands. In many countries, surface runoff of dissolved P is considered to be more important.

Soil erosion in northern Europe is generally not high, particularly on grassland soils, and average losses of P in soil erosion are probably of the same order as the figures shown in Table 8.5. In southern Europe, where soil erosion can be a serious problem, much higher losses may occur.

The modelling approach provides insight into agriculturally inevitable losses as well as losses to the environment and is therefore a useful innovative tool in the development of sustainable fertilization strategies. Further development and testing of the models under various conditions is required to quantify losses experienced by the farmers, as well as losses to the environment. At this moment, modelling studies cannot quantitatively explain the high 'losses' at medium to high soil-P levels, as shown in Table 8.6, unless immobilization is still a significant source of P 'losses'.

It is not yet possible to say if the minimum soil-test P levels compatible with crop production will also be compatible with good water quality. It is likely that the answer will be that in some lake and river catchments the two

Table 8.6. Total inevitable P 'losses' (kg P ha^{-1} year^{-1}) from the topsoil at optimum soil-P levels in the Netherlands.

Land use	Method	P loss	Standard error
Grassland	Plot of changes in soil P against P surplus*†	22	
Cultivated soils	Multiple regression analyses‡	18	9.6
Grassland	Multiple regression analyses‡	16	7.4

* P surplus = P applied − P uptake by crops.
† Mean or median values extrapolated to zero change.
‡ Ehlert et al., 1995. Loss is zero for grassland at optimum soil-P levels with broadcast P fertilizers; the figure of 16 was found when slurry was injected at 15–20 cm depth.

objectives will be compatible but that in others maintaining soil-P levels that will give good water quality may give crop yields lower than optimum. The definition of sustainable soil-test P levels, in terms of what is the minimum required for optimum crop production and the maximum acceptable for losses to water that will avoid pollution due to eutrophication, is likely to be an important area for research over the next decade. Work on sustainable P use is only beginning and brings a new discipline to bear on traditional soil-fertility work. In addition to the quantity of P to use at different soil-test P levels, more information is also required on timing of P application, the maximum that should be applied in a single dressing, the type and solubility of the P fertilizer, the nature of the catchment and the soil.

Phosphorus use in a sustainable agriculture needs to optimize crop production at minimum risk to the environment. In situations where soil-P concentrations representing a eutrophication risk are larger than those required for optimum crop production, sustainable agricultural production and water quality are compatible. At present, extractable soil-P levels for optimum crop growth across different soil types have still to be adequately identified in the majority of countries. It is well established that, once clay soils have reached a satisfactory level of extractable P, fertilizer inputs can be reduced compared with other soil types. Such estimates of soil buffer capacity are clearly required for optimum fertilizer use.

In regions where organic manures are regularly applied, the risk of soil-P accumulation is greater and upper soil-P limits may need to be introduced, although the economic implications of their implementation need to be carefully judged. Clearly, P balance sheets that take account of farm P inputs, farm P outputs and the gradual fixation of applied P into non-plant available forms in the soil are required for sustainable agriculture. Routine soil-P tests are required to monitor that such a balance is being maintained. The suitability of the wide range of destructive extraction procedures now employed in different countries requires re-evaluation and it is recommended that a core methodology for the EU should be established.

CONCLUSIONS

There is a wide range both in methods for soil analyses of P and in strategies for making P fertilizer recommendations in different countries. This makes comparison and sharing of results and experience difficult. It is evident that P recommendations can vary between countries at the same soil-P levels obtained from the same test method.

No attempt has yet been made to harmonize soil-test P methods in the EU. At present, at least ten different soil-P tests are being used in 15 countries.

Soil-testing methods for P and strategies for P recommendations were mainly developed a half-century or more ago, when soils were generally deficient or low in P. Considerable reserves of P have been built up in most agricultural soils in the EU and it is now rare to find very P-deficient soils. However, reserves of P are still being accumulated on many soils with already very high soil-P levels. While this situation does not appear to present a major problem for productivity, economically there is an increased cost of applying P fertilizer that is not needed.

However, there is increasing evidence that increasing P losses from agricultural soils are contributing to eutrophication. Therefore, it is now appropriate to review critically the soil-P test methods and the strategies used for making P fertilizer recommendations. In the first instance, this should ensure that the soil-test P levels and the P recommended should be the minimum compatible with optimum agricultural production levels. This approach would minimize the input costs and maximize profit for the farmer, while at the same time minimizing the risk of pollution due to P loss from agriculture to water.

A more harmonized approach to soil-test methods and strategies for P recommendations would help these aims by allowing pooling of information and experience.

It is appropriate that the scientific community, working with the agricultural, water and fertilizer interests, should clarify the critical parameters in P movement from agriculture to water, define what levels are sustainable and establish how these standards can best be achieved to meet the needs of agriculture and the environment.

SUMMARY

Soil analysis started 150 years ago, has become widely used in developed countries over the past 40 years and is an important tool in modern scientific farming. With regard to phosphorus (P), there are about ten different soil tests being used in European Union (EU) member states and this makes accurate comparison of results and P fertilizer recommendations difficult. There is a wide variation in the average P fertilizer use between member states from a low of 9 kg P ha^{-1} year^{-1} in Spain up to 25 in Belgium. This is partly due to differences in agricultural cropping and intensity, but it is also influenced by different recommendations and different interpretation of recommendations. For example, preliminary comparison in P fertilizer recommendations in Ireland (Morgan P) and the UK (Olsen P) indicates that there can be a twofold difference between the P recommendation for a crop in broadly similar soil and climatic conditions. This implies that over-recommendation of P fertilizer may occur in addition to the indications that in practice farmers may apply more than is recommended. This chapter extends this comparison of P recommendations to other countries. In forage-based systems, most of the P is recycled in the animal manure and net removal is of the order of 10 kg P ha^{-1} year^{-1} on intensive farms. With tillage crops, such as cereals, which are sold off the farm, the net removal of P can be as high as 40 kg P ha^{-1} year^{-1}.

There is evidence that P inputs in many developed countries are much higher than removals and the build-up in soil-P reserves is continuing. In the EU, for example, more than 2 million tonnes of P, mostly as fertilizer, is imported each year at a cost of over 2 billion European currency units (ECU). The export of P from the EU is less than 20% of imports; this indicates that EU soil-P reserves are increasing at a rate of over 1 million tonnes year^{-1}.

Phosphorus loss to water from agriculture is increasing and this is contributing to eutrophication of fresh waters (and possibly sea water), particularly in northern and alpine regions of the EU and in other regions of the world where lakes were traditionally of very high quality. There is evidence that the surplus inputs of P to soils are contributing to P loss to water. This chapter addresses strategies for sustainable P use, where inputs are related to outputs and where soil-P levels and P use are the minimum necessary to give optimum crop production.

ACKNOWLEDGEMENTS

Assistance from the following persons is gratefully acknowledged: Dr A. Giapponesi, Italy; Dr E. Sibbesen, Denmark; Dr A.H.C. Roberts, New Zealand; A.N. Sharpley, USA; B. Pommel and C. Jouany, France, O. Oenema, the Netherlands; and A.E. Johnston, UK.

REFERENCES

Anon. (1992) *Adviesbasis voor de bemesting van akkerbouwgewassen 1992–1993.* Informatie en Kennis Centrum Akker- en Tuinbouw, Afdeling Akkerbouw en Groenteteelt in de Vollegrond, Lelystad, the Netherlands.

Anon. (1994) *Adviesbasis voor de bemesting van grassland en voedergewassen.* Informatie en Kennis Centrum Veehouderij, afdeling Rundvee-, Schapen- en Paardenhouderij, Pubikatie nr. 44, Lelystad, the Netherlands.

Baker, J.M. and Tucker, B.B. (1994) *Fertilizer Recommendation Guide.* OSU Extension Facts No. 2225, Oklahoma State University, Oklahoma, pp. 1–4.

Bray, R.H. and Kurtz, L.T. (1945) Determination of total, organic and available forms of phosphorus in soils. *Soil Science* 59, 39–45.

Breeuwsma, A., Reijerink, J.G.A. and Schoumans, O.F. (1995) Impact of manure on accumulation and leaching of phosphate in areas of intensive livestock farming. In: Steele, K. (ed.) *Animal Waste and the Land–Water Interface.* Lewis Publishers, Boca Raton, pp. 239–249.

Buondonno, A., Coppola, E., Felleca, D. and Violante, P. (1992) Comparing tests for soil fertility: 1. Conversion equations between Olsen and Mehlich 3 as phosphorus extractant for 12 soils of South Italy. *Communications in Soil and Plant Analyses* 23, 699–716.

Campbell, L.C. (1994) Plant and soil analyses: an Australian perspective. *Communications in Soil Science and Plant Analysis* 25, 767–780.

COMIFER (1995) *Fertilisation phosphatée et potassique des grandes cultures.* Comité Français d'Etude et de Developpement de la Fertilisation Raisonnée, Paris, 28 pp.

Dyer, B. (1894) On the analytical determination of probably available 'mineral' plant food in soils. *Journal of the Chemical Society* 65, 115–167.

Dyer, B (1901) A chemical study of the phosphoric acid and potash content of the wheat soils of Broadbalk Field Rothamsted. *Philosophical Transactions of the Royal Society B* 184, 235–290.

Egnér, H. and Riehm, H. (1955) Die Doppellaktatmethode. In: Thon, R., Hermann, R. and Knikemann, E. (eds) *Die Untersuchung von Boden Verbandes Deutscher Landwirtschaftlicher-Untersuchungs- und Forschungsanstalten, Methodenbuch I.* Nuemann Verlag, Radebeul and Berlin. Based on: Egnér, H. (1932) Meddelande Report No. 425 från Centralanstalten for forsoksvasendet poa jordbrukssomrao det, Avdelningen Forlantbrukschemie, Stockholm, Nr. 51.

Egnér, H., Riehm, H. and Domingo, W.R. (1960) Untersuchungen über die chemische Bodenanalyse als Grundlage für die Beurteilung des Nährstoffzustandes der Boden. II Chemische Extractionsmethoden zur Phosphor und Kalumbestimmung. *Kunliga Landboukshogskolans Annaler* 26, 199–215.

Ehlert, P.A.I., Buegers, S.L.G.E. and Steenhuizen, J.W. (1995) *Changes in Phosphate Availability in Soils Due to Fertilization.* Report 51 (in Dutch), DLO–Institute of Agrobiology and Soil Fertility Research. Haren (Groningen), the Netherlands.

Finck, A. (1982) *Fertilizers and Fertilization: Introduction and Practical Guide to Crop Fertilization.* Verlag Chemie, Weinheim.

Froment, M., Dampney, P., Goodlass, G., Dawson, C. and Clarke, J. (1995) *A Review of Spatial Variation of Nutrients in the Soil.* A report to the UK Ministry of Agriculture, Fisheries and Food, London, 58 pp.

Hanotiaux, G. and Vanoverstraeten, M. (1990) *Study of the Utilisation of Mineral Phosphate Fertilizer in Western Europe.* Study sponsored by IMPHOS, World Phosphate Institute, Casablanca, Morocco.

Houba, V.J.G., Novozamsky, I. and van der Lee, J.J. (1992) Soil testing and plant analyses in Western Europe. *Communication in Soil and Plant Analyses* 23, 2029–2051.

Johnston, A.E. and Poulton, P.R. (1977) Yields on the exhaustion land and changes in the N P K contents of the soils due to cropping and manuring 1852–1975. In: *Rothamsted Experimental Station Report for 1976*, Harpenden, Part 2, pp. 53–85.

Johnston, A.E. and Poulton, P.R. (1992) The role of phosphorus in crop production and soil fertility: 150 years of field experiment at Rothamsted, United Kingdom. In: Schultz, J.J. (ed.) *Proceedings of International Workshop on Phosphate Fertilizer and the Environment.* International Fertilizer Development Centre, Tampa, Florida, USA, pp. 45–63.

Joret, G. and Herbert, J. (1955) Contribution à la determination des besoins des sols en acide phosphorique. *Annales Agronomique* 25, 233–299.

Liebig, H. von, (1872) Soil statistics and soil analysis. *Journal of the Chemical Society* 25, 318, 837.

McIntosh, J.L. (1969) Bray and Morgan soil test extractants modified for testing acid soils from different parent materials. *Agronomy Journal* 61, 259–265.

MAFF (1994) *Fertilizer Recommendations for Agricultural and Horticultural Crops.* Reference Book 209, Ministry of Agriculture Fisheries and Food, HMSO, London, 112 pp.

Mehlich, A. (1984) Mehlich 3 soil test extractant: a modification of Mehlich 2 extractant. *Communications in Soil Science and Plant Analyses* 15(12), 1409–1416.

Morgan, M.F. (1941) *Chemical Soil Diagnoisis by the Universal Soil Testing System.* Connecticut Agricultural Experimental Station Bulletin 450, Connecticut, USA.

Nelson, W.L., Mehlich, A. and Winters, E. (1953) The development, evaluation and use of soil tests for phosphorus availability. In: Pierre, W.H. and Norman, A.G. (eds) *Soil and Fertilizer Phosphorus.* Agronomy No. 4., American Society of Agronomy, Madison, Wisconsin, pp. 153–188.

Nemeth, K. (1979) The availability of nutrients in the soil as determined by electro-ultrafiltration (EUF). *Advances in Agronomy* 31, 155–188.

Oenema, O. and van Dijk, T.A. (eds) (1994) *P Losses and Surpluses in Dutch Agriculture.* Report of the Technical P Desk Study (in Dutch), Ministry of Agriculture, Nature and Fisheries, The Hague.

Olsen, S.R., Cole, C.V., Watanabe, F.S. and Dean, L.A. (1954) *Estimation of Available Phosphorus in Soils by Extraction with Sodium Bicarbonate.* Circular No. 939, US Department of Agriculture, 19 pp.

Power, V. (1992) Phosphorus maintenance requirement for silage production and availability of soil phosphorus reserves. PhD thesis, School of Botany, Trinity College Dublin.

Quemener, J. (1985) L'interprétation des analyses. *Cultivar* 184, 107–117.

Rhoades, J. (1952) *The Georgics,* Book 2, lines 346–348. Translated into English verse in the series Great Books of the Western World – No. 13 Virgil, Encyclopaedia Britannica, London.

Roberts, A.H.C., Morton, J. and Edmeades, D.C. (1994) *Fertilizer Use on Dairy Farms.* Dairy Research Corporation, Hamilton, New Zealand.

Ryan, M. and Finn, T. (1976) Grassland Productivity: 3. Effects of phosphorus on the yield of herbage at 26 sites. *Irish Journal of Agricultural Research* 15, 11–23.

Schachtschabel, P. and Köster, W. (1985) Beziehung zwischen dem Phosphatgehalt im Boden und der optimalen Phosphatdüngung in langjährigen Feldversuchen. *Zeitschrift für Pflanzenernährung und Bodenkunde* 123, 48–63.

Schüller, H. (1969) Die CAL Methode, eine neue Methode zur Bestimmung des pflanzenverfugbaren Phosphates in Boden. *Zeitschrift für Pflanzenernährung und Bodenkunde* 123, 48–63.

Sharpley, A.N., Jones, C.A. and Gray, C. (1984) Relationship among soil P test values for soils of differing pedogenesis. *Communications in Soil Science and Plant Analysis* 15, 985–995.

Sharpley, A.N., Sims, J.T. and Pierzynski, G.M. (1994a) Innovative soil phosphorus availability indices: assessing inorganic phosphorus. In: *Soil Testing: Prospects for Improving Nutrient Recommendations.* Special Publication 40, Soil Science Society of America, Wisconsin, pp. 115–141.

Sharpley, A.N., Chapra, S.C., Wedepohl, R., Sims, J.T., Daniel, T.C. and Reddy, K.R. (1994b) Managing agricultural phosphorus for protection of surface waters: issues and options. *Journal of Environmental Quality* 23, 437–451.

Sharpley, A.N., Meisinger, J.J., Breeuwsma, A., Sims, T., Daniel, T.C. and Schepers, J.S. (1996) Impacts of animal manure management on ground and surface water quality. In: Hatfield, J. (ed.) *Effective Management of Animal Waste as a Soil Resource.* Lewis Publishers, Boca Raton, pp. 1–50.

Sherwood, M. (1990) Runoff of nutrients following landspreading of slurry. In: *Environmental Impact of Landspreading of Wastes*, Proceedings of a Seminar, Johnstown Castle Centre, Wexford, Ireland. Teagasc, Wexford, p. 111.

Sibbesen, E. (1983) Phosphate soil tests and their suitability to assess the phosphate status of soil. *Journal of Science of Food and Agriculture* 34, 1368–1374.

Sibbesen, E. and Runge-Metzer, A. (1995) Phosphorus balance in European agriculture – status and policy options. In: Tressen, H. (ed.) *Phosphorus in the Global Environment.* SCOPE 54, J. Wiley and Sons, Chichester, pp. 43–57.

Sims, J.T. (1993) Environmental soil testing for phosphorus. *Journal of Production Agriculture* 6, 501–507.

Sissingh, H.A. (1971) Analytical technique of the Pw method, used for the assessment of the phosphate status of arable soils in the Netherlands. *Plant and Soil* 34, 483–486.

Skinner, R.J., Church, B.M. and Kershaw, C.D. (1992) Recent trends in soil pH and nutrient status in England and Wales. *Soil Use and Management* 8, 16–20.

Stanford, G. and deMent, J.D. (1957) A method for measuring short-term nutrient absorption by plants. 1. Phosphorus. *Soil Science Society of America Proceedings* 21, 612–617.

Teagasc (1994) *Soil Analysis and Fertilizer, Lime, Animal Manure and Trace Element Recommendations.* Teagasc, Johnstown Castle Research Centre, Wexford, 36 pp.

Tunney, H. (1990) A note on a balance sheet approach to estimating the phosphorus fertilizer needs of agriculture. *Irish Journal of Agricultural Research* 29, 149–154.

Tunney, H. (1992) Some environmental implications of phosphorus use in the European Community. In: *Phosphorus, Life and Environment from Research to Application.* Proceedings of the 4th International IMPHOS Conference 8–11 September 1992. Ghent, Belgium. World Phosphate Institute, Casablanca, pp. 347–359.

Tunney, H. (1995) *P Maintenance for Sustainable Grass Production.* Johnstown Castle Research Centre, Annual Research Report 1994, Teagasc, Johnstown Castle.

Tunney, H. and Power, V. (1995) *STRIDE Study – Lee Valley Report.* RPS Cairns, Cork.

van den Hende, A. and Cothenie, A.H. (1960) *L'estimation de la fertilité du sol par les méthodes chimiques nouvelles.* IRSIA, Brussels, 174 pp.

van der Salm, C. and Breeuwsma, A. (1995) *Phosphate Losses in Noncalcareous Sandy Soils: Comparison of Model Calculations and Measurement on Grass and Maizeland* (in Dutch). DLO Winland Staring Centre for Integrated Land, Soil and Water Research Report 404, Wageningen, the Netherlands.

van Raij, B. (1994) New diagnostic techniques, universal soil extractants. *Communications in Soil Science and Plant Analysis* 25, 799–816.

Vetter, H. and Fruchtenicht, K. (1974) Wege zür Ermittlung des Düngerbedarfs mit grösserer Treffsicherheit (Methods of determining fertilizer requirement with more accuracy). *Landwirtschaftliche Forschung* 31(1), 290–320.

Werner, W. (1987) Phosphatdungung: Was die Lufas jetzt empfehlen. *Top agrar* 8, 44–47.

Withers, P.J.A. (1995) Minimizing phosphorus and potassium fertilizer inputs on calcareous soils. In: Cook, H.F. and Lee, H.C. (eds) *Soil Management in Sustainable Agriculture.* Wye College Press, Ashford, UK, pp. 217–225.

Withers, P.J.A. and Sharpley, A.N. (1995) Phosphorus fertilisers. In: Nechcigl, J.E. (ed.) *Soil Amendments and Environmental Quality.* CRC Press, Boca Raton, Florida, pp. 65–107.

9 Sources and Pathways of Phosphorus Loss from Agriculture

A.L. Heathwaite
Department of Geography, University of Sheffield, UK

INTRODUCTION

Although eutrophication may not be considered a widespread problem in UK waters, an increasing number of noxious blooms have occurred in recent years (Heathwaite *et al.*, 1996; Rast and Thornton, 1996). Thus it would be complacent to ignore the potential environmental deterioration eutrophication may cause. Response to the problem has largely focused on tackling the effects rather than the causes. Where causes have been identified, the emphasis has been on point rather than non-point sources, in line with recent European Community (EC) legislation (EC waste-water directive; Mariën, Chapter 17, this volume). Thus the enrichment of fresh waters from agricultural sources has either escaped attention or been seen as too difficult or too costly to control. This is despite evidence to suggest that, on average, agriculture contributes 40% of the total external phosphorus (P) loading of surface waters in the UK (Withers, 1994). The relative importance of non-point agricultural sources is likely to increase in the future as point sources such as sewage and industrial discharges meet EC standards.

The environmental significance of enhanced P export from agricultural land varies from catchment to catchment. External P loads have the greatest impact on surface waters with naturally low P concentrations or those where available P is not readily bound with other chemical elements, such as iron and calcium. There are a number of controls on the magnitude and form of P export. These may be subdivided into natural controls, where the most important are:

- climate;
- soil type and drainage characteristics;

- soil P content;
- runoff incidence and hydrological pathway(s) linking land to stream;
- erosion potential;

and land management controls, including:

- land use;
- timing, form (organic/inorganic) and method of fertilizer application;
- presence or absence of grazing animals.

The proportion of land under cultivation, together with animal stocking rates, is known to influence phosphorus concentrations in the drainage network of agricultural catchments. Even though we have come some way in elucidating which forms of land management are responsible for generating high phosphorus losses, there is less evidence to define the mechanisms by which P reaches the drainage network. In particular, the relative importance of surface and subsurface hydrological pathways and the extent to which the initial form in which P is mobilized and transformed during transport remain to be quantified. This is important in the context of evaluating the bioavailability of P fractions both as they reach the drainage network and in terms of subsequent in-stream transformations. Clearly, dissolved inorganic P is directly available for biotic uptake, whereas desorption, dissolution and degradation of other P fractions may increase their bioavailability. Here the balance between dissolved and particulate P (PP) in agricultural runoff may be important. Non-point agricultural sources of P usually contribute a greater proportion of P in particulate form relative to point (e.g. sewage) inputs.

In this chapter the key sources of P export from agricultural land are identified and examined in the context of the hydrological pathways which transport mobilized P from land to stream. An evaluation of current progress in this area of P research is made, together with a discussion of recent experimental work on P export from grassland by the author and coworkers.

PHOSPHORUS FRACTIONATION

The form of P in agricultural runoff is dependent on the initial P source, the hydrological pathway it follows from land to stream, and any physical, chemical or biological transformations that take place *en route* to the drainage network. Subsequent sedimentation and in-stream transformations may further modify the form of P in fresh waters (Heathwaite, 1993). Phosphorus in runoff and receiving waters is present primarily as ions of inorganic orthophosphate or in association with organic or inorganic colloidal and particulate material. It exists in both dissolved and particulate states. Dissolved organic P (DOP) is generated from organic matter and organic biomass residues; dissolved inorganic P is released through mineralization, although some may be directly derived from agricultural sources, such as fertilizers (Heathwaite, 1993). The dissolved fraction is generally considered to be bioavailable. Particulate P clearly is potentially available given suitable conditions for its transformation. Ryding and Rast (1989) suggest that around one-third of P associated with suspended sediment is biologically available.

For the experimental results described later in the sections on surface pathways of P loss and land use and management, all samples were fractionated, using the protocol and analytical approach described in Johnes and Heathwaite (1992) and Heathwaite and Johnes (1996). Briefly, this involved persulphate digestion of filtered and unfiltered samples and subsequent analysis of these samples, together with fresh, undigested samples, following the colorimetric method of Murphy and Riley (1962). All undigested samples were analysed within 12 h of sample collection. All digested samples were analysed within 1 week of collection. This enabled the total P load to be subdivided into the contribution from total PP (TPP) and total dissolved P (TDP). Total dissolved P may be further subdivided into dissolved inorganic P (usually defined as the molybdate-reactive P (MRP) fraction, after the analytical technique) and dissolved unreactive P or DOP. Total PP reflects the contribution from both molybdate-reactive PP (MRPP) and particulate unreactive P (PUP).

SITE DETAILS

The results described below are taken mainly from research catchments in south-west England (Slapton, national grid reference (NGR) SX817475; Seale Hayne, NGR SX828721), with one study catchment in the Midlands near Ashby de la Zouch (Trent, NGR SK356197). The south-west research sites are predominantly grassland, with dairying and/or beef cattle production. The Trent catchment has mixed arable and grassland use. The site characteristics are reported in detail in the following publications: for Slapton, in Heathwaite and Burt (1991) and Heathwaite and Johnes (1996); for Seale Hayne, in Griffiths *et al.* (1995) and A.L. Heathwaite, P. Griffiths and R.J. Parkinson (unpublished results); and, for the Trent catchment, in Dils and Heathwaite (1996).

A number of different scales of approach have been used in these studies to investigate P export under different environmental conditions. For the Slapton land-use studies (Heathwaite and Johnes, 1996) and the Seale Hayne fertilizer studies (Griffiths *et al.*, 1995), a hill-slope-plot scale approach was used, with hydrologically isolated plots varying in size from 0.5 m^2 (Slapton) to 150 m^2 (Seale Hayne). At both sites, rainfall simulation was used to monitor P export under controlled conditions. For the Trent catchment, a nested approach was used, combining bound plots and field-scale and small-catchment-scale experiments (Dils and Heathwaite, 1996).

PATHWAYS OF PHOSPHORUS TRANSPORT

Mechanisms of Phosphorus Transport from Agricultural Land

The hydrological pathways transporting the various P fractions vary in time and space. Catchment controls, such as slope, proximity to the drainage network, soil texture and structure, together with management controls, such as land use and fertilizer applications, all contribute to the spatial variation in the pattern of P export. Overlying this spatial patterning, temporal controls operate through changes in the rate of biological activity and chemical reactions. Such controls are largely climate-derived and, for P transport, one of the most important factors is the balance between rainfall input and runoff generation. As a consequence, most P loss will occur during periods of high rainfall and when the land is relatively saturated; this is usually from autumn through to spring in the UK. The exceptions are summer thunderstorms, which, owing to their intensity, may initiate large P losses, which are, however, strictly localized in extent.

Surface Pathways of Phosphorus Loss

In terms of nutrient transfer, surface runoff has, in principle, a clear link from agricultural land to receiving waters. Hence source areas should be detectable and the conditions which generate runoff may be determined. It follows that the magnitude of P export via this pathway has been reasonably well documented for a range of different land uses (Heathwaite et al., 1990; Foy and Withers, 1995; Johnes, 1996; A.L. Heathwaite, P. Griffiths and R.J. Parkinson, unpublished results). However, much of this research has been conducted at the plot scale. Hill-slope- and catchment-scale studies are necessary to trace the fate of P once mobilized along this pathway (e.g. Dils and Heathwaite, 1996). This is because it is still not clear to what extent particular P fractions, once mobilized, are modified before they reach the stream or, where the P source area is some distance from the stream, whether mobilized P actually reaches it.

Both PP and dissolved P fractions may be transported in surface runoff, although the former usually predominates (but see section on land management later in this chapter). The transport of eroded material in surface runoff is particle-size-selective and hence highly effective at transporting P adsorbed on to organic-rich clay and silt-sized soil fractions. Storage or transformation of P during transit may occur and depends largely on physical factors, such as the different land uses over which surface runoff may pass down-slope and the infiltration capacity of the soil surface. Chemical (e.g. adsorption/desorption) and biological (e.g. plant uptake) factors may operate, but they are more important for subsurface transport of P (see below).

For P, it is clear that field-by-field attributes, such as runoff potential (infiltration capacity, vegetation roughness coefficient, slope, initial concentration of P in the soil), together with storm duration and magnitude and the amount of erosion, are important in determining both the likelihood and the subsequent transformation of P being mobilized in surface runoff. On this basis, it is possible to define high- and low-risk land (Lemunyon and Gilbert, 1993). High-risk land might be located adjacent to a stream, have a soil surface subject to capping or be characterized by a low infiltration capacity as a result of, for example, trampling by cattle or compaction by farm machinery. Using slurry application to such land as an example, Withers (1994) suggests that up to 20% of P applied in slurry to high-risk land may be lost in runoff, whereas the average risk is only around 3% of the application. Quantification of P loss in surface runoff is reported later in this chapter for different land uses and under different fertilizer forms.

Subsurface Transport of Phosphorus

Phosphorus export via subsurface hydrological pathways is less well documented than that of surface runoff. This may be a result of both the difficulty in measurement and the general assumption that this pathway is quantitatively less important than surface delivery. Most subsurface transport of P is assumed to be in the soluble fraction, where typical concentrations of soluble P percolating through soil are of the order of 0.1 mg l^{-1} orthophosphate (PO_4)-P, even where soil-P concentrations are high (Withers, 1994). Recent research suggests that other P fractions may also be transported via this pathway (Dils and Heathwaite, 1996). For example, for a single storm event it was found that the soluble inorganic fraction in subsurface and near-surface flow formed only 10% of the total P export in both undrained and tile-drained plots. Particulate P formed the bulk of total P mobilized for this event.

For subsurface pathways, the link between land use and P export is less clearly defined than that for surface runoff. Below the ground surface, soil characteristics and P transformations along flow pathways become relatively more important in characterizing P loss. Transformations include re-adsorption of dissolved P by soil particles (Sharpley and Syers, 1979; Sharpley et al., 1992) and are part of the dynamic and labile nature of certain P fractions (Lee et al., 1989; Owens et al., 1989; Pinay et al., 1990). Soil characteristics are important in terms of defining the rate of water movement through soil pores and in indicating the potential for chemical and biological modification of the P load. For example, sandy and peaty soils may have a high leaching potential with regard to P, owing to minimal retention through adsorption, a process which is important in soils with a higher clay content. Research also suggests that organic P compounds may be more readily transported though the soil relative to inorganic P, but this may be soil-specific, for example, in upland peaty soils (Williams and Edwards, 1993). Soil structure will influence P fractionation through its indirect control of the length of contact time between percolating water, soil water and soil particles. In this context, macropores, which exist in structured soils or develop through cracking during dry periods, enable rapid bypass flow through the soil. This will reduce the contact time between soil and percolating water. There is evidence to suggest that water moving through soil fissures may show elevated P concentrations (Dils and Heathwaite, 1996). Bypass or macropore flow will probably transport P in a form similar to that recorded in surface runoff; however, it does not require the high rainfall intensity and duration events characteristic of infiltration- or saturation-excess overland flow conditions. Thus P may be transported by macropore flow for relatively small storm events. Similar arguments exist for artificial drainage pipes, which act effectively as large, more or less permanent, macropores in the soil.

The Significance of Storm Events for Phosphorus Transport

Catchment studies have shown that P export from diffuse agricultural sources is generally low during base-flow conditions. Phosphorus export appears to be triggered by storm events, which activate a number of hydrological pathways for P transport and may also mobilize P from point sources, such as slurry stores. Storm events lead to rapid variations in the concentration of dissolved P and PP in receiving waters. They demonstrate that both the P load and P fractionation vary both between catchments and for different times of year. This variation can be used to identify potential source areas and calculate the time delay between the onset of rainfall, the generation of runoff and the receipt of storm-derived nutrients in the drainage network.

For the Trent catchment in the UK Midlands, Dils and Heathwaite (1996) showed that P fractionation in runoff from agricultural land for consecutive storms changed through the hydrologically active period from autumn to spring. The magnitude of the total P load also differed, with earlier storms showing greater P loss; although this will be in part a function of the duration and intensity of the storm event. Initial storms in early autumn were dominated by the soluble inorganic fraction (Table 9.1), where the contribution of MRP to the TDP load ranged from 87 to 99%. Soluble inorganic P may be derived from stores of mineralized P that have accumulated in the upper soil profile during summer, when little water

Table 9.1. Fractionation of phosphorus export from the Trent catchment for consecutive storm events, autumn 1994 (after Dils and Heathwaite, 1996). Figures in brackets give the concentration range recorded.

Date	Total rainfall (mm)	Duration (h)	Total P (μg P l^{-1})	TPP (μg P l^{-1})	TDP (μg P l^{-1})	MRP (μg P l^{-1})
15/09/94	54	38	1076 (109–1981)	211 (19–572)	865 (86–1735)	835 (78–1702)
2/10/94	19.4	24	513 (104–1309)	363 (67–1037)	150 (37–299)	128 (31–258)

movement and hence little P transport take place. The extent to which P accumulation occurs will depend, to some extent, on the amount of crop uptake of P. However, uptake may be moisture-limited during summer; thus the flushing of soluble P via soil macropores could take place in autumn. Subsequent storms showed increasing contribution from the particulate fraction to the total P load (Table 9.1), with, on average, up to 71% of the total P load being recorded in the TPP fraction (Dils and Heathwaite, 1996). For these later storms, overland flow was recorded, particularly at the base of the hill slope, where a saturated wedge developed as the soil became waterlogged (R.M. Dils, unpublished results).

The above results are part of an integrated catchment study that links catchment-scale measurements of stream chemistry with detailed plot studies, in order to elucidate the processes of P export. The study has been able to highlight the importance of subsurface flow pathways in P delivery from land to stream. Historically, surface runoff and associated erosion have been viewed as the dominant mode of P transport to receiving waters. Current research suggests that this may not always be the case, particularly in grassland catchments, where erosion may be limited as long as there is good ground cover with a high vegetation roughness coefficient. It is important to recognize that P is still lost from such systems, but the pathway of loss may be subsurface, especially through soil macropores and tile drains.

SOURCES OF PHOSPHORUS EXPORT

Soil Phosphorus

The rate of chemical weathering of P from rock is low – of the order of 0.01–5.0 kg P ha^{-1} year^{-1} (Newman, 1995). The wide range is indicative of the large number of environmental factors which influence the weathering rate, including rock type, particle size, temperature and water quality. The rate of chemical weathering shows a positive correlation with catchment runoff. Where runoff is high, the rate of weathering is approximately ten times greater than that at low runoff (Dethier, 1986). Thus, in catchments with a high rate of water throughput, a relatively high rate of P release through weathering might be anticipated.

Phosphorus losses from agricultural land are related to the capacity of the soil to retain P. This is regardless of the initial P source which may be naturally released through the mineralization of organic matter or artificially applied in the form of fertilizers and manures. Some soils – for example, sandy soils – have a limited P retention capacity; thus subsurface P losses of artificially applied P may be high. Phosphorus export is still likely from soils with a high P retention capacity if they are prone to surface runoff and erosion.

Adsorption is the main mechanism of P retention in soils in the range pH 4–7.5. Above or below this pH range, calcium- and metal (aluminium/iron)-complexing reactions are important. Phosphorus adsorbed on to hydrated non-crystalline oxides (especially iron) is thought to be the most bioavailable fraction (Sonzongni *et al.*, 1982), although organically bound P is quantitatively more important in soils (Harrison, 1987) and the concentration of organic P in soil solution is usually greater than dissolved inorganic P (Heathwaite, 1993). Soil-P concentrations usually decrease significantly with depth. For example, in the Trent catchment study (Dils and Heathwaite, 1996), the mean total P concentration at the soil surface was over 115 μg P g^{-1}, whereas that at 150 cm depth was around 40 μg P g^{-1}. Regardless of soil depth, the water-extractable fraction formed only a very small proportion (< 0.5%) of total soil P; hence leaching losses of soil P under natural conditions are low, but this depends on the extent to which the soil is saturated with respect to P. It is important to establish the point at which the capacity of soil to adsorb P becomes saturated, because this will influence the potential for P export in drainage waters. The degree of saturation is dependent on the concentration of P in soil solution, because P becomes weakly held as the adsorption capacity of the soil approaches saturation. Recent Dutch research suggests that only 30% saturation is sufficient for downward displacement of P in sandy soils in the Netherlands (Uunk, 1991). In such conditions, the potential for P export via subsurface hydrological pathways may be high.

With respect to the magnitude of P losses from agricultural land, it is the residual soil-P content which is critical rather than, for example, the annual fertilizer application rate. Recent research suggests that in some regions, by disregarding livestock inputs and incorrectly calculating crop P requirements, a P surplus (inputs > outputs) is operated and thus soil-P concentrations may be increasing. This problem is further enhanced by the limited use of soil-P tests to determine manure applications. Even where soil tests are used to determine manure application rates, they tend to be calculated on the basis of nitrogen requirements, which consequently grossly exceed P requirements. As a result, there is evidence to suggest that the residual level of P in soils has increased – for example, in Northern Ireland (Foy and Withers, 1995) – although UK-wide the extent of such increases appears to be restricted in area at present (Withers, 1994). If concentrations of P in the soil have built up – for example, as a result of excess fertilizer applications – the potential for P export via subsurface pathways may be increased. The timescales involved are important. For the medium-term time-scale (years to decades), the amount of P present in the soil is usually greater than the annual uptake by plants. This implies that the magnitude of P input to agricultural systems is less important than the form of P and the nature of P transformations in soil. In the longer term (decades to centuries), the balance between inputs and outputs of P and the consequences for P cycling may be more important (Newman, 1995).

Land Use and Land Management Practice

In addition to the type of land use, a number of related land management practices are important in influencing both the magnitude and the form in which P may be lost from agricultural land. These include: the presence or absence of grazing animals, the timing of ploughing and related operations, the amount and timing of fertilizer application, the type of manure and inorganic fertilizer used and the extent of crop cover and root activity (Heathwaite, 1993). These controls are discussed in greater detail below.

Land use
The export of P from agricultural land shows large temporal and spatial variation. However, a number of land-use trends are discernible which may influence the spatial extent and magnitude of losses from land with a high potential to contribute P to receiving waters. These include: an increased area of winter cereals, the increased use of slurry-based livestock systems and a greater area of underdrained land (Withers, 1994). The two former trends are likely to contribute P via surface runoff pathways, while the latter may increase the rate and magnitude of P loss via subsurface flow. This is significant, because drainage systems create a direct link between the land and the stream.

Different land-use types present a different degree of risk in terms of P loss. In general, cereal crops appear to maintain a better P balance than, for example, root crops (Withers, 1994). For arable land, the main pathway of P loss is usually soil erosion, with P predominantly removed in particulate form. The process is storm-event-driven (see earlier). Olness *et al.* (1975) report a high fraction of PP export in the range 67–97% for cropped catchments in Oklahoma. The bioavailability of P transported in this form will depend on the degree of adsorption on to soil particles and the quality of the receiving medium, where desorption may increase the bioavailability of P and sedimentation may build up a future source of P.

For grassland systems, higher rates of P mineralization are usually recorded relative to long-term arable land. Brookes *et al.* (1984) give figures for P mineralization around 20–40 kg P ha^{-1} year^{-1}, with up to 24% of the total organic P fraction being present in the biomass. However, even for unfertilized permanent grassland, Crisp (1966) reports P export of the order of 0.4 kg ha^{-1} year^{-1}. Where there is good vegetation cover, most P export from grassland may occur via subsurface runoff.

Research at Slapton focused on P export in surface runoff from different land-use types representative of the catchment. The land uses studied were: permanent grass: (i) lightly grazed (less than four dairy cows per hectare); and (ii) heavily grazed (> 15 dairy cows per hectare); grass ley; cereal; and land ploughed and rolled but with no vegetation cover. Table 9.2 gives the results for the grassland subject to different grazing intensities. The experiments, reported in detail in Heathwaite and Johnes (1996), demonstrated the significance of grazing animals in generating high P export from overgrazed land. Here, total P export was at least 16 times that of lightly grazed and ungrazed grass ley. Most of the P (> 80%) was mobilized in the particulate (PP + DOP) fraction. There was a strong correlation with suspended sediment delivery, with P export increasing in proportion to the magnitude of the suspended sediment load (Heathwaite *et al.*, 1990). Similar results are reported by Ryding and Forsberg (1979) for a study of six Swedish catchments of varying land use.

For the Slapton study, it appeared that overgrazing, by removing much of the vegetation cover, compacting the soil surface through trampling and adding P directly in excreta, created conditions in which surface runoff was a significant hydrological pathway, carrying with it a high nutrient load. The extent of overgrazing, as investigated in this study, may cover whole fields but is more representative of, for example, feeding and watering areas, where localized poaching usually occurs. However, where such areas

Table 9.2. Phosphorus export in surface runoff from grassland subject to different grazing intensities, Slapton catchment, UK. Phosphorus export was calculated under rainfall-simulated conditions for a 4-h storm of 12.5 mm hr^{-1} rainfall intensity. All losses reported in mg P m^{-2} for the 0.5 m^2 plots used. The results are means of two runs for each grazing intensity. Figures in brackets are sample ranges. (Modified from Heathwaite and Johnes, 1996.)

Grazing intensity	MRP	PP + DOP
Ungrazed	0.6 (0.5, 0.7)	1.4 (0.7, 3.4)
Lightly grazed (< 4 stock per hectare)	0.6 (0.3, 0.9)	7.0 (5.4, 8.6)
Heavily grazed (> 15 stock per hectare)	42.0 (39.2, 44.8)	249.0 (222, 276)

are located close to the drainage network, they obviously represent major 'point' sources of P export within fields generally assumed to be delivering non-point nutrient export.

Land management: fertilizers and livestock waste
Two aspects of land management are especially important in influencing the potential export of P from agricultural land. These are, first, livestock production systems, and in particular the inherent problem of waste disposal, and, second, the supplement of natural soil-P levels with fertilizers in both inorganic and organic form, where the problem may be largely brought about because soil-test P is rarely used to determine manure-application rates. In the west of the country, where much of this research is focused, the specialization and intensification of agricultural land use has led to a concentration of large livestock production systems. The increasing number of freshwater pollution incidents traced to agricultural sources suggests that the rate of intensification has not been matched by measures to effectively utilize the increased production of livestock manure. Thus, in many instances, the number of animals on a farm is not related to the area of land available for manure application, and excess manure accumulates. In order to avoid overloading limited storage facilities (and the risk of prosecution for point-source pollution), manures may have to be applied to land at times when the risk of runoff is high, for example, towards the end of the winter, when land may be waterlogged and highly susceptible to compaction from vehicles (Parkinson, 1993).

There is a link between livestock production systems and the use of inorganic P fertilizers in the UK. Current use of inorganic P fertilizer in the UK is around 190,000 tonnes of P per year and has remained fairly constant since the Second World War (Haydon, 1991). The reliance on imported P fertilizers has led to wide price variations as the market fluctuates. This has probably curbed inorganic fertilizer use. Long-term trends suggest that the average annual application of inorganic P fertilizer to arable land has increased slightly, whereas that to grassland has decreased – although the area of land receiving P fertilizer has increased (Withers, 1994). To counter this, there has been a trend towards greater reliance on organic manures as a source of P. This is despite the inherent problems of availability, handling and variable nutrient content. Organic fertilizers and livestock systems in general may be becoming increasingly important as non-point sources of P in receiving waters.

In terms of the potential loss of P applied as fertilizer, the key controls are fertilizer form and the application rate and method, particularly whether the fertilizer is incorporated into the soil or simply applied to the surface. Incorporating organic manure applications into the soil has been shown to reduce soluble P loss from arable fields (Withers, 1994). It is generally thought that organic fertilizers, such as slurry and farmyard manures, penetrate deeper into soils than the equivalent amount of inorganic fertilizers; hence only a small proportion should be mobilized in surface runoff. Harrison (1987), however, suggests that the addition of organic manures or excreta from grazing animals to soils increases the mobility of soil organic P. Such subsurface movement of organic fertilizers is highly dependent on microbial cycling (Brookes *et al.*, 1984; Withers, 1994). Contrary to other research, A.L. Heathwaite, P. Griffiths and R.J. Parkinson (unpublished results) recorded soluble-P concentrations in surface runoff from slurry-amended fields up to 20–30 mg l^{-1} PO_4-P. In grassland systems where the vegetation cover has been damaged by, for example, cattle grazing, Heathwaite *et al.* (1990) recorded high P losses in particulate form.

The transport of nutrients in runoff from grassland was examined at the Seale Hayne research site, using controlled plot experiments. The four 30 m × 5 m plots, which were replicated three times, received different forms of surface-applied inorganic and organic fertilizers (Fig. 9.1). The experimental set-up is described in detail in A.L. Heathwaite, P. Griffiths and R.J. Parkinson (unpublished results) and Griffiths *et al.* (1995). One plot in each replicated set acted as a control and received no fertilizer. The following fertilizers were added: cattle slurry at an application rate of 50 m^3 ha^{-1} (69 kg P ha^{-1}), cattle manure at an application rate of 50 t ha^{-1} (118 kg P ha^{-1}) and granular inorganic fertilizer as calcium phosphate (100 kg P ha^{-1}). These application rates are high (see other chapters in this volume) but conform with the current Ministry of Agriculture, Fisheries and Food (MAFF) guidelines given in the *Code of Good Agricultural Practice* (MAFF, 1991). Phosphorus loss in surface runoff from the plots was monitored under controlled conditions, using a rainfall-simulation approach. Phosphorus concentrations in subsurface flow were found to be below the limit of detection of the analytical technique used (10 µg P l^{-1}).

The concentration of total P in surface runoff from plots receiving fertilizer, regardless of fertilizer form, was significantly greater ($P < 0.05$) than that derived from control plots (Fig. 9.2). Phosphorus was primarily mobilized in the dissolved inorganic fraction (MRP), with the exception of plots receiving manure, where over 75% of the P exported was in particulate form (A.L. Heathwaite, P. Griffiths and R.J. Parkinson, unpublished results). A significant concentration of 'particulate' P was recorded in runoff from the inorganic fertilizer treatment. This was the result of the transport of undissolved granules of fertilizer in surface runoff, which were analytically retained in the 'particulate' fraction when the samples were filtered prior to analysis.

The results shown in Fig. 9.2 imply that significant losses of surface-applied P may occur if the surface runoff hydrological pathway is activated as a result of storm events following fertilizer application. This appears to be the case regardless of the form of fertilizer used. Unlike other research (e.g. Sharpley and Syers, 1979; Withers, 1994), in the experiments described above dissolved inorganic P forms a major fraction of the P loss.

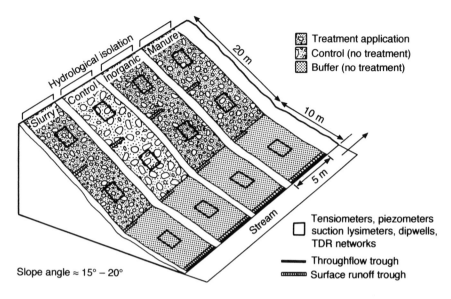

Fig. 9.1. Experimental plot layout to evaluate P export from land receiving different forms of P fertilizer, Seale Hayne research site (after A.L. Heathwaite, P. Griffiths and R.J. Parkinson, unpublished results).

This could be related in part to the grass land use of these experiments, which may increase the relative importance of dissolved P fractions by trapping and retaining any particulate material that is mobilized. The high bioavailability of the soluble P fraction may cause serious environmental deterioration if it reaches the drainage network. Management strategies

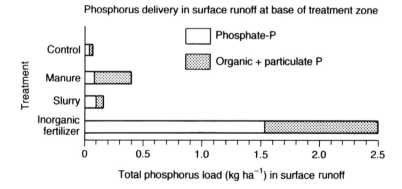

Fig. 9.2. Phosphorus export from land receiving different fertilizer treatments. The treatments used were: slurry, farmyard manure, inorganic fertilizer and control (no fertilizer). The results are means of three replicate treatments.

designed to prevent such scenarios were also investigated in this series of experiments. The results are discussed in the following section. Longer-term studies of the experimental plots showed that residues of surface-applied manure and slurry remained on the soil surface for long periods of time relative to the inorganic fertilizer granules, which dissolved rapidly. This suggests than organic fertilizers may remain a potential non-point source of P for a considerable period of time. Hence management to reduce the risk of their loss is important.

MANAGEMENT

Controlling Sources of Phosphorus Export from Agricultural Land

Research to date has gone some way in identifying high-risk land uses and land-management operations. Further work is needed to fully understand the pathways of P transport and assess the environmental significance of the various forms of P mobilized and their subsequent transformations. Inputs of P to the drainage network from agricultural land commonly occur as a result of storm events, which are unpredictable in time and space. Management strategies need to be devised to cope with such spasmodic events – although it must be accepted that it is neither feasible nor economic to control all non-point sources of P export. Hence management needs to focus on identifying and controlling 'hot spots' of P loss to reduce the risk of water-quality deterioration (Withers, 1994). The main emphasis should be to identify key source areas within a catchment and, more critically, their sensitivity to control. Thus some soils will be more vulnerable than others and some land uses will generate more P than others (Sharpley and Withers, 1994). For soils, the P index system (Withers and Heathwaite, 1994) may be used to get field-by-field estimates of potential P export and hence the risk of P loss. However, this index does not take into account catchment characteristics, notably hydrological parameters and the nature of the drainage network. However, it does help in understanding how sound application strategies for fertilizers and organic manures may reduce the build-up of topsoil P and how land management may help reduce the erosion of arable land and the runoff of manures from grassland.

Phosphorus losses from agricultural catchments cannot be fully quantified without a complete P budget for the study catchment. A number of different models exist to address this need. At their simplest, they encompass the export coefficient model (Johnes, 1996; Johnes and Heathwaite, 1997), which enables some scaling up from the field or hill-slope scale to the catchment scale by defining land use and fertilizer inputs as the key determinants of P loss from land and hence P concentrations in receiving waters. The model assumes that a given land use exports a relatively constant P load per unit area of land. This clearly may not be the case where P concentrations in soils have built up over time as a result of excess P inputs – for example, in Northern Ireland (Foy and Withers, 1995). Despite such limitations, basic models of P budgets operating at the catchment scale should enable the identification of the location of key source areas of P in the catchment and some assessment of their likely magnitude and potential contribution to the P load of receiving waters. This simplistic level of modelling gives no indication of the processes responsible for mobilizing P in the first place, nor does it give information on any subsequent transformation of the P fractions during transit. This level of detail requires more complex (and data-hungry) physically based models (Johnes, 1996).

Riparian Land Use: the Significance of Livestock

In many grassland catchments, livestock are often grazed on riparian land owing to its proximity to watering sites and the unsuitability of such land for other uses unless it is drained. Where this is the case, there is virtually no buffer between the land and the stream. This means that little transformation or trapping of P exported from the land is possible before it enters the stream. Runoff from grazed riparian areas may contain high concentrations of P in various fractions – especially around feeding and watering areas (Heathwaite, 1993; Dils and Heathwaite, 1996). Consequently, riparian land needs to be carefully managed to control P losses. The high risk of P loss from such zones offers a good opportunity for effective nutrient export control, using a number of land-management options. These options commonly involve the introduction of buffer zones of various widths and designs and the control of fertilizer and livestock inputs within and close to such zones.

Buffer zones

The *Code of Good Agricultural Practice* (MAFF, 1991) recommends leaving a 10 m buffer strip between agricultural land and adjacent watercourses. The success of this approach depends on the mechanisms by which P is transported from land to stream (Muscutt *et al.*, 1993). If the major P fraction is particulate, or at least surface runoff-derived, buffer zones need to trap and retain sediment-associated P. In this case, factors such as the roughness coefficient of the vegetation in the buffer zone are critical. Vegetation characteristics will vary seasonally and it will be important to tally maximum vegetation trapping efficiency with the main periods of P export. Unfortunately, P export in surface runoff will be primarily associated with periods of high rainfall, commonly in winter months. This coincides with periods where vegetation cover and trapping efficiency may be least.

The experiments described above, which monitored the export of P following fertilizer application (A.L. Heathwaite, P. Griffiths and R.J. Parkinson, unpublished results), were extended to examine the efficiency of a buffer zone in reducing the magnitude of P loss. A 10 m buffer zone was established between the different P fertilizer treatments and the stream receiving runoff from the monitored hill slope. The buffer zone received no fertilizer. The experimental layout is shown in Fig. 9.1. Surface runoff was collected at the base of the buffer strip below the fertilizer zone at the same time so that the differences in P loss could be compared. All samples were fractionated and the magnitude of P loss was recorded, using the methodology described earlier.

Comparison of P loss from the fertilizer zone (Fig. 9.2) with that from the buffer strip (Fig. 9.3) demonstrates the reduction in the magnitude of P loss for the various fertilizer treatments imposed. No data were available for P export from the base of the manure buffer zone. For slurry, only a small reduction (10%) in the total P exported was recorded once surface runoff from the upper fertilizer treatment had passed through the buffer zone – but note that the magnitude of loss from slurry was small relative to the other treatments. The reduction was largely in the particulate fraction. For the inorganic fertilizer treatment, a 98% reduction in the total P load was recorded at the base of the buffer zone. There was no significant difference ($P < 0.05$) between P export from the buffer zones below the fertilizer treatments and the control where no fertilizer was either applied or transported from up-slope. Hence losses of surface-applied P fertilizer were not significant after transit through the buffer zone. It appears, therefore, that the incorporation of an unfertilized grass sward between the hill-slope and the stream was effective in trapping surface-delivered P, although this depended on the form of fertilizer applied. Such a zone would be of limited value in retaining P transported in subsurface flow. However, for this study at least, P loss in subsurface pathways is thought to be minimal. This is not the case for all catchments, particularly where

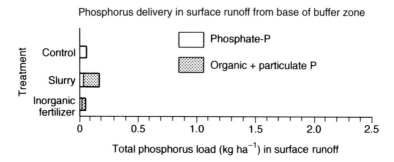

Fig. 9.3. Phosphorus export from a 10 m buffer zone installed between the treatment zones described in Fig. 9.2 and the receiving stream.

macropores or subsurface drainage systems exist (see earlier and Dils and Heathwaite, 1996).

CONCLUSIONS

Phosphorus export from agricultural land is a major source of water-quality deterioration in UK surface waters. For agricultural catchments, such as the Slapton catchment described earlier, around 85% of the total P load reaching the drainage network is thought to be derived from non-point sources (P.J.A. Withers, unpublished). Some receiving waters are more sensitive than others to P influx, but most show some change in their trophic status as a result of past and present land-management strategies. Understanding the nature of P export from land and its potential impact on receiving waters requires an integrated catchment-scale approach based on sound knowledge of the processes of P export from different land uses and the mechanisms of transport from land to stream. In the above review and in the experimental results described for the research sites in south-west England, it has been shown that land management is particularly important in governing the magnitude of P export from different land uses. Thus, where land is overgrazed or fertilizers poorly placed, high nutrient export may result. Where such land has close hydrological connectivity with the drainage network, for example, in riparian zones, there is little scope for modification of the nutrient load or form, in which case it will enter the watercourse. The experimental results described above demonstrate that even a narrow buffer of 10 m width can reduce the risk of P export in surface runoff from land receiving certain fertilizer types. Control of P export in subsurface hydrological pathways may, however, be strictly limited and more careful control of the sources of P export will be needed.

REFERENCES

Brookes, P.C., Powlson, D.S. and Jenkinson, D.S. (1984) Phosphorus in the soil microbial biomass. *Soil Biology and Biochemistry* 16, 169–175.

Crisp, D.T. (1966) Input and output of minerals for an area of Pennine moorland: the importance of precipitation, drainage, peat erosion and animals. *Journal of Applied Ecology* 3, 327–348.

Dethier, D.P. (1986) Weathering rates and chemical flux from catchments in the Pacific Northwest, USA. In: Colman, S.M. and Dethier, D.P. (eds) *Rates of Chemical Weathering of Rocks and Minerals*. Academic Press, New York, pp. 503–530.

Dils, R.M. and Heathwaite, A.L. (1996) Phosphorus fractionation in hillslope hydrological pathways contributing to agricultural runoff. In: Anderson, M. and Brookes, S. (eds) *Advances in Hillslope Processes*. John Wiley and Sons, Chichester, pp. 229–252.

Foy, R.H. and Withers, P.J.A. (1995) The contribution of agricultural phosphorus to eutrophication. *Proceedings of the Fertilizer Society* 365, 32 pp.

Griffiths, P., Heathwaite, A.L. and Parkinson, R.J. (1995) The transport of nutrients following applications of manure, slurry, and inorganic fertilizer to sloping grassland. In: Cook, H.F. and Lee, H.C. (eds) *Soil Management in Sustainable Agriculture.* Wye College Press, Ashford, Kent, pp. 510–519.

Harrison, A.F. (1987) *Soil Organic Phosphorus: A Review of World Literature.* CAB International, Wallingford, UK, 257 pp.

Haydon, P.R. (1991) Fertilizer statistics 1991. In: *Fertilizer Review 1991.* Fertilizer Manufacturers Association, Peterborough, pp. 1–2.

Heathwaite, A.L. (1993) The impact of dissolved nitrogen and phosphorus cycling in temperate ecosystems. *Chemistry and Ecology* 8, 217–231.

Heathwaite, A.L. and Burt, T.P. (1991) Predicting the effect of land use on stream water quality. In: Peters, N.E. (ed.) *Sediment and Stream Water Quality in a Changing Environment.* IAHS Press, Wallingford, pp. 209–218.

Heathwaite, A.L. and Johnes, P.J. (1996) The contribution of nitrogen species and phosphorus fractions to stream water quality in agricultural catchments. *Hydrological Processes* 10, 971–983.

Heathwaite, A.L., Burt, T.P. and Trudgill, S.T. (1990) The effect of land use on nitrogen, phosphorus and suspended sediment delivery to streams in a small catchment in southwest England. In: Thornes, J.B. (ed.) *Vegetation and Erosion.* John Wiley & Sons, Chichester, pp. 161–178.

Heathwaite, A.L., Johnes, P.J. and Peters, N.E. (1996) Trends in nutrients. *Hydrological Processes* 10(2), 263–293.

Johnes, P.J. (1996) Evaluation and management of the impact of land use change on nitrogen and phosphorus load delivered to surface waters: an export coefficient modelling approach. *Journal of Hydrology* (in press).

Johnes, P.J. and Heathwaite, A.L. (1992) A procedure for the simultaneous determination of total nitrogen and total phosphorus in freshwater samples using persulphate microwave digestion. *Water Research* 10, 1281–1287.

Johnes, P.J. and Heathwaite, A.L. (1997) Modelling the impact of land use change on water quality in agricultural catchments. *Hydrological Processes* 11, 1–19.

Lee, D., Dillaha, T.A. and Sherrard, J.H. (1989) Modelling phosphorus transport in grass buffer strips. *Journal of Environmental Engineering* 115, 409–427.

Lemunyon, J.L. and Gilbert, R.G. (1993) The concept and need for a phosphorus assessment tool. *Journal of Production Agriculture* 6, 483–486.

MAFF (1991) *Code of Good Agricultural Practice for the Protection of Water.* Ministry of Agriculture, Fisheries and Food, London, 80 pp.

Murphy, J. and Riley, J.D. (1962) A modified single solution for determination of phosphate in natural waters. *Analitica Chimica Acta* 27, 31–36.

Muscutt, A.D., Harris, G.L., Bailey, S.W. and Davies, D.B. (1993) Buffer zones to improve water quality: a review of their potential use in UK agriculture. *Agriculture, Ecosystems, and Environment* 45, 59–77.

Newman, E.I. (1995) Phosphorus inputs to terrestrial systems. *Journal of Ecology* 83, 713–726.

Olness, A.E., Smith, S.J., Rhoades, E.D. and Menzel, R.G. (1975) Nutrient and sediment discharge from agricultural watersheds in Oklahoma. *Journal of Environmental Quality* 4, 331–336.

Owens, L.B., Edwards, W.M. and Van Keuren, R.W. (1989) Sediment and nutrient losses from an unimproved, all-year grazed watershed. *Journal of Environmental Quality* 18, 232–238.

Parkinson, R.J. (1993) Changes in agricultural practice. In: Burt, T.P., Heathwaite, A.L. and Trudgill, S.T. (eds) *Nitrate: Processes, Patterns and Management*. Wiley, Chichester, pp. 321–339.

Pinay, G., Decamp, H., Cahuvet, E. and Fustec, E. (1990) Functions of ecotones in fluvial systems. In: Naiman, R.J. and Decamp, H. (eds) *The Ecology and Management of Aquatic–Terrestrial Ecotones*. Parthenon, Paris, pp. 141–170.

Rast, W. and Thornton, J.A. (1996) Trends in eutrophication research and control. *Hydrological Processes* 10(2), 295–313.

Ryding, S.-O. and Forsberg, C. (1979) Nitrogen, phosphorus, and organic matter in running waters: studies from six drainage basins. *Vatten* 1, 46–58.

Ryding, S.-O. and Rast, W. (eds) (1989) *The Control of Eutrophication in Lakes and Reservoirs*. Man and the Biosphere Series, Vol. 1, Parthenon Publishing Group, Paris, 314 pp.

Sharpley, A.N. and Syers, J.K. (1979) Phosphorus inputs into a stream draining an agricultural watershed II: amounts and relative significance of runoff types. *Water, Air and Soil Pollution* 11, 417–428.

Sharpley, A.N. and Withers, P.J.A. (1994) The environmentally sound management of agricultural phosphorus. *Fertilizer Research* 39, 133–146.

Sharpley, A.N., Smith, S.J., Jones, O.R., Berg, W.A. and Coleman, G.A. (1992) The transport of bioavailable phosphorus in agricultural runoff. *Journal of Environmental Quality* 21, 30–35.

Sonzongni, W.C., Chapra, S.C., Armstrong, D.E. and Logan, T.J. (1982) Bioavailability of phosphorus inputs to lakes. *Journal of Environmental Quality* 11, 555–563.

Uunk, E.J.B. (1991) Eutrophication of surface waters and the contribution from agriculture. In: *Proceedings of the Fertilizer Society*. The Fertilizer Society, Peterborough, UK, 56 pp.

Williams, B.L. and Edwards, A.C. (1993) Processes influencing dissolved organic N, P, and S in soils. *Chemistry and Ecology* 8, 1–21.

Withers, P.J.A. (1994) *The Significance of Agriculture as a Source of Phosphorus Pollution to Inland and Coastal Waters in the UK*. ADAS Bridgets, Winchester, UK, 57 pp.

Withers, P.J.A. and Heathwaite, A.L. (1994) Modelling non-point phosphorus loss from agricultural land. In: *Modelling the Fate of Agrochemicals and Fertilizers in the Environment*, European Society for Agronomy, Venice, 3–5 March 1994.

10 Hydrological and Chemical Controls on Phosphorus Loss from Catchments

H.B. Pionke, W.J. Gburek, A.N. Sharpley and J.A. Zollweg
USDA-ARS, Pasture Systems and Watershed Management Research Laboratory, University Park, PA 16802-3702, USA

INTRODUCTION

Phosphorus (P) contamination of surface water is of growing concern in many areas of the world. When in excess, the role of P in generating unwanted algal production in freshwater lakes and reservoirs is well documented. Most of the strategies designed to limit P entry to susceptible lakes and reservoirs are land or land-management based. Much of this P control strategy has focused on agricultural land uses and managements because agriculture is a dominant land use in many catchments.

A land-based control strategy presumes the linkage between P source (land) and impact areas (freshwater bodies) to be simple and direct. This typically is not true. First, not all P contributes to the problem; the bioavailability of the inflow P depends on its chemical form and the nature of the receiving water. Inflow P ranges from dissolved and desorbable sediment P (most bioavailable) to crystalline mineral and refractory organic P associated with sediment (least bioavailable). Most of the dissolved P (DP), especially orthophosphate, is readily available. The methods for estimating bioavailable P (BAP) of exported suspended sediment (subsequently referred to as sediment) range from algae to resin (chloride (Cl), iron (Fe))-, to chemical (sodium hydroxide (NaOH))-based extractions and provide results that are useful as indices. However, these measures may overestimate the true algal-available P in a small, rapidly sedimenting reservoir and underestimate it in a large, partly or periodically anoxic reservoir. Second, a catchment or basin is a collection of P sources, storages and sinks tied together by a flow framework. The position of the P sources, storage and sinks relative to the primary flow pathways and each other defines the key linkages from source to impact area. Clearly, flow and proximity to flow are critical. All else being equal, the sources, storages and sinks closest to the primary flow pathways will have the largest and quickest impact. The nature of the sources, storages or sinks is also important. Flow considerations being equal, their intensity, capacity and efficiency will affect the extent of their impact. Practically, P sinks are not important unless water or sediment containing P is transferred out of the catchment. Third, the importance of intensity, capacity and efficiency of sources and storages applies to flow and erosion (sediment) as well as P. Sources with low P contents and high runoff and/or erosion potentials can be major P sources, whereas those with high P concentrations but low runoff and erosion potentials may be minor P sources. For a hill-land catchment in the northeastern USA, Gburek and Pionke (1995) and Zollweg et al. (1995) provide good examples of this situation.

The focus of this chapter will be on the small catchment, a step above the farm and field scale. It will establish the critical P source areas in the context of the catchment flow system. This critical source area is a special condition of the critical zone, defined earlier by Pionke and Lowrance (1991) as 'a bounded area or volume within which one or a set of related processes dominate to provide excessive production (source), permanent removal (sink), detention (storage) or dilution of NO_3-N'. When applied to P, the focus is more on an area than a volume and less on sinks and dilution, due to the differences between nitrate (NO_3) and P. Our goals are: (i) to describe some basic controls on P losses from critical source areas; (ii) to describe and evaluate a simulation-based approach to delineate critical P source areas in the context of a small catchment; and (iii) to examine these ideas in terms of remediation and controlling P export from catchments. We shall focus on a small agricultural hill-land catchment located in east-central Pennsylvania, USA.

METHODS

The following describes the simulation methodology and basis of application. The simulation results were compared with measured runoff and DP response to storms on the experimental catchment. Earlier results show this simulation methodology to be useful and applicable to this catchment (Zollweg et al., 1995).

Runoff Generation Using the Soil-Moisture-Based Runoff Model

The soil-moisture-based runoff model (SMoRMod) developed by Zollweg (1994) has been shown to successfully simulate both the long-term daily hydrograph and individual storm hydrographs for small to medium non-winter storms on small (< 200 ha) New York and Pennsylvania catchments (Zollweg et al., 1995, 1996). It is a physically based, spatially distributed model of catchment processes, using climatic variables as input and requiring topography, land use and soil distribution as data layers. It includes the hydrological processes of infiltration, soil-moisture redistribution, groundwater flow and surface-runoff generation. It divides the catchment into homogeneous small rectangular cells, for which calculations are performed.

The soil-moisture submodel simulates daily variations in soil-moisture status and groundwater conditions for each cell over the catchment and estimates a daily stream flow at the catchment outlet. For each daily time step, water can be added to or removed from the soil, split into the root-zone layer, the subroot-zone layer and the subsoil plus bedrock layer by the processes of evapotranspiration, precipitation, interflow, percolation and surface runoff.

The runoff-generation submodel computes the storm hydrograph from the rainfall amount, rainfall intensity and initial soil-moisture conditions. This submodel provides a variable source area (VSA) response (Ward, 1984), where storm runoff is produced from expanding and contracting zones of filled storage, due to high water-tables or soil moisture. Surface runoff due to rainfall in excess of soil infiltration capacity is also computed.

The outflow hydrograph for the storm event combines the runoff amounts generated with the times for these to reach the catchment outlet. From overland flow velocities and travel times calculated for each cell, based upon the slope and land use (SCS, 1975), the travel times along the optimal flow paths are summed to generate total travel times to the catchment outlet. Organized by equal travel times, the total runoff delivered to the outlet over time becomes the storm hydrograph.

Dissolved Phosphorus Generation Using the Soil-Moisture-Based Runoff Model

The SMoRMod was modified to combine P input from the land surface with the generated surface runoff and route this mix to the catchment outlet, using the time–area approach previously described. The DP in surface runoff was computed from Bray P for both cropland and forest–grassland soils, using the linear relationships presented in Daniel et al. (1994). The Bray P values for each field within the Brown catchment were obtained by analysing multiple soil samples collected from the top 15 cm during May 1985. The storm runoff is diluted in the stream by subsurface flow, which is assigned a DP concentration of 0.007 mg l^{-1} (Pionke et al., 1988). Baseflow rate is simulated by the long-term hydrological submodel of SMoRMod.

Erosion and Sediment Bioavailable Phosphorus Generation

Erosion was computed for each model cell using the soil loss equation (soil loss = RKSLCP) developed by Wischmeier and Smith (1965). Cells with no surface runoff were designated as non-contributing. The soil loss terms were determined directly from maps and available data. The sediment BAP was computed directly as enrichment ratio (ER) × Bray P × soil loss, where ER is computed as: $\ln(\text{ER}) = 1.21 - 0.16 \ln(\text{soil loss})$ (Sharpley, 1985). Bioavailable P was computed as the sum of sediment BAP and DP.

STUDY CATCHMENT

The 25.7 ha Brown catchment is typical of small, first-order, upland agricultural catchments in the north-eastern USA. It is located approximately 40 km north of Harrisburg, Pennsylvania, within the Susquehanna River Basin (Fig. 10.1). The climate is temperate and humid. Precipitation is approximately 1100 mm year^{-1} and stream flow about 450 mm year^{-1}. Land use is almost entirely agricultural. Figure 10.2 shows land-use distribution and elevation with respect to mean sea level. The universal transverse Mercator (UTM) geographical coordinate system (zone 18) is used as a reference. A small amount of forest and a non-harvested grass strip borders the lower portion of the stream channel. The typical crop rotation is maize, small grain and hay, with about 35% of the catchment being in maize in any year. The soils are all shaly or channery silt loams that range in depth from 25 to 120 cm but are otherwise hydrologically similar. The Bray P concentration, expressed in terms of 1 cm depth, ranged from 6 kg ha^{-1} cm^{-1} in permanent grass and woods to 36 kg ha^{-1} cm^{-1} in the intensively cropped areas.

The topographic data were derived from a US Geological Survey topographical map augmented by a local survey. Soil properties and related hydrological characteristics were taken from the Soil Conservation Service (SCS, 1969). Agricultural field distribution and associated land use were determined from our own survey.

RESULTS AND DISCUSSION

Basic Controls on Phosphorus Losses from Critical Source Areas

The critical source area, as defined earlier, can be critical in hydrological, erosional, chemical and P-use terms. The source areas of surface runoff and erosion within the catchment provide the underlying control on P export. Clearly, without surface runoff, neither eroded soil, DP nor sediment BAP will be exported and thus that area cannot be a P critical source area.

Fig. 10.1. Brown catchment – location.

Surface-runoff controls
Surface runoff in our study catchment can be generated by two processes. One is by generation of precipitation excess, which occurs when the rainfall rate exceeds the soil infiltration capacity (Smith and Williams, 1980). Accumulated over the storm, this excess becomes surface runoff. This process is typically controlled at the soil scale, and the infiltration curve very much depends on the properties of the surface and shallow soil. The other is the conversion of precipitation to surface runoff where potential soil-

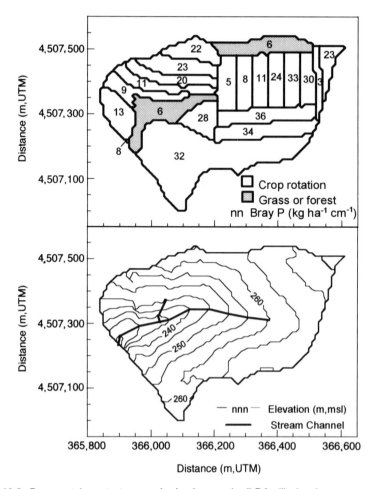

Fig. 10.2. Brown catchment – topography, land use and soil-P fertility levels.

water storage is filled. This control is at the catchment scale and results from the close proximity of the water-table to the land surface, primarily in near-stream zones (Ward, 1984). Here, soil properties are much less important, because the control is lack of water storage, not infiltration capacity. In fact, many of these soils exhibit high infiltration rates when not saturated. In our study catchment, the VSA system is the dominant control on where and how much surface runoff is generated. The VSA control is common to humid, temperate, hill-land regions (Dunne and Black, 1970) and is well known, with work having been done in Canada, the UK, New Zealand, the USA and Scandinavian countries.

Erosion controls

Erosion in the Brown catchment is typically very low, but has been severe when a certain combination of conditions is met, such as very intense storms on already wetted bare soils. Also, much of the gully erosion occurs in concentrated flow zones within the VSA rather than up on the steeper slopes. In terms of the soil loss equation parameters, the cover (C) factor appears critical, particularly in the VSA. The C factor ranges over two orders of magnitude from permanent grass to row crops (Musgrave, 1947). The soil loss equation, although instructive, estimates total erosion at the soil scale over the long term (Wischmeier and Smith, 1965). In a critical-area context, we are more interested in those erosion source areas that deliver most of the sediment BAP to the catchment scale. For 56 sediment-bearing storms in the Brown and encompassing 7.4 km^2 Mahantango Creek research catchment (Pionke and Kunishi, 1992), nearly all the sediment exported was clay and fine silt (< 5 µm diameter). However, the source soils ranged from 29 to 53% sand. Thus, the conversion of soil to exported sediment is mostly due to the much reduced export of sand. This sand fraction consists of primary particles, not sand-sized aggregates. The aggregates found in these soils are observed to be unstable and break down to primary particles when wetted. We concluded that most sand is left behind where erosion occurs, with the rest being deposited during transport to the catchment scale.

To relate eroded P to its soil-source P contents, Menzel (1980) proposed expressing this as an ER. Summarizing his own and Menzel's work, Sharpley (1992) showed: (i) the eroded materials to be enriched in both BAP and clay relative to its soil source; and (ii) the BAP and clay contents in the eroded materials to be highly related. This implies that the critical source area for sediment may need to be redefined in terms of supplying the fine sediments being exported from the catchment. In the Brown catchment, which is small, well drained and without apparent storages for fine sediments *en route*, we hypothesize a direct and quick connection between critical sediment source areas and the catchment scale. Here, the much used sediment delivery ratio (SDR), showing sediment loss per unit land area to decrease with increasing catchment size (Walling, 1977), reflects sand deposition *en route* from source to catchment. Thus, the SDR is not an issue and does not alter sediment BAP export from the Brown catchment.

Phosphorus use and chemical controls

The P use and chemical controls can vary, depending on the chemical and mineralogical nature of the soil and sediment system under study.

The soil-P content or fertility level is typically the dominating control in determining P critical source areas from P use and chemical perspectives. For our catchments, we have examined and considered P exposure, soil properties and vegetation as controls on BAP export. We have concluded these to be important, but their effects were largely overshadowed by P fertility level over the longer term and at larger scales.

Increased P fertility level of a soil translates into increased DP in runoff and BAP in sediment (Romkens and Nelson, 1974; McCallister and Logan, 1978; Sharpley, 1992; Daniel et al., 1994; Pote et al., 1996). The relationship between DP and sediment or soil BAP is curvilinear (McCallister and Logan, 1978; Wolf et al., 1985; Sharpley, 1992), with DP increasing much more rapidly than sediment or soil BAP as P fertility increases to some limit established by the solubility of Fe, aluminium (Al) or calcium (Ca) phosphorus compounds, depending on the mineralogy. The relationship can be linearly approximated (Pote et al., 1996), especially at the lower P fertility levels (McCallister and Logan, 1978; Wolf et al., 1985; Daniel et al., 1994).

Surface placement or exposure of manure or P fertilizer to surface runoff can greatly increase DP concentrations in surface runoff (Timmons et al., 1973; Sharpley et al., 1994, 1996). The question relative to selecting P critical source areas is: will exposure be sufficiently long-term and spatially extensive to control BAP export at the catchment scale? We concluded no, because here surficial P applications tend to be incidental and local and the effects transitory. However, the timing of the runoff-generating storm following application is critical. Sharpley (1980) and Westerman and Overcash (1980), respectively, have shown the DP and total P concentration in runoff to decrease greatly within a relatively few days following surface application. Apparently, diffusion and infiltration redistribute surface-applied P into the soil profile. In most field situations, dew and small rainfall events occur much more often than major runoff events soon after application. Where P is surface-applied concurrently over most of the critical surface-runoff areas and quickly followed by a major runoff event, P exposure may well control P export from the catchment for that storm event. However, the contribution from this single event needs to be assessed in terms of the long-term P losses and impact downstream.

Soil properties are clearly important in defining P critical source areas. These properties can greatly affect runoff (where precipitation excess is the cause), erosion, P bioavailability and the P distribution between sediment and surface runoff. In terms of P chemistry, the primary control in our catchments is on the relationship between solution, sorbed and the remaining sediment BAP. Based on our own unpublished work and that of Sharpley (1983), the changes in isotherm shape and buffering capacity seem mostly to depend on the clay content, although we expect some lesser effects due to Fe content (Wolf et al., 1985). However, where all soils are developed from the same parent material and subject to similar P fertilizer-management strategies as in this catchment, the soil-P chemistry and mineralogy are not expected to be differentiating criteria for defining critical source areas.

In our catchments, the surface-runoff critical areas (VSA) are generally grassed. This results from a strong conservation ethic among local farmers and because these areas are often too wet to plant in spring. These grassed riparian zones have two other effects. One, they act as buffer strips and are effective in removing the sand fraction from erosion-containing through flows. Two, vegetation can be a source of DP and may set the lower limit on P export from the catchment. Schreiber (1990), Sharpley (1981), Sharpley and Smith (1991), and Gburek and Broyan (1974) examined the DP concentration in leachate from simulated rainfall on soybeans, cotton, sorghum and orchardgrass. The overriding effect appeared to be the extent of vegetative leaching, with an initially high DP concentration in leachate (0.06–0.27 mg l^{-1}) rapidly declining to a much lower (0.015–0.1 mg l^{-1}) and stable concentration at storm's end. For orchardgrass (Gburek and Broyan, 1974), the ending values for a 12 and 150 mm rainfall were 0.05 and 0.035 mg l^{-1} DP, respectively. Because the channel in the Brown catchment is neither a sediment source nor storage, these extended leaching values may represent the lower DP concentration limit achievable in surface runoff by improved P management or remediation.

Critical Source Delineation in the Context of a Small Catchment

The surface runoff, erosion, DP and sediment BAP critical areas were delineated for the Brown catchment when subject to a 21 mm rainfall. These delineations were based on simulations applied to this 25 April 1992 storm for the particular set of initial conditions existing then. For a different-size storm under different initial conditions, the storm yields could change greatly and the areal extent of the response might expand (larger storm) or contract (smaller storm). However, the landscape pattern of surface runoff should change little. The erosion and P loss patterns may change more, but only in response to major changes in key parameters, such as C factor or P fertility levels.

The simulations of surface runoff and DP in storm flows from the Brown catchment were found to reasonably approximate storm data collected from this catchment during 1983–1985 and in 1992 (Zollweg *et al.*, 1995). The erosion and sediment BAP simulations were not tested or compared against collected data and are used for demonstration only.

The results show surface runoff and erosion to be generated from very limited areas within the Brown catchment (Fig. 10.3). Most surface runoff originates near the stream channel, with 98% being produced from about 14% of the catchment area. There is some surface runoff generated along slope breaks that parallel the channel. Basically, this is a VSA hydrological system, where the location of surface runoff is controlled by catchment (water-table position)- rather than soil (infiltration)-scale processes. The erosion critical source areas are even more limited, located primarily in the lower catchment, where a sloped maize field intersects the hydrologically active area. Very little erosion occurs elsewhere in the surface-runoff critical area because of the excellent vegetative cover.

The critical source areas of surface runoff and erosion establish the outer boundaries for DP and sediment BAP loss (Fig. 10.4). Within these boundaries, most of the DP is lost from cropland, where the Bray P concentrations in soil are highest, within and about the ephemeral upper stream channel. In contrast, the much lower Bray P values associated with the permanently grassed area around the lower channel negate the effect of the high surface runoff on DP export. The sediment BAP is more localized, primarily in cropland near the lower channel, because of the high erosion rate. Combined as BAP loss (Fig. 10.4), there are two very small areas that contribute most of the BAP total. Based on the simulation, DP

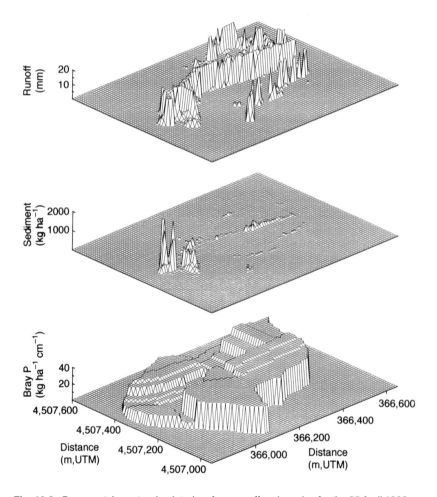

Fig. 10.3. Brown catchment – simulated surface runoff and erosion for the 25 April 1992 storm.

(68 g) and sediment BAP (77 g) losses from the catchment are about equal. In terms of area, 98% of the sediment BAP loss is from 6% of the catchment and most of the DP loss is from 11% of the catchment. About 20–30% of the BAP originates from 1% of the catchment area.

This analysis shows that a very small portion of the catchment is the source of most of the exported BAP. It also shows where these BAP losses are concentrated within the catchment. With this simulation approach, it is possible to better target monitoring, research and remediation programmes and more realistically explore the impact of alternative land use and P-management options.

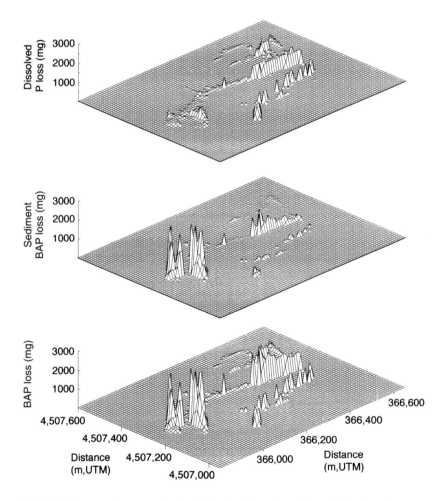

Fig. 10.4. Brown catchment – simulated dissolved P, sediment bioavailable P and bioavailable P losses for the 25 April 1992 storm.

Implications for Remediation and Control of Phosphorus Export from the Catchment

This section provides additional observations and interprets key catchment processes and responses in terms of remediation. These interpretations also serve to summarize key points developed in the chapter. Remediation is defined broadly to include assessment and prevention, not just the application of remedial practices to critical source areas.

Delineation and evaluation of critical source areas
Typically, small areas within the total catchment are the source of most BAP export. The base control for these BAP critical source areas is where surface-runoff occurs. In VSA-controlled catchments, the location of surface-runoff critical areas is predictable and follows a pattern, even when hydrological conditions change. If BAP loss from the catchment is a major issue, the critical source areas can be identified for remediation. In addition, optional remediation strategies and tactics can be tested via simulation. Knowing where surface runoff is least likely to occur can also be important for remediation. These might be the best areas for land-based disposal of P if required. However, this becomes more complicated with manurial P sources because of the nitrogen (N) component, a situation that will be addressed last.

Delineation of critical source times and events
Analogous to critical source areas, there can be critical time or storm periods when relatively little time or few storms account for most of the BAP lost from the catchment. Using data from the 7.4 km^2 Mahantango Creek research catchment (Pionke *et al.*, 1996), which contains the Brown catchment, nearly 70% of the DP exported occurred during the 10% of time dominated by the larger runoff events (Fig. 10.5). This increased to about 90% BAP exported when the sediment BAP was included. These data represent the mean DP by flow interval for over 1000 observations (samples collected from two to three times weekly over 9 years) and are used instead of the Brown catchment data, which indicated similar patterns but with fewer observations (about 240 over $3\frac{1}{2}$ years). Of the 1000 samples, 109 were taken during storm hydrographs, with 62 of these included in the highest flow period ($> 300 \, \text{l s}^{-1}$) shown in Fig. 10.5. These 62 storm events, averaging seven per year, controlled BAP loss. We have not yet determined the seasonality or similarity of these storms. The remedial implications of a time period or an event class being critical are several. First, we can probably ignore the sources of base flow and the small storm events, which together export most water but very little of the BAP. Secondly, we can develop design or index storms to represent the range in size, intensity and initial conditions for the 62 controlling events. These can be used to do

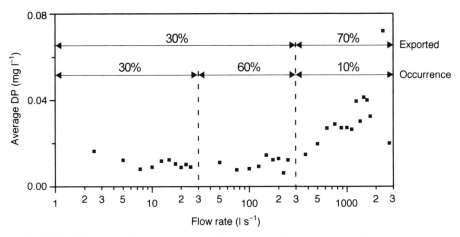

Fig. 10.5. Mahantango Creek research catchment – dissolved P concentration, occurrence and export summarized for 1984 to 1993.

simulations, as well as to establish the location and dimensions of critical source areas for these critical storms.

Critical process or characteristics at source areas that can be managed
At the catchment scale, relatively few processes and parameters in source areas control BAP export and even fewer are efficiently or readily manageable. The amount and location of surface runoff are not typically or readily managed on site. Erosion is more manageable, particularly by manipulating the C factor directly or indirectly through a conservation practice, such as modified tillage. Dissolved P is managed by controlling the P fertility level in surface-runoff zones, and sediment BAP is managed by controlling erosion and/or P fertility level in the primary erosion zones. In terms of remediation, how might we simply examine, contrast and identify potential control strategies? In Fig. 10.6, sediment, runoff, sediment BAP and DP in catchment outflow are compared for this purpose. These curves represent conceptual, not data-derived, equations. The parameters used are the sediment-to-solution ratio, expressed as sediment concentration, and the concentration ratio of sediment BAP to DP, which is defined as r. For high sediment concentrations, i.e. above 10,000 mg l^{-1}, erosion controls BAP export, and clearly erosion control is the primary remedial candidate. For low sediment concentrations, i.e. below 100 mg l^{-1}, most of the BAP is dissolved and the main control will be to reduce the P fertility level in the surface-runoff critical zones, unless the surface runoff volume can be greatly reduced, a much less likely option. Between 100 and 10,000 mg l^{-1} sediment, curve selection, representing the different sediment BAP-to-DP ratios, exerts control on whether water- or sediment-phase P dominates.

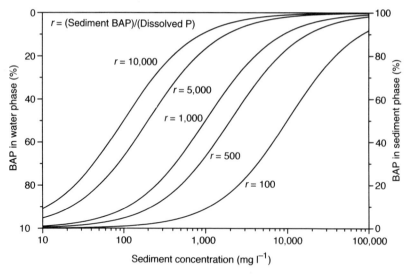

Fig. 10.6. Implications for controlling BAP export based on sediment and BAP concentrations in catchment outflow.

The change in sediment BAP/DP lumps the effects of soil properties and P fertility levels together. Assuming P equilibrium, this ratio (r) will decrease as the P fertility level is increased, simply due to the curvilinearity of the isotherm. Consequently, the DP concentration will increase more rapidly than sediment BAP as more P is added to the soil or sediment. Thus, as the P fertility level is raised (e.g. decrease r from 5000 to 500 at 1000 mg l^{-1} sediment), the BAP export will be dominated by water and not sediment, which implies a very different remediation strategy. Conversely, a remediation strategy implemented to greatly reduce P fertility level in soil may eventually elevate erosion control to a realistic remedial approach if further BAP-loss control is needed. The Brown catchment averages about 200 mg l^{-1} sediment in storm runoff and an r value of about 5000 (298 mg kg^{-1} of 0.1 N NaOH-extractable P in sediment ÷ 0.053 mg l^{-1} DP), which puts 50% of BAP in the water phase. Thus, if a remedial control strategy is needed for the Brown catchment, it should include the reduction of P soil-fertility levels in the critical surface-runoff areas in order to be effective.

Unified remedial strategies where nitrate losses must also be controlled
The selection of remediations for controlling BAP export from catchments should not cause or aggravate other water-quality problems. Bioavailable-P control strategies based on reducing surface-runoff losses either by increasing the infiltration rate or by placing most P in non-runoff zones may well increase NO_3 recharge to groundwater. This is an issue where the application of manure or organic materials is the primary source of BAP

excess. Thus, remedial approaches in these circumstances must be developed and selected to optimize BAP control relative to achieving these other nutrient-control objectives.

CONCLUSION

Control of BAP export from a catchment must be examined at the catchment, as well as farm and field, scale. There are two reasons. First, hydrologically related processes that dominate at the catchment scale can control which farms or fields contribute most of the exported BAP. Second, catchment elements that exist between field and catchment outlets may alter the timing, amount and concentration of BAP exported. These elements can include wetlands, channels, reservoirs and flow confluences. Although the impact of these elements is important, it is catchment-specific and beyond the scope of this chapter. Instead, we delineated major BAP source areas from a catchment-scale hydrological perspective, which are not predictable using a field- or farm-scale approach. Only after hydrological and erosion source areas were established was it appropriate to bring in soil, chemistry and P-use information. After doing this, we found that nearly all of the BAP exported from this catchment originated from less than 10% of the land area. To tie off-site impacts of BAP export to field use and management, it will be necessary to integrate field use and management with the dominant catchment-scale processes, as was done here.

ABSTRACT

This chapter explores the controls on loss of bioavailable phosphorus (BAP) from catchments. It establishes the concept of critical BAP source areas in the context of a flow framework. The hydrological, erosion, chemical and phosphorus (P)-use controls on BAP loss from critical source areas are identified and discussed in terms of BAP export from hill-land agricultural catchments located in the humid north-eastern USA. The application of the concept and its importance are demonstrated for an experimental Pennsylvania catchment, using a simulation method based on data collected from this catchment. The results show most exported BAP to originate from small and predictable hydrological source areas within the catchment. These ideas, simulations and additional data are summarized and used to show that most of the BAP loss is concentrated in space and time and by process, so that relatively small areas, few storm events and few processes control BAP export from catchments. These implications are discussed in terms of remediation strategies and options likely to be most effective.

REFERENCES

Daniel, T.C., Sharpley, A.N., Edwards, D.R., Wedepohl, R. and Lemunyon, J.L. (1994) Minimizing surface water eutrophication from agriculture by phosphorus management. In: *Nutrient Management*, supplement to *Journal of Soil and Water Conservation* 49, 30–38.

Dunne, T. and Black, R.D. (1970) Partial area contributions to storm runoff in a small New England watershed. *Water Resources Research* 6, 1296–1311.

Gburek, W.J. and Broyan, J.G. (1974) A natural non-point phosphate input to small streams. In: Loehr, R.D. (ed.) *Proceedings of Agricultural Waste Management Conference*, Cornell University, Rochester, New York State, March 1974. Cornell University, Ithaca, New York, pp. 39–50.

Gburek, W.J. and Pionke, H.B. (1995) Management strategies for land-based disposal of animal wastes: hydrologic implications. In: Steele, K. (ed.) *Animal Waste and the Land–Water Interface*. Lewis Publishers, New York, pp. 313–323.

McCallister, D.L. and Logan, T.J. (1978) Phosphate adsorption–desorption characteristics of soils and bottom sediments in the Maumee River basin of Ohio. *Journal of Environmental Quality* 7, 87–92.

Menzel, R.G. (1980) Enrichment ratios for water quality modeling. In: Knisel, W.J. (ed.) *CREAMS, a Field Scale Model for Chemicals, Runoff, and Erosion from Agricultural Management Systems*. USDA Conservation Research Report 26, Washington DC, pp. 486–492.

Musgrave, G.W. (1947) Quantitative evaluation of factors in water erosion, a first approximation. *Journal of Soil and Water Conservation* 2, 133–138.

Pionke, H.B. and Kunishi, H.M. (1992) Phosphorus status and content of suspended sediment in a Pennsylvania watershed. *Soil Science* 153, 452–462.

Pionke, H.B. and Lowrance, R.R. (1991) Fate of nitrate in subsurface waters. In: Follett, R.F., Keeney, D.R. and Cruse, R.M. (eds) *Managing Nitrogen for Groundwater Quality and Farm Profitability*. SSSA Special Publication, Soil Science Society of America, Madison, Wisconsin, pp. 237–257.

Pionke, H.B., Hoover, J.R., Schnabel, R.R., Gburek, W.J., Urban, J.B. and Rogowski, A.S. (1988) Chemical–hydrologic interactions in the near-stream zone. *Water Resources Research* 24, 1101–1110.

Pionke, H.B., Gburek, W.J., Sharpley, A.N. and Schnabel, R.R. (1996) Flow and nutrient export patterns for an agricultural hill-land watershed. *Water Resources Research* 32, 1795–1805.

Pote, D., Daniel, T.C., Sharpley, A.N., Moore, Jr, P.A., Edwards, D.R. and Nichols, D.J. (1996) Relating extractable phosphorus in a silt loam to phosphorus losses in runoff. *Soil Science Society of America Journal* 60, 855–859.

Romkens, M.J.M. and Nelson, D.W. (1974) Phosphorus relationships in runoff from fertilized soils. *Journal of Environmental Quality* 3, 10–13.

Schreiber, J.D. (1990) Estimating soluble phosphorus from green crops and their residues in agricultural runoff. In: *Small Watershed Model (SWAM) for Water, Sediment, and Chemical Movement*. ARS-80, USDA-Agricultural Research Service, Washington, DC, pp. 77–95.

SCS (1969) *Soil Survey Interpretations for Mahantango Creek Watershed Pennsylvania*, Vol. 1. US Department of Agriculture Soil Conservation Service, Harrisburg, Pennsylvania, 113 pp.

SCS (1975) *Urban Hydrology for Small Watersheds.* Technical Release No. 55, SCS Engineering Division, USDA-Soil Conservation Service, Washington DC.

Sharpley, A.N. (1980) The effect of storm interval on the transport of soluble phosphorus in runoff. *Journal of Environmental Quality* 9, 575–578.

Sharpley, A.N. (1981) The contribution of phosphorus leached from crop canopy to losses in surface runoff. *Journal of Environmental Quality* 10, 160–165.

Sharpley, A.N. (1983) Effect of soil properties on the kinetics of phosphorus desorption. *Soil Science Society of America Journal* 47, 462–467.

Sharpley, A.N. (1985) The selective erosion of plant nutrients in runoff. *Soil Science Society of America Journal* 49, 1527–1534.

Sharpley, A.N. (1992) Indexing phosphorus availability for soil productivity and water quality. In: Yost, R. (ed.) *Phosphorus Decision Support Systems.* Texas A&M Press, College Station, Texas, pp. 91–107.

Sharpley, A.N. and Smith, S.J. (1991) Effect of cover crops on surface water quality. In: Hargrove, W.L. (ed.) *Cover Crops for Clean Water.* Soil and Water Conservation Society, Ankeny, Iowa, pp. 41–50.

Sharpley, A.N., Chapra, S.C., Wedepohl, R., Sims, J.T., Daniel, T.C. and Reddy, K.R. (1994) Managing agricultural phosphorus for protection of surface waters: issues and options. *Journal of Environmental Quality* 23, 437–451.

Sharpley, A.N., Meisinger, J.J., Breeuwsma, A., Sims, T., Daniel, T.C. and Schepers, J.S. (1996) Impacts of animal manure management on ground and surface water quality. In: Hatfield, J. (ed.) *Effective Management of Animal Waste as a Soil Resource.* Lewis Publishers, Boca Raton, Florida.

Smith, R.E. and Williams, J.R. (1980) Simulation of the surface water hydrology. In: Knisel, W.J. (ed.) *CREAMS, a Field Scale Model for Chemicals, Runoff, and Erosion from Agricultural Management Systems.* USDA Conservation Research Report 26, Washington DC, pp. 13–64.

Timmons, D.R., Burwell, R.E. and Holt, R.F. (1973) Nitrogen and phosphorus losses in surface runoff from agricultural land as influenced by placement of broadcast fertilizer. *Water Resources Research* 9, 658–667.

Walling, D.E. (1977) Natural sheet and channel erosion of unconsolidated source material (geomorphic control, magnitude, and frequency of transfer mechanisms). In: Shear, H. and Watson, A.E.P. (eds) *Proceedings of the Workshop on the Fluvial Transport of Sediment-associated Nutrients and Contaminants.* International Joint Commission of the US and Canada Research Advisory Board and Pollution Land Use Activities Reference Group, Windsor, Ontario, pp. 11–33.

Ward, R.C. (1984) On the response to precipitation of headwater streams in humid areas. *Journal of Hydrology* 74, 171–189.

Westerman, P.W. and Overcash, M.R. (1980) Short-term attenuation of runoff pollution potential for land-applied swine and poultry manure. In: *Livestock Waste – a Renewable Resource.* Proceedings of the 4th International Symposium on Livestock Wastes, American Society of Agricultural Engineers, St Joseph, Michigan, pp. 289–292.

Wischmeier, W.H. and Smith, D.D. (1965) *Predicting Rainfall-Erosion Losses from Cropland East of the Rocky Mountains.* Agriculture Handbook 282, USDA-Agricultural Research Service, Washington DC.

Wolf, A.M., Baker, D.E., Pionke, H.B. and Kunishi, H.M. (1985) The use of soil test P measurements for estimating labile P, soil solution P, and algae-available P in noncalcareous agricultural soils. *Journal of Environmental Quality* 14, 341–348.

Zollweg, J.A. (1994) Effective use of geographic information systems for rainfall-runoff modeling. PhD dissertation, Cornell University, Ithaca, New York State.

Zollweg, J.A., Gburek, W.J., Sharpley, A.N. and Pionke, H.B. (1995) GIS-based modeling of phosphorus output from a northeastern upland agricultural watershed. In: *Proceedings of an International Symposium on Water Quality Modeling.* American Society Agricultural Engineers, 2–5 April 1995, St. Joseph, Michigan pp. 251–258.

Zollweg, J.A., Gburek, W.J. and Steenhuis, T.S. (1996) SMoRMod – a GIS-integrated rainfall-runoff model applied to a small northeast US watershed. *Transactions of the American Society of Agricultural Engineers* 39, 1299–1307.

11 Movement of Phosphorus from Agricultural Soil to Water

B. Pommel[1] and J.M. Dorioz[2]
[1]INRA Laboratoire d'Agronomie, 78850 Thiverval Grignon, France; [2]INRA Institut de Limnologie, 74203 Thonon les Bains, France

INTRODUCTION

Phosphorus (P) concentrations in Lac Léman (Lake Geneva, Western Europe's largest lake) increased almost tenfold between 1950 and 1975 (CIPEL, 1984) and symptoms of eutrophication became obvious. During this period, the population increased substantially on both the French and the Swiss sides of the lake, while intensive industrial, urban and agricultural activities became concentrated in the plains along the various tributaries that flow into the lake. In the 1970s, attention was focused on reducing traditional point-source discharges in the Lac Léman basin. Consequently, the P concentrations in the lake have fallen substantially, but they remain above the goal of 30 µg l^{-1} soluble orthophosphate (SOP) (CIPEL, 1992). Now, management attention is shifting to non-point sources of P; however, as in other areas (Novotny and Olem, 1994), it is difficult to define the origins and magnitude of non-point, especially agricultural, diffuse sources of P.

Effective management of agricultural diffuse sources of P requires knowledge of the origins and magnitude of P pools, an understanding of the impact that hydrology and land use have on the storage, transport and export of P and the effects on the receiving water body. An agricultural subcatchment of Lac Léman was studied to elucidate the origins of P and the processes of transfer, storage and export. This pilot catchment, which is representative of the French portion of the Lac Léman basin (Fig. 11.1), was studied at different scales in order to understand how the transferred P is modified before it reaches the lake.

Changes of P speciation form and bioavailability in water and sediment during transport are briefly discussed.

METHODS FOR WATERSHED MONITORING

The watershed has gentle and regular slopes (< 2% in cultivated areas). The land use consists of 40% cultivated fields (cereals, mainly winter wheat), 50% hay meadows and wetland associated with a small forest area (10%). Soils are predominantly clay loam and are derived from glacial deposits of compacted calcareous sand/clay mixtures. At the outlet of the main agricultural subwatershed (15 ha), the annual average discharge is about $6 \, l \, s^{-1}$ and storm flows are brief (24 h) and sometimes violent (maximum output (Q_{max}) $150 \, l \, s^{-1}$). The stream is dry in summer (Hamid et al., 1989). The stream waters from the agricultural areas are rich in calcium carbonate ($CaCO_3$) and discharge into a 2 ha wetland, which discharges into the Redon River (Pilleboue and Dorioz, 1986). The monitoring of watershed included three sample points for water in order to describe the origin of P

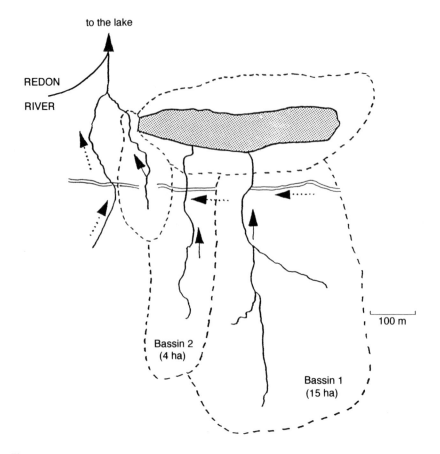

Fig. 11.1. Pilot watersheds in the agricultural area of Perrignier in the Redon basin.

exported from the agricultural areas. The monitoring programme provided a continuous record of P loads and concentrations at the outlet of the 15 ha agricultural subwatershed and at the outlet of the wetland. During 1 year, water was sampled every 30 min. Each sample was analysed during 12 storm-flow events and with flow-proportional average samples during low-flow periods (Dorioz et al., 1991; Dorioz and Ferhi, 1994).

Complementary sampling was done on composite water at the outlet of the smaller agricultural subwatershed (4 ha) and on sediment from representative locations along the streams and the Redon River.

Analyses of P in water were according to standard methods (AFNOR, 1982; CIPEL, 1984) for total P (TP), total dissolved P (TDP), molybdenum reactive P (MRP) and total suspended solids (TSS). On the size fraction of sediment less than 200 µm, TP and sodium bicarbonate ($NaHCO_3$)-extractable P (Olsen et al., 1954) were determined. The immediately exchangeable (1 min) P and P-fixation capacity of the soils and sediments were determined using ^{32}P (Tran et al., 1988). Interstitial waters obtained by centrifugation were analysed for TDP (AFNOR, 1982).

RESULTS AND DISCUSSION

Overview of Mass Balance of Phosphorus in the Watershed

The total annual P loads (Table 11.1) measured at the 15 ha subwatershed outlet varied greatly, from 0.45 to 0.8 kg P ha^{-1} year^{-1}, according to the hydrological conditions of the year. These outputs were very large compared with the natural inputs but very small considering the total amounts of P which were added by agriculture and rain within the watersheds. The TP load exported at the outlet was less than 2% of the fertilizer applied. Half of the TP was TDP and 30% was MRP. The mean value of annual P loads was relatively small compared with other data obtained in fertilized and tilled watersheds of Lac Léman by CIPEL (1988) using a similar sampling programme (Table 11.2).

Table 11.1. Phosphorus mass balance in a 15 ha agricultural subwatershed in the area of Perignier in the Redon basin, France (mean values for 3 years).

	Total P (TP) (kg P year^{-1})	Soluble P (TDP) (%)
Rainfall	7	90
Background level	< 0.1	> 95
Fertilization	500	–
Load at the outlet	8.4	50

Table 11.2. Agricultural diffuse-sources loading values in Lac Léman basin area (based on data from CIPEL, 1988; Pilleboue, 1987; Blanc, 1994; Jordan-Meille, 1994).

	Range in total P (kg ha^{-1} year^{-1})	Range in total P concentration (annual mean value) (mg l^{-1})
Rain	0.2–1	0.02–0.06
Spring water and phreatic groundwater	–	0.01–0.02
Undisturbed watershed (wooded areas or natural grasslands)	0.01–0.3	0.020–0.1
Agricultural watershed (cultivated and fertilized)	0.5–2	0.1–0.5
Vineyards	6.9–10.2	–

Phosphorus in Soils and Sediments

Table 11.3 summarizes the P content of stream sediments in the Redon watersheds. River sediments downstream from the point-source discharges contained an average of 800 mg TP kg^{-1} dry weight, while the sediments downstream from natural and agricultural areas had lower TP contents, averaging 415 and 600 mg TP kg^{-1}, respectively. Sediments at the outlet of the wetland area averaged 1450 mg TP kg^{-1}. The TP content of agricultural soils varied between 510 and 1200 mg TP kg^{-1}, whereas samples of the riverbank soils averaged 410 mg TP kg^{-1} (Dorioz et al., 1997). The soils and sediments that have lower TP contents show little ability to provide TP to the surrounding environment because of low values of immediately exchangeable P and Olsen P (Table 11.3). On the contrary, the riverbottom sediments downstream in the Redon River show a high propensity for exchanging TP, as evidenced by high values of immediately exchangeable P and Olsen P.

The index of P-fixation capacity, a relative measure of the ability of sediment and soil particulates to take up P from the surrounding environment (Tran et al., 1988), is highest in river-bottom sediments not impacted by point-source discharges and lowest in river-bottom sediments downstream in the Redon River (Table 11.3).

Weekly Variations in Phosphorus Export

The weekly fluxes at the outlet of the 15 ha watershed varied greatly, from 0.005 to 0.3 kg TP. Total P was lost during the storm-flow periods in both particulate and soluble forms. Approximately 90% of particulate P (PP) and 60% of annual TDP load were exported in the six largest storm flows. During low-flow periods with stable water discharge, the TP was very low (< 0.02 mg P l^{-1}) and mostly (90%) as TDP. The average concentration in

Table 11.3. Comparison of phosphorus in sediments (< 200 μm sieve) at different locations in the pilot watershed: average values for total P content, Olsen P, exchangeable P and fixation capacity for P in sediments (from Dorioz et al., 1997).

Locations	Total P ($mg\ kg^{-1}$)	Olsen P ($mg\ kg^{-1}$)	Immediately exchangeable P ($mg\ kg^{-1}$)	Fixation-capacity index (% maximum)	Number of samples
Forested and natural grassland	415	4	1.1	90	3
Agricultural area	600	10	2.8	93	6
Outlet of wetland	1450	8	2.0	100	1
Downstream in Redon River	800	45	8.0	50	4

TDP and MRP exceeded the standard quality levels for Lac Léman only during the storm-flow periods.

Pattern of Export and Origin of Phosphorus

Figure 11.2. summarizes the pattern of P export during storm flows, at the outlet of the 15 ha subwatershed. Particulate P and suspended sediment peaks appeared during the rising limb of the hydrograph. Sometimes the peaks occurred at the very beginning of the storm flow, with a long lag effect, indicating that PP came mainly from sediments stored in the ditches and streams. Generally, the lag effect was smaller and PP originated from land runoff and erosion, the concentration reaching very high levels, up to 1.2 mg PP l^{-1}.

Soluble P (TDP) generally showed a single peak, which coincided with the PP peak and reached 0.06 mg P l^{-1} (curve b). This peak could easily be explained by mixing of runoff with soil water, solubilization of PP or exchange during transport. During some storm flows studied in detail by Dorioz and Ferhi (1994) and Pilleboue (1987), another important peak of TDP and MRP was observed. It took place during the decreasing limb of the hydrograph (curve a) and corresponded with a maximum for nitrate (NO_3^-) and chloride (Cl^-) concentrations. This latter peak indicated movement of TDP in subsurface water flow.

Although runoff was the main origin for P, it occurred only under certain conditions. In the absence of these conditions, the load exported was very low and consisted of sediment and soluble P from rain. Sometimes, the TP exported was less than the inputs in rain.

In this agricultural area, runoff usually occurred from tilled areas of bare soil only after development of surface crusting, a process that reduces

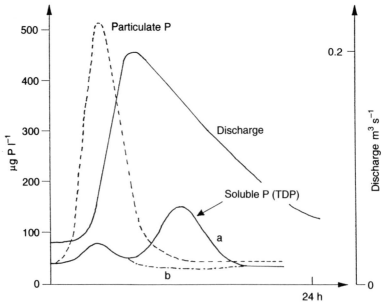

Fig. 11.2. Phosphorus export during a typical storm flow.

the infiltration capacity of many soils in the temperate zone of Europe (Boiffin *et al.*, 1988). In addition, surface compaction by farm machinery during sowing and harvesting can further reduce infiltration capacity (Vansteelant *et al.*, 1997). In the winter and spring, soil-moisture levels are high and crusting is well developed, so that the potential for surface runoff is great. Low soil-moisture levels and relatively undeveloped soil crusting and cracking of soils during the summer months result in low runoff potential.

Evolution of Phosphorus in the Landscape

To evaluate the impact on water quality of these diffuse P loads generated by agricultural watersheds, it is necessary to take into account not only concentration and fluxes at the outlet but also attenuation during transfer to the receiving lake.

The changes of P concentration and speciation from a 3 ha wetland through which the two agricultural watersheds discharged was studied. A very significant decrease in the concentration of TP, soluble P and fluxes was observed (Fig. 11.3). Over the period of a year, about 70% of the input was stored in this wetland. Retention affected MRP, TDP and PP, which indicated that numerous processes were involved, including sedimentation of PP, biological uptake of MRP by macrophytes and sorption of P on to particles and particularly on to iron hydroxides precipitated by iron bacteria. Release processes from the wetland, for example erosion, desorption and dissolution, were less than retention processes and were noticeable only in a small area near the outlet (Dorioz and Ferhi, 1994) during some storm flows. At the outlet of the wetland, the mean TDP and MRP concentration was very small. The wetland also induced transformation of PP, which was less available at the outlet than at the inlets, for the same P content but with a very high fixation capacity. This change coincided with a drastic change in the nature of the sediment, which was richer in $CaCO_3$ ($\times$ 3), total iron ($\times$ 5) and organic matter ($\times$ 4) at the outlet.

This type of sediment with high fixation capacity, transported into the polluted Redon River, contributed to the trapping of soluble P discharged by point sources (Pilleboue and Dorioz, 1986; Dorioz et al., 1989). The maximum P retention efficiency was observed during low-flow periods, especially when they occurred after a high storm-flow event which had renewed the sediment of the river bed (Dorioz and Orand, 1996). The sediments at the outlet of the Redon River had a large content of TP and a high exchangeability, but a low fixation capacity (Table 11.3) resulted from P retention by sediment. Perhaps sediment from agricultural land could play the same role and this phenomenon could partly attenuate point-source impact (Dorioz et al., 1997).

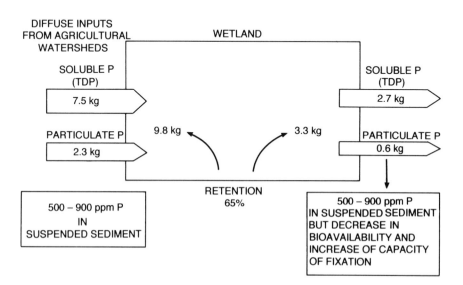

Fig. 11.3. Phosphorus balance in the experimental wetland.

CONCLUSION

The impact of a watershed on a lake depends on many parameters: the TP load, the bioavailability of PP, the fixation capacity of associated suspended matter and the distribution of P inputs over the annual cycle. These factors control the use of P inputs by the lake ecosystem (Labroue, 1995).

Consequently, managers who have the responsibility for implementing P reduction plans for lakes must know the origins of P within the watersheds feeding the lake, as well as the amount, speciation and properties of the P ultimately delivered to the lake. An understanding of the origin and speciation of P requires a knowledge of how P is transferred from its origins through the watershed to the lakes.

The transfer of P through watersheds to lakes must be considered as a discontinuous process (Verhoff et al., 1982) driven by: (i) hydrology; and (ii) different components of the landscape, particularly agricultural practices and land management.

Consequently, plans to decrease P losses to water must take into account the possibilities of reducing the various sources of P. This may mean changes in agricultural systems, land use and tillage practices to decrease diffuse pollution, together with improving or restoring natural filters, such as wetlands, or by implementing, when possible, artificial filter systems.

Success will probably depend on the capacity to develop plans that are compatible with the diversity of the agricultural systems.

REFERENCES

AFNOR (1982) *Dosage des orthophosphates, polyphosphates et du phosphore total.* NFT 90–93, Essais des eaux, Paris, 10 pp.

Blanc, P. (1994) Apports au Léman par les retombées atmosphériques. In: *Rapport de la Commission Internationale pour la Protection des Eaux du Léman contre la Pollution, campagne 1993.* CIPEL, Lausanne, pp. 129–142.

Boiffin, J., Einberk, M. and Papy, F. (1988) Influence des systèmes de culture sur les risques d'érosion par ruissellement concentré. I Analyse des conditions de déclenchement de l'érosion. *Agronomie* 8, 663–673.

CIPEL (1984) *Le Léman: synthèse 1957–1982.* Commission Internationale pour la Protection des Eaux du Léman contre la Pollution, Lausanne, 650 pp.

CIPEL (1988) *Rapport d'étude sur les pollutions diffuses.* CIPEL, Lausanne, 88 pp.

CIPEL (1992) *Rapport sur les études et recherches entreprises dans le bassin lémanique.* Commission Internationale pour la Protection des Eaux du Léman contre la Pollution, Lausanne, 119 pp.

Dorioz, J.M. and Ferhi, A. (1994) Pollution diffuse et gestion du milieu agricole: transferts comparés de phosphore et d'azote dans un petit bassin versant agricole. *Water Research* 28, 395–410.

Dorioz, J.M. and Orand, A. (1996) Dynamique du phosphore dans un bassin du Lac

Léman: application à la détermination de l'origine des flux. *Environment, Technique* 153, 61–64.

Dorioz, J.M., Pilleboue, E. and Ferhi, A. (1989) Dynamique du phosphore dans les bassins versants: importance des phénomènes de rétention dans les sédiments. *Water Reseach* 23, 147–158.

Dorioz, J.M., Orand, A., Pilleboue, E. and Blanc, P. (1991) Prélèvement et échantillonage dans les petits bassins versants ruraux. *Revue Française des Sciences de l'Eau* 4, 211–238.

Dorioz, J.M., Pelletier, J. and Benoit, P. (1997) Propriétés physico-chimiques et biodisponibilité potentielle du phosphore particulaire selon l'origine des sédiments, dans un bassin du Lac Léman (France). *Water Reseach* (in press).

Hamid, S., Dray, M., Ferhi, A., Dorioz, J.M., Normand, M. and Fontes, J.C. (1989) Etude des transferts d'eau à l'intérieur d'une formation morainique dans le bassin du Léman: transfert d'eau dans la zone non saturée. *Journal of Hydrology*, 109, 369–385.

Jordan-Meille, L. (1994) Etude sur les origines et les transferts de nutriments dans un bassin versant rural du bassin lémanique: cas particulier du phosphore. Travail de diplome, Institut de Limnologie, ST 10–94, EPFL, INRA, Thonon-les-Bains, 90 pp.

Labroue, L. (1995) Cycle des nutriments: l'azote et le phosphore. In: Pourriot, R. and Meybeck, M. (eds) *Limnologie générale*. Paris, Masson, pp. 727–764.

Novotny, V. and Olem, H. (1994) *Water Quality: Prevention, Identification and Management of Diffuse Pollution*. Van Nostrand-Reinhold, New York, USA.

Olsen, S.R., Cole, C.V., Watanabé, F.S. and Dean, L.A. (1954) *Estimation of Available Phosphorus in Soil by Extraction with Sodium Bicarbonate*. Circular no. 939, USDA, Washington, 19 pp.

Pilleboue, E. (1987) Origines, bilans et mécanismes de transfert de l'azote et du phosphore d'un bassin versant vers un lac. Thèse de doctorat, Université de Paris VI, 220 pp.

Pilleboue, E. and Dorioz, J.M. (1986) Mass-balance and transfer mechanism of phosphorus in a rural watershed of Lac Léman, France. In: Sly, P.G. (ed.) *Sediment and Water Interactions: Proceedings of 3rd Sediment/Freshwater Symposium*. Springer, New York, pp. 91–102.

Tran, S.T., Fardeau, J.C. and Giroux, M. (1988) Effects of soils properties on plant available phosphorus determined by the isotopic dilution phosphorus-32 method. *Soil Science Society of America Journal* 52, 1383–1390.

Vansteelant, J.Y., Trévisan, D., Dorioz, J.M., Perron, J.Y. and Roybin, D. (1997) Conditions d'apparition du ruissellement dans les cultures annuelles de la région lémanique, relation avec le fonctionnement des exploitations agricoles. *Agronomie* 17, 17–34.

Verhoff, F.H., Melfi, D.A. and Yackish, S.H. (1982) An analysis of total-P transport in a river system. *Hydrologia* 91, 241–252.

12 Losses of Phosphorus in Drainage Water

P.C. Brookes,[1] G. Heckrath,[1] J. De Smet,[2] G. Hofman[2] and J. Vanderdeelen[2]

[1]Soil Science Department, IACR-Rothamsted, Harpenden, Hertfordshire, UK; [2]Faculty of Agricultural and Applied Biological Sciences, University of Ghent, Belgium

INTRODUCTION

Phosphorus (P) may be lost from agricultural land to water by several processes. These include erosion, surface runoff and subsurface flow (leaching). As most soils have a very high absorption capacity for P, usually far exceeding the quantities of P added as manures or fertilizers, it has long been considered that leaching losses of P from soil to water are negligible in most cases. While this is certainly true in terms of economic losses to the farmer, the concentrations of P required to trigger eutrophication in fresh water are extremely small (as low as 0.02–0.035 mg P l^{-1}). Indeed, the quality of surface waters throughout Europe, in terms of risk of eutrophication, is a major current concern, because of the increasing concentrations of P. In much of the work reviewed here, it is difficult to precisely relate P loss to soil-P concentrations, in either the plough layer or the subsoil, simply because such data are either not given or insufficient.

Here we attempt to assess the significance of the losses of P in drainage water by subsurface flow as a contribution towards eutrophication. We also assess some of the possible mechanisms involved in the losses. Recent studies of P losses in drainage waters from the Broadbalk Experiment at Rothamsted and from heavily manured soils in West Flanders, Belgium, will be presented.

PHOSPHORUS USE IN AGRICULTURE

Phosphorus is an essential element for plant nutrition and is generally added to agricultural soils as inorganic fertilizer or as organic manures. Unlike nitrogen (N), P is strongly fixed by soil particles, so that the efficiency of use of P fertilizers by crops is rather small. As practically no adverse effects of high soil-P concentrations on plant growth have been observed, farmers have frequently added P in amounts which exceed crop removal. This is particularly so in regions with intensive cattle or pig breeding. As a consequence, P has accumulated with these repeated P applications (e.g. Sibbesen, 1989; Behrendt and Boekhold, 1993). Many intensively farmed soils have more than twice the total P content in the topsoil now compared with the content 50 years ago, before chemical fertilizer was applied (Tunney, 1992). It has long been known that P may move from agricultural soils to the aquatic environment by processes such as erosion and surface runoff (Miller *et al.*, 1982; Isermann, 1990; Vighi *et al.*, 1991). However, losses of P from soil in drainage water were previously considered to be of little significance, other than in a few specialized cases, such as poorly drained soils high in organic matter (Sharpley *et al.*, 1994).

The process of P transport in subsurface runoff, therefore, has generally received little attention. This can be attributed to the fact that P in drinking-water has no known adverse effects on human health, and also that other sources of P loss have usually been the major contributors to eutrophication of surface waters (Driescher and Gelbrecht, 1988). There are reports in the literature, though, which demonstrate that P in subsurface runoff might well reach concentrations which have clear environmental implications (Ryden *et al.*, 1973). More recently, strong concern about this problem has been raised, especially in the Benelux countries, mainly in regions with a history of long-term applications of organic manures (Breeuwsma *et al.*, 1995; Lookman *et al.*, 1995).

In the European Union (EU), the average annual application of inorganic P fertilizer for 1989 ranged from 25 kg P ha^{-1} in Belgium to 9 kg P ha^{-1} in Spain and Portugal (Table 12.1). In addition, in some EU countries, large amounts of animal manures are applied. For example, the total P (TP) inputs per hectare, including inorganic fertilizer, are now at a mean level of 80 kg year^{-1} in the Netherlands (Tunney, 1992) and 75 kg year^{-1} in Flanders, Belgium (VLM, 1993). These inputs are far above crop removal of P, so that P continues to accumulate in the soil profile. Tunney (1992) calculated that the P status of agricultural soils in the EU has increased by 45 million tonnes during the last 30 years, equivalent to more than 10 kg P ha^{-1} year^{-1}.

Phosphorus is usually considered to be the nutrient which limits eutrophication in freshwater ecosystems. A critical P concentration for triggering eutrophic effects in lakes can be as low as 0.02–0.035 mg l^{-1} (Vollenweider, 1975). However, factors such as weather conditions, depth of

Table 12.1. Average chemical fertilizer consumption in the countries of the EU in 1989 per unit of agricultural area (from Tunney, 1992).

Country	kg P ha^{-1}
Belgium/Luxemburg	25
Germany	22
France	21
Italy	18
Netherlands	17
Denmark	16
Greece	15
Ireland	12
United Kingdom	10
Portugal	9
Spain	9

the lake and turnover time also play a role in the initiation of the eutrophication process. A concentration of 0.15 mg TP l^{-1} was proposed by Uunk (1990) as a maximum permissible P concentration in fresh water. The movement of even relatively small quantities of P from the terrestrial to the aquatic environment can therefore pose a threat to water quality.

PHOSPHORUS LOSSES FROM AGRICULTURE TO DRAINAGE WATER

Phosphorus Losses to Drainage Water in Soils Given Inorganic Fertilizer

Until the 1980s, the problem of P leaching from soil to water was generally not considered important. Cooke (1976) regarded P concentrations in drainage water to be small (up to about 1 mg P l^{-1}) and not related to fertilizer use but to the nature and pH of the soil parent material and to weather conditions, i.e. factors that the farmer cannot control. He quoted others who reported average concentrations of molybdate-reactive P (MRP) in drainage water in the Netherlands of less than 0.1 mg P l^{-1} from marine clay soils and 0.04 mg P l^{-1} from river clay soils. Similarly, Zwerman *et al.* (1972) reported P concentrations of less than 0.01 mg MRP l^{-1} in drainage water from tile drains under moderate applications of fertilizer (about 30 kg P ha^{-1}). They concluded that MRP concentrations in drainage water were not influenced significantly by fertilizer or management variables, e.g. presence or absence of cover crops or crop-residue incorporation. Baker *et al.* (1975) believed that annual P losses in drainage water were negligible, because subsoils fix much of the P leached from the plough layer. Sharpley and Menzel (1987) also considered that losses of P in subsurface drainage water were small with applications of fertilizer at recommended rates. They quoted concentrations of P in subsurface drainage waters from unfertilized and P-fertilized arable and grassland soils in North America ranging from undetectable to 0.064 mg P l^{-1}, except for three soils where it ranged around 0.2 mg P l^{-1}.

More recently, Haygarth and Jarvis (1995) found significant seasonal fluctuations in P leaching (mainly as MRP) from grass monoliths in lysimeters. There were clear maxima in the spring (> 0.1 mg P l^{-1}) for each of the four soil types studied, all given 40 kg P ha^{-1}. The spring peaks were thought to be due to interactions between effects of hydrological wetting and drying cycles with soil biological processes, thus promoting the release of P to the groundwater. There was no relationship between P leaching and Olsen P concentrations of the soils.

Howse et al. (Chapter 19.10, this volume) measured P losses in drainage water in the Brimstone experiment on a clay soil in Oxfordshire, UK, and reported that concentrations of MRP ranged from 0.03 to 0.50 mg P l^{-1} between 1990 and 1995. Total P in the drainage water was more variable (0.06 to 1.31 mg P l^{-1}), and the concentration often increased with drainage flow rate, as a result of an increased transport of P in particulate form. Three to four weeks after applying superphosphate at 17 or 33 kg P ha^{-1} to wet soil in December 1994, the MRP concentrations in the drainage water were considerably greater than before or after fertilizer application. The two application rates resulted in different soluble P losses during the drainage period 1994–1995. Nevertheless, losses only amounted to 38 and 61 g P ha^{-1} 100 mm^{-1} drain flow for the plots given the lower and higher fertilizer rates, respectively.

A comprehensive study of the P concentrations in drainage waters at experimental stations in south-east England was made by Williams (1976). At Saxmundham, the soil is a calcareous sandy clay loam, pH 8.0, overlying an impervious calcareous clay. Although the Olsen P concentration of the soil differed between grass/lucerne (15 mg P kg^{-1}) and arable cropping (35 mg P kg^{-1}), the average P concentrations in drainage waters at 60–80 cm depth were very similar, at around 0.02 mg P l^{-1}. At Broom's Barn, the mainly arable soils investigated range from sandy loams to clay loams. Between 1972 and 1974, in soils below 45 mg Olsen P kg^{-1} soil, phosphate concentrations in drainage waters were frequently undetectable, and the maximum was 0.08 mg P l^{-1} (Williams, 1976). However, drainage water from fields where mean soil-P concentrations varied between 46 and 100 mg Olsen P kg^{-1} contained up to 0.5 mg P l^{-1}. In another catchment, at Rothamsted, on a silty clay loam, up to 1.45 mg P l^{-1} was found in drainage water. Williams (1976) offered no explanation for these differences in P concentrations in drainage water. Recent work (Heckrath et al., 1995), showing a rapid increase in P loss to drainage water above a critical Olsen P concentration, offers an explanation, as we shall show later.

Phosphorus Losses to Drainage Water in Organically Manured Soils

Most of the P present in pig slurry is present as inorganic P (Gerritse, 1976). Of the TP in organic form (~20%), about 10–15% is present in the liquid phase. Rolston *et al.* (1975) considered that, although some of the organic P compounds appeared quite mobile, they were only of minor importance in determining P movement. However, in column leaching studies on a sandy soil, to which pig slurry was applied at the equivalent of 1000 t ha^{-1}, Gerritse (1981) found that high-molecular-weight organic P compounds were eluted shortly after a chloride tracer, reaching up to 1.5 mg P l^{-1}. In contrast, inorganic P concentrations remained below 0.1 mg l^{-1}. Nevertheless, it was suggested that the mobility of P added to soils in pig slurry was mainly governed by orthophosphate, although the model that was developed to describe the processes involved in P leaching was not very accurate.

Several authors have reported on significant P accumulation below the plough layer after organic manure applications (e.g. Campbell and Racz, 1975; Unwin, 1981; Spallacci and Boschi, 1985). For example, Johnston (1976) showed that, where farmyard manure (FYM) was applied each year to the Barnfield experiment, there was appreciably more enrichment of the 23–30 cm and 30–46 cm soil layers, compared with inorganic fertilizer-treated soils. In addition, some slight enrichment was observed below 46 cm. Similarly, Johnston (1976) reported much P enrichment in both the 30–46 and 46–62 cm layers on the FYM-treated plots of the market-garden experiment on a sandy loam, compared with plots receiving inorganic P fertilizer. This suggests enhanced mobility of manure-derived P through soil, almost certainly due to movement of some P in organic forms.

Phosphorus movement after long-term pig-slurry applications to sandy soils was also reported by Vetter and Steffens (1981). With annual applications between 185 and 660 kg P ha^{-1}, they show large increases in double-lactate-extractable P down to 90 cm soil depth. Up to 13% of P applied accumulated in the 60–90 cm layer, with a clear tendency for greater P movement on sandy soils with high organic-matter content. Phosphorus movement was much more pronounced after pig-slurry application, compared with similar rates of mineral fertilizer. However, in other work, 7% of the TP enrichment in a heavy loam soil occurred below 60 cm depth following mineral fertilizer treatments (Vetter and Steffens, 1981). Unfortunately, no details of application rates are given.

After 4 years of applying up to 440 kg P ha^{-1} year^{-1} in pig slurry to a heavy loam on an opencast mining reclamation site, Laves and Thum (1984) found a significant increase in ammonium acetate-extractable P down to 80 cm depth. There was a linear relationship between TP inputs and the P accumulation, as measured by ammonium acetate, in the soil below the plough layer. Unlike superphosphate, FYM applications over 4 years raised sodium biocarbonate (NaHCO$_3$)-extractable P concentrations significantly in the 30–60 cm layer of a silty clay (Meek *et al.*, 1979). At the maximum rate of 1336 kg P ha^{-1} year^{-1}, the P concentrations were up to eight times higher compared with an unfertilized control. At the end of the high-manurial treatment, TP in soil solution in the 30–60 cm layer reached 4 mg l^{-1}, indicating transport of inorganic or organic P. However, other authors who showed an increase in P below the plough layer concluded that slurry might have moved directly into the subsoil, either because of dry summer conditions (Unwin, 1981) or because soils were water-saturated (McAllister and Stevens, 1981).

Breeuwsma *et al.* (1995) considered that, in the Netherlands, leaching of P was a problem in areas with intensive livestock farming, where high manure loads coincided with shallow water-tables and sandy soils. In the catchment they studied, more than 80% of the agricultural soils were saturated with P due to past manure applications. Percentage P saturation was defined as: 100(P extracted by oxalate)/(P sorption capacity). The mean estimated P leaching rate was 2.5 kg P ha^{-1} year^{-1}. They reported total P concentrations in drainage water under manured soils in the Netherlands of more than 1 mg P l^{-1} and recognized this amount as a major environmental problem.

Although there is evidence for P enrichment in deeper soil layers, only a few studies in the older literature have confirmed elevated P concentrations in drainage water or shallow groundwater. This was mainly explained by sufficient sorption capacity in the subsoil. However, in a drained sandy soil, Lookman (1995) found an abrupt decrease in P sorption capacity below the plough layer, suggesting that colloidal iron (Fe) and aluminium (Al) hydroxides may have been leached from the subsoil. Together with a tendency for P to be translocated from the plough layer in a colloidal form (Lookman, 1995), this may also indicate possible limitations in P retention in the subsoil of sandy soils.

Phosphorus Losses to Drainage Water in Organic Soils

Losses of P from organic soils to drainage water can be several orders of magnitude larger than from mineral soils and can have a marked effect on water quality. Miller (1979) reported very large losses of MRP in tile drainage water from three organic soil sites in Ontario. The MRP concentrations in the drainage water, measured between 1972 and 1975, ranged from about 2 to 18 mg P l^{-1}, giving a maximum estimated annual P loss by drainage of 37 kg P ha^{-1}. Cogger and Duxbury (1984) compared the behaviour of P in eight organic soils and measured annual leaching losses between 1 and 30 kg ha^{-1}. These large amounts of P in the drainage water were attributed to the very high fertilizer-P applications and especially to the low P-fixation capacity of these organic soils, due to a lack of Fe and Al hydroxides. Phosphorus does not tend to accumulate with fertilization in arable-farmed organic soils. Kuntze and Scheffer (1979) reported 10 to 20 times higher P leaching rates from organic compared with mineral soils.

PHOSPHORUS LEACHING FROM BROADBALK

Experimental Site

For more than 100 years, soils from the Broadbalk continuous wheat experiment at Rothamsted have received annually either no P, P in FYM at c. 40 kg P ha^{-1} or P as superphosphate at 35 or 17 kg P ha^{-1}. Further details of the experiment are given by Johnston (1969). The soil is a silty clay loam of the Batcombe series overlying clay with flints and is maintained at about pH 7.5. The soil Olsen P concentrations in the plough layer (0–23 cm) of the single replicate now range from less than 10 to about 100 mg P kg^{-1} soil (air-dry basis). The different soil-P concentrations, other than in the plot receiving no P (the nil plot), are mainly caused by the different P off-takes by wheat given different rates of N fertilizer. Tile drains were installed along the middle of each plot in 1849 (1884 on the FYM plot) at a depth of 65 cm. In 1993 the old drains were replaced (see Heckrath et al., 1995, for details).

Drainage-water samples were collected at the drain outlets of plots selected to cover a wide range of Olsen P concentrations in the plough layer on five occasions between October 1992 and February 1994. The water samples were analysed for TP before filtration, and for MRP and total dissolved P (TDP) after filtration. Total particulate P (TPP) was calculated as the difference between TP and TDP, and dissolved organic P (DOP) as the difference between TDP and MRP (Heckrath et al., 1995).

Phosphorus in the Drainage Water

Total P concentrations above 1 mg P l^{-1} were measured on several plots at all dates other than October 1992. These concentrations are higher than most others reported (e.g. Cooke, 1976; Bottcher et al., 1985; Sharpley and Menzel, 1987). Similarly, MRP values, of which the largest ranged between 1.04 and 1.83 mg P l^{-1}, were very high compared with those measured by the above authors.

Relatively, the largest P fraction in drainage water on all plots, apart from the nil plot, was MRP. It ranged from 66 to 86% of the total P in the drainage water for different plots and drainage events, being a rather smaller fraction of the TP after the new drains were installed than before. Total particulate P accounted for 8–15% in waters from the old drains and 23–35% from the new drains, respectively. Dissolved organic P, averaged across plots and time, was 4.5%, excluding the nil and FYM plot, the latter averaging 11%. Similar proportions of these different P forms were reported by Culley *et al.* (1983). Phosphorus in drainage waters from the nil plot showed more variability between different sampling times. With a maximum TP concentration of 0.2 mg TP l^{-1}, MRP ranged from none to 50%, DOP from 2 to 40% and TPP from 44 to 66%, being generally the largest fraction.

Relationship Between Soil Olsen Phosphorus Concentrations and Phosphorus in Drainage Water

Below about 60 mg Olsen P kg^{-1} soil, the TP concentrations in the drainage water were small (< *c.* 0.15 mg P l^{-1}). Above *c.* 60 mg Olsen P kg^{-1} soil, there was a rapid linear increase in the P concentration in drainage water up to the maximum Olsen P concentration of about 100 mg P kg^{-1} soil. A simple split-line model described the relationship well for all the drainage events, and exponential models did not result in better fits (Fig. 12.1). In an analysis of parallelism, split lines, fitted to data from each sampling date, showed no evidence for separate initial slopes or different change points ($P < 0.01$). The fit was slightly improved with MRP rather than TP (Fig. 12.2a). Nevertheless, for both TP and MRP, the Olsen P change points were estimated to be almost the same, at 57 mg P kg^{-1} soil for TP and 56 mg P kg^{-1} soil for MRP.

Significance of the Broadbalk Results

As far as we know, the Broadbalk data are unique, in that they show such a clear relationship between soil-P concentration in the plough layer and P concentration in the drainage water above 60 mg Olsen P kg^{-1} soil. Hanway and Laflen (1974) reported a positive relationship between acid-fluoride-extractable P at depth and soluble inorganic P in drainage water. Sharpley *et al.* (1977) found that 0.1 M sodium chloride (NaCl)-extractable P at drain depth (40–50 cm) was closely related to TP load in the drainage water, but not to MRP concentration in the drainage water.

Between 1983 and 1988, about 65% of agricultural soils in England and Wales were between Agricultural Development and Advisory Service (ADAS) indices 2 and 3 (16–46 mg Olsen P kg^{-1} soil) (Skinner *et al.*, 1992). Thus, if a soil Olsen P concentration of 60 mg P kg^{-1} soil does eventually prove to be a critical level, the risk of leaching from most soils will be small. However, about 20% of arable soils were at or above ADAS index 4 (i.e. > 46 mg Olsen P kg^{-1} soil). Thus, if soil-P concentrations increase in the future, there may be an enhanced risk of leaching of P. This could occur, for example, in areas of high livestock density or sewage-sludge application or very high inorganic fertilizer use.

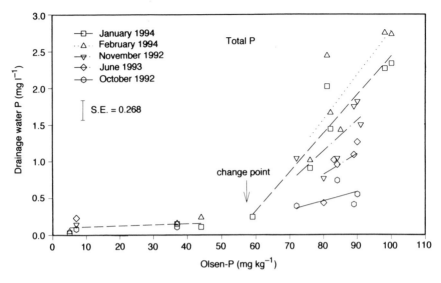

Fig. 12.1. Relationship between total P (TP) in drainage water and Olsen P from different Broadbalk plots. The split-line model shown fitted a common initial line and change point (at 57 mg Olsen P kg^{-1} soil) but separate changes in slope in the steep section for the five drainage events; 90% of the variance was accounted for (modified from Heckrath et al., 1995).

Phosphorus Movement from the Plough Layer to Drainage Water

The texture of the subsoil on Broadbalk varies, with more than 55% clay in some places but only about 35% in others (Avery and Bullock, 1969). Water flows through only a relatively small proportion of the subsoil at Rothamsted, the remainder of the water being immobile (Lawes et al., 1882; Addiscott et al., 1978). Recently, Goulding and Webster (1992) have provided evidence that nitrate from fertilizers can be carried from the soil surface into the drains through preferential pathways, i.e. solutes may interact little with the bulk of the subsoil.

A subsoil containing 35–55% clay should adsorb P strongly, and that from Broadbalk has been shown to adsorb substantial quantities of P (D. Thomas, 1995, Rothamsted, personal communication). Even if P is carried through preferential pathways, should it not be adsorbed by the surfaces over which it moves? One possibility is that the P moves in organic or other forms that are not susceptible to sorption, as seems likely in plots that receive FYM. Another possibility arises if the preferential pathways occur in the same place each year. For example, cracks may open up along the same plane of weakness. If the water carries P over the same surfaces for many years, localized P saturation may occur and the sorption of P cease, with the result that P is carried through to the drains.

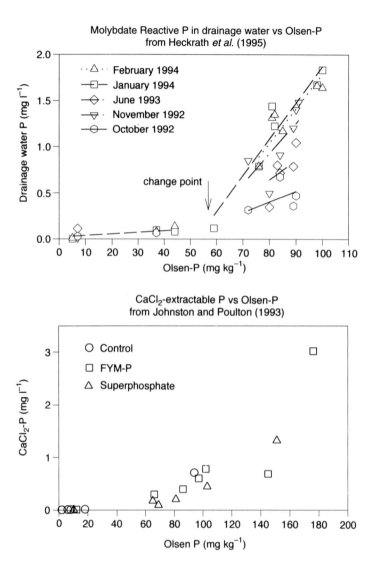

Fig. 12.2. (a) Relationship between molybdate-reactive P (MRP) in drainage water and Olsen P from different Broadbalk plots. The split-line model shown fitted a common initial line and change point (at 56 mg Olsen P kg^{-1} soil) but separate changes in slope in the steep section for the five drainage events; 92% of the variance was accounted for. (b) Relationship between P extracted by 0.01 M CaCl$_2$ and Olsen P (modified from Johnston and Poulton, 1993).

The Significance of the Change Point

We need to know if the change point at about 60 mg Olsen P kg^{-1} soil occurs in other soils and differs between soil types. It could be an important indicator of the risk of P movement in drainage water. Unfortunately, there are few experiments like Broadbalk from which such data could be obtained. The expense and time required to set up new, suitable field experiments also preclude this possibility. However, there is evidence in the scientific literature that an alternative approach may exist. Thus, the data from Johnston and Poulton (1993) may be replotted to show the relationship between Olsen P and calcium chloride (CaCl$_2$)-soluble P over a wide range of soil P concentrations (Fig. 12.2b). Remarkably, the graph is extremely similar in shape to that showing the relationship between P in drainage water and soil Olsen P (Figs 12.1 and 12.2a). Thus, the amount of CaCl$_2$-extractable P was extremely small until about 60 mg Olsen P kg^{-1} soil. There was then a rapid increase in the amount of CaCl$_2$-extractable P, which, again, was linearly related to soil Olsen P concentrations. The change point at 60 mg P kg^{-1} soil was precisely the point at which P began to be detected in the drainage water of Broadbalk. Being a weak electrolyte, CaCl$_2$ extracts reflect only the intensity of P in soils and therefore a readily leachable P pool (Sharpley *et al.*, 1977). Thus the ratio between, for example, CaCl$_2$-extractable P and Olsen P might be a useful predictor of the soil-P concentrations at which there is significant risk of P movement from heavy, rather calcareous soils to water. This needs to be tested on different soils given different rates of P fertilization.

PHOSPHORUS LEACHING IN SOILS OF WEST FLANDERS, BELGIUM

The soils studied were loamy sand to sandy loam soils which received regular inputs of various amounts of pig slurry. In contrast to the Broadbalk experiment, the P content of the total soil profile, with the base of the profile taken to be the depth of the mean highest groundwater-table, was considered.

In non-calcareous light-textured soils, the phosphate sorption capacity (PSC) is mainly governed by the quantity of reactive, amorphous, microcrystalline Fe and Al in the soil profile (Van der Zee and Van Riemsdijk, 1988). Van der Zee *et al.* (1990a) calculated the PSC of the soil profile using the oxalate-extractable Fe and Al (expressed as mmol kg^{-1}):

$$\text{PSC} = \frac{(\text{Fe}_{ox} + \text{Al}_{ox})}{2} \text{ (mmol P kg}^{-1}) \qquad \text{equation (1)}$$

Then, they determined the phosphate saturation degree (PSD), i.e. the ratio between oxalate-extractable P and PSC:

$$\text{PSD} = 100 \times \frac{P_{ox}}{\text{PSC}} \quad (\%) \qquad \text{equation (2)}$$

Based on the Langmuir equation, PSD might be related to the partitioning of P between the solid and the solution phase in soils (Van der Zee et al., 1990b):

$$\text{PSD} = \frac{\gamma Kc}{1 + (\gamma Kc)} \qquad \text{equation (3)}$$

where:

> PSD = phosphate saturation degree of the soil profile between the soil surface and the groundwater table
> γ = the ratio between total fixed P and reversibly adsorbed P, rated at a value equal to 3
> K = Langmuir adsorption constant, measured as 1.13 l mg^{-1}
> c = MRP concentration in the groundwater at the reference level (mg P l^{-1})

Using equation (3) allows us to determine critical levels of P saturation in soils in terms of maintaining acceptable P concentrations in the groundwater. For example, it was found that a concentration of 0.1 mg MRP l^{-1} in groundwater, which is sometimes chosen as the maximum acceptable P concentration in waters in view of environmental protection, was reached at a PSD of only 24%. Due to field heterogeneity, this phosphate saturation criterion should be interpreted with caution. Taking 15 samples per field, a field was considered to be phosphate-saturated with a probability of 97.5% if the field average PSD of the soil profile exceeded 30% (Van der Zee et al., 1990b). This implies that P is leached to the groundwater table long before the PSC is reached.

Phosphorus losses to tile-drainage water were investigated during the winter seasons 1993–1994 and 1994–1995 on 19 artificially drained fields which received various amounts of slurry in the past. The P concentrations in the drainage water in the two seasons ranged from less than 0.1 mg MRP l^{-1} to more than 1.0 mg MRP l^{-1} (Fig. 12.3). Similar P concentrations in the drainage waters were found in both seasons, indicating that high P losses may continue for several years once elevated soil-P concentrations are reached.

With 15 soil samples taken per field, the PSD of the soil profile was

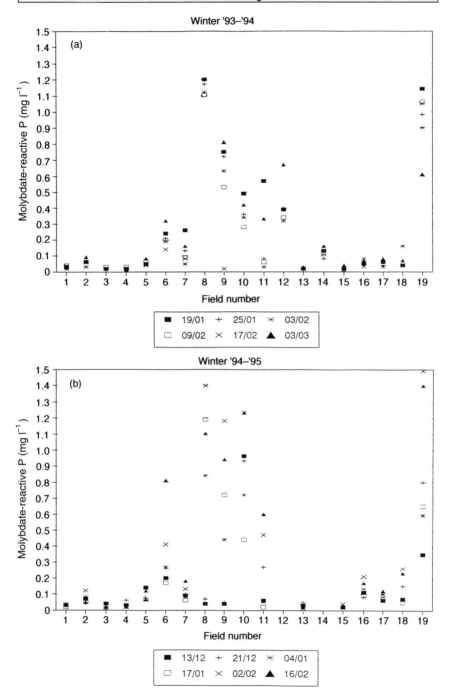

Fig. 12.3. Concentrations of molybdate-reactive phosphate in the drainage water from 19 different fields during the winter seasons (a) 1993–1994 and (b) 1994–1995.

determined according to Van der Zee *et al.* (1990b). Figure 12.4 shows the concentrations of MRP in the drainage water averaged over the two seasons and sampling times in relation to the PSD of the soil profile of the 19 fields studied. Based on equation (3), the solid line represents the relationship between the PSD of the soil profile and calculated P concentration in the soil water. The dashed line, in comparison, represents the best fit of our data to this model, with only the *K* value installed as a variable. For the sandy loam soils investigated, the ratio of total fixed P to adsorbed P was, indeed, found to be equal to 3 (De Smet *et al.*, 1995). Fitting the field data therefore resulted in an almost identical relationship between PSD and P concentration compared with the prediction based on equation (3) indicating that the proposed model is also valid for the sandy loam soils. A somewhat lower *K* value of 0.93 l mg^{-1} was found, instead of 1.3 l mg^{-1}. This means that the sandy loam and loamy sand soils have a somewhat lower capacity to adsorb P than the acid sandy soils for which the model was designed.

A PSD below about 35% resulted in P concentrations less than 0.1 mg P l^{-1}. The higher the PSD, the higher the P losses and the higher the

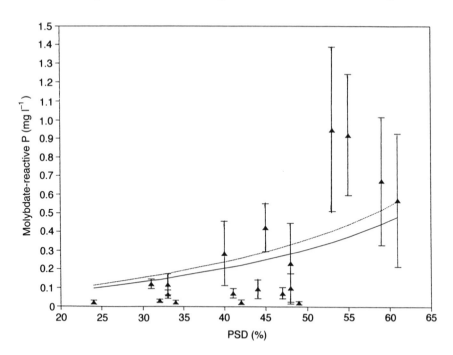

Fig. 12.4. Relationship between the molybdate-reactive P concentration in the drainage water and the phosphate saturation degree (PSD) of the soil profile. Standard deviations are shown for MRP averaged over all sampling times of the two seasons. The solid line represents the theoretical relationship according to Van der Zee *et al.* (1990b). The dashed line is the theoretical model fitted to the field data.

variability in P concentrations. This was due to variability within and between fields, caused by differences in depth of drainage, fluctuations in groundwater levels and differences in P distribution down the soil profiles. Phosphate saturation degrees higher than 50% resulted in an almost exponential increase in P losses, and concentrations of more than 0.5 mg P l^{-1} were measured. Such high P loads have very damaging efffects on the aquatic ecosystem.

CONCLUSIONS

The P moving down the soil profile in drainage water can vary widely, from negligible, almost undetectable amounts to a few mg P l^{-1} in arable and grassland soils and even higher concentrations in organic soils. Although the quantities are unimportant in economic terms to the farmer, being at most only a few kg P ha^{-1}, such quantities of P may have significant effects on water quality.

There is evidence that accumulation of P in the soil profile may lead to increased losses of P in drainage water. In sandy soils, determination of the PSC and the PSD of the total soil profile to the groundwater table is a good indicator of the risk of P losses by drainage. This was tested on various fields in the sandy loam soil region of West Flanders, Belgium.

The Broadbalk experiment shows that, in clay loams, until soil Olsen P concentrations reach a certain critical concentration, little P moves from the plough layer. After this critical soil-P concentration, P loss to drainage water is significant.

The concentrations of soil P which may be allowed to accumulate in soil before the risk of leaching increases needs evaluation in different soil types and under different managements.

The mechanisms of P movement in drainage water need evaluation. In sandy soils, P may move down a P-saturated subsoil. In clay soils, this does not seem possible. Preferential flow is a more likely mechanism.

Simple indicators are required to predict the soil-P concentrations at which the risk of leaching becomes significant.

REFERENCES

Addiscott, T.M., Bolton, J. and Rose, D.A. (1978) Chloride leaching in the Rothamsted drain gauges: influence of rainfall pattern and soil structure. *Journal of Soil Science* 29, 305–314.

Avery, B.W. and Bullock, P. (1969) The soils of Broadbalk: morphology and classification of the Broadbalk soils. In: *Report of the Rothamsted Experimental Station for 1969*, Part 2. Rothamsted Experimental Station, Harpenden, UK, pp. 63–81.

Baker, J.L., Campbell, K.L., Johnson, H.P. and Hanway, J.J. (1975) Nitrate, phosphorus, and sulfate in subsurface drainage water. *Journal of Environmental Quality* 4, 406–412.

Behrendt, H. and Boekhold, A. (1993) Phosphorus saturation in soils and groundwaters. *Land Degradation and Rehabilitation* 4, 233–243.

Bottcher, A.B., Monke, E.J., Beasley, D.B. and Huggins, L.F. (1985) Sediment and nutrient losses from subsurface drainage. In: El-Swaify, S.A., Moldenhauer, W.C. and Lo, A. (eds) *Soil Erosion and Conservation.* Soil Conservation Society of America, Ankeny, Iowa, pp. 622–633.

Breeuwsma, A., Reijerink, J.G.A. and Schoumans, O.F. (1995) Impact of manure on accumulation and leaching of phosphate in areas of intensive livestock farming. In: *Proceedings of Conference: Animal Waste and the Land–Water Interface.* ARWC, Arkansas, pp. 1–11.

Campbell, L.B. and Racz, G.J. (1975) Organic and inorganic P content, movement and mineralization of P in soil beneath a feedlot. *Canadian Journal of Soil Science* 55, 457–466.

Cogger, C. and Duxbury, J.M. (1984) Factors affecting phosphorus losses from cultivated organic soils. *Journal of Environmental Quality* 13, 111–115.

Cooke, G.W. (1976) A review of the effects of agriculture on the chemical composition and quality of surface and underground waters. In: *Agriculture and Water Quality.* Ministry of Agriculture, Fisheries and Food Technical Bulletin No. 32, HMSO, London, UK, pp. 5–57.

Culley, J.L.B., Bolton, E.F. and Bernyk, V. (1983) Suspended soils and phosphorus loads from a clay soil: I. Plot studies. *Journal of Environmental Quality* 12, 493–498.

Driescher, E. and Gelbrecht, J. (1988) Phosphat im unterirdischen Wasser. 1. Mitteilung: Zum Vorkommen von Phosphat im Grundwasser – eine Literaturübersicht. *Acta Hydrophysica* 32, 213–235.

Gerritse, R.G. (1976) Phosphorus compounds in pig slurry and their retention in the soil. In: Voorburg, J.H. (ed.) *Utilization of Manure by Land Spreading.* Commission of the European Communities, Luxemburg, pp. 257–266.

Gerritse, R.G. (1981) Mobility of phosphorus from pig slurry in soils. In: Hucker, T.W.G. and Catroux, G. (eds) *Phosphorus in Sewage Sludge and Animal Waste Slurries.* D. Reidel Publishing Company, Dordrecht, the Netherlands, pp. 347–366.

Goulding, K.W.T. and Webster, C.P. (1992) Methods for measuring nitrate leaching. *Aspects of Applied Biology* 30, 63–70.

Hanway, J.J. and Laflen, J.M. (1974) Plant nutrient losses from tile-outlet terraces. *Journal of Environmental Quality* 3, 351–356.

Haygarth, P. and Jarvis, S.C. (1995) Phosphorus losses from grassland soils in the UK. In: *Proceedings of the Second International IAQW Specialised Conferences and Symposium on Diffuse Pollution,* Brno and Prague, Czech Republic, 13–18 August 1995. IAQW, Brno and Prague, Czech Republic, Part 1, pp. 287–291.

Heckrath, G., Brookes, P.C., Poulton, P.R. and Goulding, K.W.T. (1995) Phosphorus leaching from soils containing different P concentrations in the Broadbalk experiment. *Journal of Environmental Quality* 24, 904–910.

Isermann, K. (1990) Share of agriculture in nitrogen and phosphorus emissions into the surface waters of Western Europe against the background of their eutrophication. *Fertilizer Research* 26, 253–269.

Johnston, A.E. (1969) The soils of Broadbalk: plant nutrients in Broadbalk soils. In:

Report of the Rothamsted Experimental Station for 1969, Part 2. Rothamsted Experimental Station, Harpenden, UK, pp. 93–115.

Johnston, A.E. (1976) Additions and removals of nitrogen and phosphorus in longterm experiments at Rothamsted and Woburn and the effect of residues on total soil nitrogen and phosphorus. In: *Agriculture and Water Quality*. Ministry of Agriculture, Fisheries and Food Technical Bulletin No. 32, HMSO, London, UK, pp. 111–144.

Johnston, A.E. and Poulton, P.R. (1993) The role of phosphorus in crop production and soil fertility: 150 years of field experiments at Rothamsted, United Kingdom. In: Schultz, J.J. (ed.) *Phosphate Fertilizers and the Environment*. International Fertilizer Development Center, Muscle Shoals, Alabama, pp. 45–63.

Kuntze, H. and Scheffer, B. (1979) Die Phosphatmobilität im Hochmoorboden in Abhängigkeit von der Düngung. *Zeitschrift für Pflanzenernährung und Bodenkunde* 142, 155–168.

Laves, D. and Thum, J. (1984) Phosphor- und Kalium-Akkumulation nach hohen Güllefugat-Gaben auf einem Kipplehmstandort. *Archive für Acker- und Pflanzenbau und Bodenkunde* 28, 279–286.

Lawes, J.B., Gilbert, J.H. and Warington, R. (1882) On the amount and composition of drainage water collected at Rothamsted. III. The quantity of nitrogen lost by drainage. *Journal of the Royal Society of England, 2nd Series* 18, 43–71.

Lookman, R. (1995) Phosphorus behaviour in excessively fertilized soils. PhD thesis, Katholieke Universiteit Leuven, Louvain, Belgium.

Lookman, R., Vandeweert, N., Merckx, R. and Vlassak, K. (1995) Geostatistical assessment of the regional distribution of phosphate sorption capacity parameters (Fe_{ox} and Al_{ox}) in northern Belgium. *Geoderma* 66, 285–296.

McAllister, J.S.V. and Stevens, R.J. (1981) Problems related to phosphorus in the disposal of slurry. In: Hucker, T.W.G. and Catroux, G. (eds) *Phosphorus in Sewage Sludge and Animal Waste Slurries*. D. Reidel Publishing Company, Dordrecht, the Netherlands, pp. 383–396.

Meek, B.D., Graham, L.E., Donovan, T.J. and Mayberry, K.S. (1979) Phosphorus availability in a calcareous soil after high loading rates of animal manure. *Soil Science Society of America Journal* 43, 741–744.

Miller, M.H. (1979) Contribution of nitrogen and phosphorus to surface drainage water from intensively cropped mineral and organic soils in Ontario. *Journal of Environmental Quality* 8, 42–48.

Miller, M.H., Robinson, J.B., Coote, D.R., Spires, A.C. and Draper, D.W. (1982) Agriculture and water quality in the Canadian Great Lakes basin: III. Phosphorus. *Journal of Environmental Quality* 11, 487–492.

Rolston, D.E., Rauschkolb, R.S. and Hoffman, D.L. (1975) Infiltration of organic phosphate compounds in soil. *Soil Science Society of America Proceedings* 39, 1089–1094.

Ryden, J.C., Syers, J.K. and Harris, R.F. (1973) Phosphorus in runoff and streams. *Advances in Agronomy* 25, 1–45.

Sharpley, A.N. and Menzel, R.G. (1987) The impact of soil and fertilizer phosphorus on the environment. *Advances in Agronomy* 41, 297–324.

Sharpley, A.N., Tillman, R.W. and Syers, J.K. (1977) Use of laboratory extraction data to predict losses of dissolved inorganic phosphate in surface runoff and tile drainage. *Journal of Environmental Quality* 6, 33–36.

Sharpley, A.N., Chapra, S.C., Wedepohl, R., Sims, J.T., Daniel, T.C. and Reddy, K.R. (1994) Managing agricultural phosphorus for protection of surface waters: issues and options. *Journal of Environmental Quality* 23, 437–451.

Sibbesen, E. (1989) Phosphorus cycling in intensive agriculture with special reference to countries in the temperate zone of Western Europe. In: Tiessen, H. (ed.) *Phosphorus Cycles in Terrestrial and Aquatic Ecosystems.* University of Saskatchewan, Saskatoon, Canada, pp. 112–122.

Skinner, R.J., Church, B.M. and Kershaw, C.D. (1992) Recent trends in soil pH and nutrient status in England and Wales. *Soil Use and Management* 8, 16–20.

Spallacci, P. and Boschi, V. (1985) Long-term effects of the landspreading of pig and cattle slurries on the accumulation and availability of soil nutrients. In: Williams, J.H., Guidi, G. and L'Hermite, P. (eds) *Long-term Effects of Sewage Sludge and Farm Slurries Applications.* Elsevier Applied Science Publishers, London, pp. 33–44.

Tunney, H. (1992) Some environmental implications of phosphorus use in the European Community. In: *Phosphorus, Life and Environment. From Research to Application. Proceedings of the 4th International IMPHOS Conference,* Ghent, Belgium. World Phosphate Institute, Casablanca, Morocco, pp. 347–359.

Unwin, R.J. (1981) Phosphorus accumulation and mobility from large applications of slurry. In: Hucker, T.W.G. and Catroux, G. (eds) *Phosphorus in Sewage Sludge and Animal Waste Slurries.* D. Reidel Publishing Company, Dordrecht, the Netherlands, pp. 333–343.

Uunk, E.B.J. (1990) Impact of manure and fertilizer usage on surface waters: load, effect and control. In: Del Castilho, P. (ed.) *Animal Manure: Problems and Solutions.* Dutch Royal Chemical Society, s'Gravenhage, the Netherlands, pp. 49–75.

Van der Zee, S.E.A.T.M and Van Riemsdijk, W.H. (1988) Model for long-term phosphate reaction kinetics in soil. *Journal of Environmental Quality* 17, 35–41.

Van der Zee, S.E.A.T.M, Van Riemsdijk, W.H. and de Haan, F.A.M. (1990a) *Het protokol fosfaatverzadigde gronden. Deel I: Toelichting.* Vakgroep Bodemkunde en Plantevoeding, Agricultural University of Wageningen, Wageningen, the Netherlands, 69 pp.

Van der Zee, S.E.A.T.M, Van Riemsdijk, W.H. and de Haan, F.A.M. (1990b) *Het protokol fosfaatverzadigde gronden. Deel II: Technische uitwerking.* Vakgroep Bodemkunde en Plantevoeding, Agricultural University of Wageningen, Wageningen, the Netherlands, 25 pp.

Vetter, H. and Steffens, G. (1981) Phosphorus accumulation in soil profiles and phosphorus losses after the application of animal manures. In: Hucker, T.W.G. and Catroux, G. (eds) *Phosphorus in Sewage Sludge and Animal Waste Slurries.* D. Reidel Publishing Company, Dordrecht, the Netherlands, pp. 309–327.

Vighi, M., Soprani, S., Puzzarini, P. and Menghi, G. (1991) Phosphorus loads from selected watersheds in the drainage area of the northern Adriatic Sea. *Journal of Environmental Quality* 20, 439–444.

VLM (1993) *Mestactieplan.* Flemish Land Society, Brussels, Belgium, 108 pp.

Vollenweider, R.A. (1975) Input–output models with special reference to the phosphorus loading concept in limnology. *Schweizerische Zeitschrift für Hydrologie* 37, 53–84.

Williams, R.J.B. (1976) The chemical composition of rain, land drainage and borehole water from Rothamsted, Brooms Barn, Saxmundham and Woburn

Experimental Stations. In: *Agriculture and Water Quality.* Ministry of Agriculture, Fisheries and Food Technical Bulletin No. 32, HMSO, London, UK, pp. 174–200.

Zwerman, P.J., Grewling, T., Klausner, S.D. and Lathwell, D.J. (1972) Nitrogen and phosphorus content of water from tile drains at two levels of management and fertilization. *Soil Science Society of America Proceedings* 36, 134–137.

13 Sustainable Phosphorus Management in Agriculture

G. Bertilsson[1] and C. Forsberg[2]

[1]*Environmental Manager, Hydro Agri, Landskrona, Sweden;* [2]*Institute of Limnology, Uppsala University, Sweden*

INTRODUCTION

Sustainability in the absolute sense is hard to define and even more difficult to achieve. Our goal here is limited to defining some obvious problems and pointing to steps which will lead towards improved sustainability for both agriculture and the environment in society as a whole.

One important part concerns the use of phosphorus (P) to soils and crops, another the wider circulation of P in society. Important sources for recycling are animal manure and urban waste. A problem with manure occurs when there is a local surplus in relation to the P requirement of the agricultural land. In general, P in animal manure is recycled efficiently. Sweden has the favourable situation that, on average, only half of the P uptake by crops is covered by manure. For urban waste, the situation is totally different. At present, there is practically no recycling. Both technology and systems need fundamental development efforts.

RECYCLING PHOSPHORUS FROM URBAN WASTE

Urban sanitary systems in modern society are far from optimal for saving and recycling natural resources, e.g. P. Instead, the present flux of P from agriculture to the urban society results in accumulations in sludge, organic wastes and ash, and, as a long-term consequence, irreversible losses to fresh and marine waters and water pollution. Fluxes and accumulations are illustrated here with some data from Sweden (Fig. 13.1.)

In Sweden, chemical precipitation of P in sewage is standard, and more

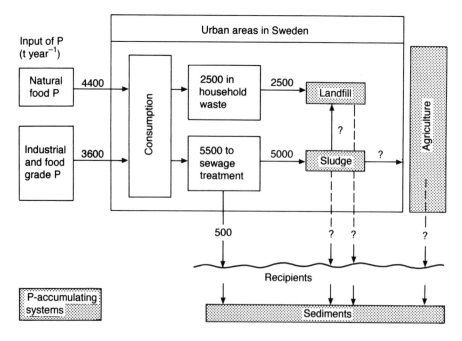

Fig. 13.1. Lack of effective recycling systems for phosphorus between urban areas and agriculture/industry has resulted in accumulation of P in sludge deposits/landfills and also in neighbouring farmlands, where P has been spread in amounts exceeding those removed by crop harvest.

than 90% of the urban population is connected to this treatment (Balmér and Hultman, 1988). More than 90% of the sewage P may be bound in sludge. However, recycling of sludge back to agriculture is not effectively developed and most sludge is deposited as a waste.

The agricultural sector, especially the food-processing industries, feel a great uncertainty about both known and unknown contaminants in the sludge and thus the use of sludge in agriculture has diminished during the last decade. In the mid-1980s 60% of the sludge was recycled back to agriculture (Bollmark, 1991) while a few years later only 35% could be spread. Recently, the use of 'pure' sludge, from waste-treatment plants in non-industrial areas, has increased a little.

The total input of P to urban areas in Sweden is about 8000 tons (about 1 kg P person^{-1} year^{-1}), which, after consumption, will be accumulated to a high degree in sludge deposits and landfill, together with household wastes. Varying amounts will be spread on neighbouring agricultural areas and unknown amounts are lost by non-point losses to water. As an example, in 1990 the total accumulation of P in the Stockholm region was estimated to be 80% of the import (Forsberg and Rengefors, 1993). The long-term effects of sludge deposition are unknown, and uncertainty prevails concerning the rate and size of future loss. This problem was already being addressed in 1973 (Stumm, 1973).

The annual accumulation of P in Sweden as a whole during the period 1975–1985 was estimated to be 90% of the import of 60,000 tons (Karlsson, 1989). The use of fertilizer P is now only one-third of the 1975 figure and the overall balance is much improved. However, the one-way flux of P from agriculture to waste deposits continues.

Sludge Deposits

The use of sludge in Swedish agriculture is, in accordance with regulations, limited by the risk for P accumulation. A maximum of 22 kg P ha^{-1} year^{-1} may be applied. On average, the municipal waste distributed on agricultural land would give 1.9 kg P ha^{-1}. There is agricultural capacity to utilize the P in sludge in Swedish conditions.

For landfills, etc., as well as for uncontrolled diffuse depositions around farmyards, in urban areas, etc., the situation is different. Here, the accumulation is uncontrolled as far as amount is concerned.

The accumulation and loss of P to or from a P-accumulating system, such as a sludge deposit, may be described as von Bertalanffy growth (Odum, 1983):

$$Q_t = J/k(1-e^{-kt}) + Q_0 e^{-kt}$$

where

J = flow to a storage (amount per time)
k = rate constant (the relative rate of change)
Q_0 = available amount of P at time $t = 0$

In systems without P inflow, k can be looked upon as a rate constant for P export. In laboratory experiments where artificial rainwater was passed on through iron (Fe)- or aluminium (Al)-precipitated sludge, the maximum release of P at 20°C was 95% during anaerobic conditions and 20–30% during aerobic conditions. The release rate (k values) varied between 0.02 and 0.3, where the lowest values were obtained for small leaching volumes during aerobic conditions. A k value of 0.02 means a decrease of 2% per day of the stored, mobile P. In an aerobic environment, the k values increased to 0.2–0.3. Oxygen status and leachate volume were the main controlling factors (Rydin, 1996).

The results demonstrate that increasing amounts of deposited sludge must increase the risk of future P pollution to groundwaters and surface waters. This further emphasizes the need for new management systems, permitting effective recycling of P from urban areas back to agriculture.

Future Perspective

The long-term perspective is that waste P will become transformed into a new resource and recycled in society, and that the loading to fresh and marine water will be substantially reduced. In the short-term perspective of 10–20 years, more municipal sludge will be recycled from urban areas back to agriculture and industry. Harmful substances in sewage have been reduced. New techniques have improved the handling, transport and distribution on farmland. New methods and systems have been developed for collecting and separating important plant nutrients from sludge, and other elements (metals) are separated and recovered from sludge and ash. There is a great international demand for technologies in this field.

New Methods for Recycling

The awareness of the need for recycling of nutrients from urban areas to agricultural land has increased. At a local level, there is a political interest in developing recycling activities, such as changing the sanitation system from standard toilets to source (urine)-separating ones. The goal is to separate the nutrient-rich part of the waste before it is mixed with other waste flows and in this way facilitate recycling.

Prerequisites for Recycling: Hygiene and Total Environmental Benefit

Recycling in itself has a positive value in transforming a waste into a resource. However, some problems must be solved. One is hygiene, where at least three elements cause concern: sanitary safety, known or unknown organic compounds, heavy metals, etc. In particular, uncertainties about the effect of organic compounds and heavy metals have affected the development in Sweden.

Another aspect to consider is the total environmental effect of recycling measures. For instance, if energy use and emissions for processing and transporting waste outweigh what is saved by the recycling of nutrients, the benefit of recycling can be questioned. A total life-cycle analysis (LCA) is necessary for successful development.

New Recycling Research in Sweden

Several research projects are under discussion on sustainable sanitation systems in urban areas. New systems and processes are proposed for recovery and separation of P and other elements from sewage treatment processes, sludge and ashes.

For toilet waste, new systems involving urine separation and subsequent dual waste treatment are a possibility under preliminary study. However, in the following discussion, it is assumed that the prevailing urban waste-handling system based on water transportation will be used for a long time. It is a costly and time-consuming process to replace the present toilets with something else. The following projects are discussed.

1. Recycling technologies of resources from human wastes and criteria for product quality. The aim is to provide a fundamental background for recycling of resources and to evaluate alternatives, for instance:
- recovery and reuse of nutrients in enriched streams;
- biological P removal;
- treatment of waste solutions by ion exchange or absorption processes;
- concentrated waste streams handled locally.

2. Beneficial use of sludge from sewage treatment.
The aim is to recycle P and organic material from sewage sludge. The sludge is heated at a low pH to separate inorganic and organic pools. From these pools, organic sludge can be used for soil improvement, biofuel or activated carbon. The inorganic part can be used for recycling the coagulant, reuse of P and a heavy-metal residue, which perhaps can be used by the metal industry. The process will also produce liquid carbon, which can be used to improve nitrogen (N) reduction in sewage treatment or for biogas production (Göransson and Karlsson, 1994).

3. New technique for recycling nutrients in sludge. The aim is to develop a new and very promising technique for manufacturing pellets from sludge, when sludge quality permits recycling on agricultural lands. The vision is that sludge will be handled and used as fertilizers, easy to store and apply. There are a lot of questions to be answered, for instance energy for drying and sanitary safety.

TOWARDS SUSTAINABLE USE OF FERTILIZER PHOSPHORUS

The use of P in many West European countries is declining. Some of the reasons and causes may be highlighted by the example of Sweden.

The use of P in Sweden from 1920 to today is shown in Fig. 13.2. With some fluctuations, the use increased up to 1973, after which it declined fairly rapidly. The fertilizer recommendations were based on field experiments. They were based on agricultural economics, as well as soil chemistry. In order to optimize production, it was necessary to apply P not only for the crop but for the soil. It was necessary to apply more P than that exported by crop products, because P fixation had to be considered.

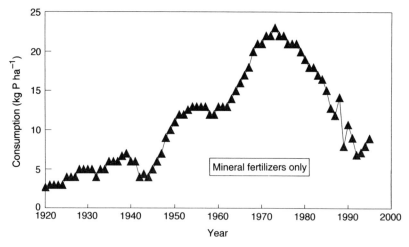

Fig. 13.2. Average application of mineral P fertilizer in Sweden.

In addition, P application often exceeded recommendations. Manure was seen by many farmers (but not the advisers) as a 'premium' on top of fertilizer needs. The favourable effects of P application had also become deeply rooted in the agricultural community.

What happened in 1974? A sharp increase in P price caused a reassessment. It was realized that the soil-P status was sufficient in many cases and that the P fertilization rates could be reduced. Subsequently, during the 1980s environmental questions became prominent. Agriculture was accused of overuse of fertilizers. Lake eutrophication resulting from P enrichment was widely discussed and brought to people's attention. Fertilizer taxes were introduced, which increased the price per unit N and P by 30–40%. Gradually, the fertilizer P application rates were reduced from an average of 25 kg P ha^{-1} in 1973 to 8 in 1993.

In concept, we had an enrichment phase up to the 1970s, followed by a replacement phase by the end of the 1980s. The enrichment was necessary to develop Swedish agriculture. The soils were low in P, so that fertilizer P additions caused dramatic yield increases and at the same time increased the soil fertility, measured both by yield potential and soil tests.

Phosphorus Fixation and Replacement

Several recent investigations show that the P status of soils with an enrichment history can be maintained by replacement (McCollum, 1991; Junck *et al.*, 1993). However, other investigations show replacement to be insufficient to maintain the soil-P status (Saarela, 1995).

In work in Sweden (G. Bertilsson, 1995, unpublished results), 11 soils of different origin, P status and fertilization history were treated with: zero P, P based on crop removal, and P removal plus 20 kg P ha^{-1}.

The experiment (microplots) was started in 1981 and was cropped each year, predominantly with spring cereals. Yield development and soil status (P-AL (ammonium lactate extraction), according to modified Egnér–Riehm) was monitored. Some of the results are given in Fig. 13.3. The yield measurements showed no response for added P if the P-AL was above 10 (Fig. 13.3).

The soil-P (available P) changes are shown in Table 13.1. For zero P, the P-AL decreased by an average of 0.36 units per year, for replacement plus 20 P it increased by 0.33, whereas for replacement there was on average no detectable change since the start in 1981. About 30% of the soil-P balance (negative for zero P, positive for replacement plus 20 P) was reflected in the P-AL analysis.

The soil analysis in general did not contradict the hypothesis that replacement maintained the soil-P status during the period. However, two soils showed a declining tendency, and it is clear that there are no universal truths in this matter.

The selection of soils allows another aspect to be covered. Five of the soils came from soil fertility experiments, with a preceding treatment of replacement plus 15 kg P annually since 1957. Thus, these five soils had been subject to an enrichment period in the field-scale soil fertility

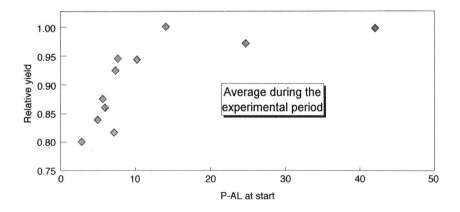

Fig. 13.3. Yield response to fertilizer.

Table 13.1. Soil-P changes in 11 soils during 13 years: P-AL at start and linear regression coefficients (change in P-AL per year).

Soil	P-AL	Regression coefficients		
		No P	Repl.	Repl. + 20
1	7.0	−0.33	0.13	0.36
2	5.2	−0.21	−0.01	0.23
3	8.2	−0.38	−0.07	0.26
4	11.0	−0.57	−0.21	0.08
5	6.8	−0.27	0.06	0.20
6	3.1	−0.14	−0.01	0.19
7	23	−0.70	0.13	0.50
8	6.7	−0.17	0.02	0.31
9	6.0	−0.16	0.00	0.18
10	14.5	−0.59	−0.17	0.08
11	38	−0.43	0.62	1.34
Average		−0.36	0.04	0.33

experiments during 1957–1980, and the conclusion of that work was that replacement is not sufficient (Jansson, 1975). Then the later experimental work – 1981–1993 described above – showed that replacement was sufficient on the same soils after the enrichment. There are explanations: the microplot experiment starts from an elevated P level, where the soil pools have been enriched, the yield level is raised and consequently the replacement dose is higher. The behaviour of the soil, the fixation, might have been changed by the enrichment period from 1957 to 1980. It has been shown that previous P additions affect the fixation (Barrow, 1974; Nielsen, 1994).

There is some scientific support for a replacement policy. However, situations differ (replacement would not have worked for Sweden in 1920), soils differ and crops differ.

The Effect of Crop and Soil Phosphorus Level

Crops vary in their ability to extract and utilize soil P. Examples of demanding crops are potatoes and onions. It is quite evident that the demands of these crops cannot normally be met by replacement. If such crops occur often in the rotation, a continuous P accumulation might take place even at proper P fertilization. There are two ways of improving P utilization and improving sustainability in this case.

- Rotational measures with intermediate crops utilizing the residual P.
- Improved fertilizer efficiency (products and application techniques).

Techniques for Improved Phosphorus Utilization

If replacement works, the long-term efficiency of P is 100% and there is not much room for improvement. However, techniques for improved P utilization may either give the same yield from a smaller soil-P pool or increase the agricultural output of the system. Work is going on with placement of P or multinutrient fertilizers, additions to fertilizers to reduce P fixation and foliar P fertilization.

Rotations and agricultural systems should be adapted to utilize residual P from demanding crops, such as potato. If potatoes are grown too often, a continued enrichment will result. This can be a problem in regions with a high density of crops with high demands of P fertilization. Improved fertilizer techniques might help.

It is important that the agricultural extension and advice are accepted by the farmer. A fertilizer-use adjustment will be for his/her long-term benefit. The concept of replacement cannot be applied rigorously. There are varying soils and varying crops to consider. Local research and experience should always be the base.

CONCLUSION

The concept of replacement is put forward as a guide for fertilizer use, which might serve as a starting-point, to be adjusted according to local data concerning soils and crop responses. Replacement has a sustainability component, although there are questions concerning resources, environment and soils. In most cases, P replacement would be a step in the right direction.

The amount of fertilizer P needed for replacement depends on P recycling. If P from recycling increases, the need for fertilizer P additions decreases. If P recycling conditions were ideal, there would be no need for fertilizer P once the system had been brought into a steady state.

However, the separation of agriculture and urban settlements in modern society makes it very difficult and expensive to recycle P. Even where the P in sewage is efficiently precipitated and contained in the sludge phase, as in Sweden, there are difficulties, mainly contamination with heavy metals. Until a new P management philosophy and new management systems have been developed, P will be accumulated in urban regions. From a national and global point of view, it can be concluded that development of improved P management systems is needed for: decreasing P accumulation in different ecosystems, decreasing P export from land to water, decreasing eutrophication of lakes and marine waters, decreasing the irreversible P losses to marine sediments and saving P mineral resources.

In general, there is great hope for improvement in the wealthier and more technically advanced economies. Techniques and concepts are ready for use. In reality, the change toward lower P inputs is quite rapid.

The challenge is much greater in the developing economies. Those countries are in the middle or at the start of the 'enrichment phase'. The issues concern both economy and ecology. There is a great need for methods which lower the need for continued P enrichment but still give satisfactory yields.

REFERENCES

Balmér, P and Hultman, B. (1988) Control of phosphorus discharges: present situation and trends. *Hydrobiologia* 170, 305–319.

Barrow, N.J. (1974) Effects of previous additions of phosphate on phosphate adsorption by soils. *Soil Science* 118, 82–90.

Bollmark, L. (1991) *Phosphorus Flows in Swedish Agriculture* (in Swedish). 1, SLU repro, Department of Soil Science, Agricultural University, Uppsala.

Forsberg, C. and Rengefors, K. (1993) Anrikning av fosfor i stadsregionen – en fosforbudget för Stockholm 1990. (Engl. summary: Accumulation of phosphorus in urban regions – a phosphorus balance for Stockholm 1990). *Vatten* 49, 81–86.

Göransson, J. and Karlsson, I. (1994) Beneficial use of sludge from sewage plants and water works. In: Hahn, H.H. and Klute, R. (eds) *Chemical Water and Wastewater Treatment.* Springer, Berlin, pp. 341–352.

Jansson, S.L. (1975) Longterm soil fertility studies (in Swedish). *Kungliga Skogs-och Lantbruksakad. Tidskr.* Suppl. 10.

Junck, A., Claassen, N., Schultz, V. and Wendt, J. (1993) Pflanzenverfugbarkeit der Phosphatvorräte ackerbaulich genutzter Böden. *Zeitschrift für Pflanzenernährung. Bodenkunde* 156, 397–406.

Karlsson, G. (1989) *Dynamics of Nutrient Mass Transport – a River Basin Evaluation.* Linköping Studies in Arts and Science 40, University of Linköping, Linköping, Sweden.

McCollum, R.E. (1991) Buildup and decline in soil phosphorus: 30-year trends on a typic Umprabuult. *Agronomy Journal* 83, 77–85.

Nielsen, J.D. (1994) Crop recovery of fertilizer P from soils low in soluble P. *Acta Agricultura Scandinavica* 44(2), 84–88.

Odum, H.T. (1983) *Systems Ecology: An Introduction.* John Wiley and Sons, New York.

Rydin, E. (1996) Experimental studies simulating potential phosphorus release from municipal sewage sludge deposits. *Water Research* 30, 1695–1701.

Saarela, I. (1995) P balances – consequences of reduced fertilization. Paper presented at Congress 1995 of Scandinavian Association of Agricultural Scientists, Reykjavik.

Stumm, W. (1973) The acceleration of the hydrochemical cycling of phosphorus. *Water Research* 7, 131–144.

14 Phosphorus Requirements for Animal Production

P.B. Lynch[1] and P.J. Caffrey[2]

[1]Teagasc, Moorepark Research Centre, Fermoy, Co. Cork, Ireland; [2]Department of Animal Science and Production, University College, Dublin, Ireland

INTRODUCTION

Phosphorus (P) nutrition of animals was concerned with improving productivity and fertility through adequate dietary supplementation, but the emphasis has now changed to minimizing the level in diets in order to reduce excretion of excess P in manure. The problem of excess P is most acute in intensive pig and poultry systems, where the inability of the monogastric to utilize phytate P results in a high percentage of ingested P being excreted. Most pigs and poultry are now produced in large units which utilize mainly purchased feed, and this, combined with high densities of animals in particular geographical locations, causes the quantity of P in animal manure to exceed the requirements of land-based agriculture in many regions of Western Europe and elsewhere. Phosphorus excretion from ruminants is seldom a problem, as it is generally recycled on the farm of origin. The contribution of intensive animal production to the overall P balance in agriculture is sometimes overestimated. Tunney (1990) calculated that pig and poultry production contributed less than 10% to the total annual P build-up in Irish soils.

BIOLOGICAL ROLE

Phosphorus has more known functions in the animal body than any other mineral element (NRC, 1980; Jongbloed, 1987). With calcium (Ca), it plays a major role in the formation of bones and teeth, as well as in eggshells. It is a component of nucleic acids, which control cell multiplication, growth and differentiation, and in combination with other elements it has a role in the maintenance of cellular osmotic pressure and the acid–base balance. Energy transfer processes in all living cells involve interconversion of the phosphate-containing nucleotides, adenosine diphosphate (ADP) and adenosine triphosphate (ATP), and thus P participates in all biological events. Other roles include its presence in phospholipids, where it functions in cell-wall structure, fatty acid transport and protein and amino acid formation.

Symptoms of dietary deficiency are well documented and may be summarized as follows (Khan, 1994). Initially, blood plasma P falls, but this is buffered by withdrawal of Ca and P from the skeleton, loss of appetite, reduced production (growth, milk yield, egg yield), abnormal appetite, reduced fertility and abnormal bone development. Reduced appetite is especially significant in cattle (McDowell, 1992).

PHOSPHORUS ABSORPTION AND METABOLISM

Phosphorus is absorbed mainly from the upper small intestine or duodenum as orthophosphate (NRC, 1980; Breves and Schroder, 1991). The amount absorbed varies with source, Ca:P ratio, intestinal pH, lactose intake and dietary levels of Ca, P, vitamin D, iron, aluminium, manganese, potassium, magnesium and fat. The greater the need, the more efficient the absorption. In dairy cows, the absorptive capacity of the gut for Ca and P is thought to be higher in late than in early lactation, and parasite infestation can also reduce absorption (Meschy, 1994). While the P requirement for bone growth is greatest in young animals, some incorporation of P into bone occurs at all ages. Phosphorus may be withdrawn from bones to maintain blood plasma levels during periods of dietary deprivation. Plasma P level is inversely related to plasma Ca level and is regulated by parathyroid hormone and thyrocalcitonin. In monogastrics, excess P is excreted via the kidney, which is the major organ controlling levels of the mineral. Reabsorption from the kidney, in the pig, can be as high as 99% in a deficiency situation (Khan and Jongbloed, 1994). In ruminants, excess P is largely excreted in faeces (McDowell, 1992).

Dietary Ca plays a major role in P utilization (NRC, 1980; Jongbloed, 1987). The optimum Ca:P ratio for pigs has been suggested by Peo (1991) to lie between 1.0 and 1.3:1, with outside limits of 0.9 and 1.6:1. It appears that dietary Ca and P levels can vary widely provided the Ca:P ratio is close to optimal. Detrimental effects of excess Ca are greater where the level of P is marginal or inadequate.

Data reviewed by the NRC (1988) did not support the common view that a particular dietary Ca:P ratio is important in ruminants.

A beneficial effect of high dietary vitamin D on utilization of P, when present at low levels in the diet, has been reported by Edwards (1991) and Mohammed *et al.* (1991).

Phosphorus is relatively non-toxic, although high levels do interfere with metabolism of Ca and result in bone loss comparable to the effects of insufficient dietary Ca (McDowell, 1992).

PHOSPHORUS IN FEEDS

Apart from the effect of excess on the environment, formulation of feeds to contain high levels of P is, usually, expensive, and for this reason diets for meat animals are formulated to support optimal growth and feed-conversion efficiency. The dietary requirement, based on these criteria, is lower than for development of maximum bone strength and bone mineralization, and higher levels of P are advisable for breeding stock where soundness and longevity are important (Mahan, 1990). Reduced bone strength, in meat animals, may on occasion cause problems in processing of carcasses. Feeding a diet where available P was 10% below estimated requirement resulted in no apparent leg or skeletal problems, a small reduction in live weight and similar yields of boneless meat, but there was an increased incidence of rib separations and broken femurs during processing of broilers (Moran, 1994). Omission of supplemental P in diets fed in the later stages of growth had a similar effect.

The utilization of plant P by animals has been reviewed by Jongbloed (1987). It is estimated that 60–70% of P in feedstuffs of plant origin is in the form of phytate, a mixed Ca–magnesium–potassium salt of phytic acid (hexaphosphoinositol ($C_6H_{18}O_{24}P_6$), which contains 28% P) (Edwards, 1991), which is largely unavailable to monogastrics. Phytic acid forms complexes with several minerals, such as Ca, zinc, manganese and iron, and with protein (Nelson, 1967).

Phytate P must be solubilized and hydrolysed by the enzyme phytase before it can be absorbed. While animals do not possess significant amounts of phytase, many plants contain some activity. Among the cereals, rye is especially rich in the enzyme, with maize having little activity and wheat and barley being intermediate, while cereal by-products, e.g. brans and pollards, have greater phytase activities than the whole grain (Pointillart, 1994).

The role of intestinal phytase (and also intestinal alkaline phosphatase, which can hydrolyse phytate) is thought to be a minor one in the case of pigs and poultry (Pointillart *et al.*, 1984; Pointillart, 1994). Dietary supplementation with 1,25-dihydroxycholecalciferol (1,25-$(OH)_2 D_3$) is thought to stimulate intestinal phytase activity (Edwards, 1993). Phytate P is well utilized by ruminants, being hydrolysed by rumen microflora. Caecal fermentation results in hydrolysis of phytate in non-ruminants, but the extent to which this P is absorbed is not clear (Jongbloed, 1987). Phytate

Table 14.1. Total and non-phytate P content of common feedstuffs (data from Jongbloed and Kemme, 1990; NRC, 1994).

	Jongbloed and Kemme, 1990		NRC, 1994	
Feedstuff	Total P%	Non-phytate P	Total P%	Non-phytate P
Maize	0.32	0.11	0.28	0.08
Barley	0.44	0.16	0.36	0.17
Wheat	0.41	0.12	0.31	0.13
Sunflower meal	1.16	0.27	1.00	0.16
Wheat pollard	1.25	0.24	0.85	0.30
Soybean meal	0.66	0.26	0.65	0.27

may also affect trace mineral nutrition through its ability to form complexes with these elements (Suttle, 1984).

The apparent digestibility of P in feedstuffs of animal origin is high, 70–90% (Jongbloed and Kemme, 1990).

Total and non-phytate P content of common feed ingredients is shown in Table 14.1. The data of Jongbloed and Kemme (1990) showed that, while digestibility of P in feedstuffs is related to content of non-phytate P, significant differences exist, e.g. for wheat.

INORGANIC PHOSPHORUS SOURCES

The main sources of supplemental P for animal feeds are the Ca phosphates (dicalcium phosphate, monocalcium phosphate, defluorinated rock phosphate, bone-meal, guano-origin phosphates), ammonium phosphates (monoammonium phosphate, diammonium phosphate, ammonium polyphosphate), sodium phosphates (monosodium phosphate, disodium phosphate, sodium tripolyphosphate) and phosphoric acid. There is considerable variation in the biological availability of P from sources within each feed type, especially the Ca phosphates (Peeler, 1972; NRC, 1980; Khan, 1994). Dicalcium phosphate constitutes up to 60% of the feed phosphate used in Western Europe (Thomsen, 1994). Rendered animal products, such as meat and bone-meal, contain P of high biological value but may be variable in P content (Khan, 1994).

Bioavailability of dietary P may be assessed by balance studies, bone and blood parameters and *in vitro* tests of solubility (Gueguen, 1994).

Table 14.2. Phosphorus content of live pigs of typical slaughter weights.

Source	Breed	Sex	Pig wt (kg)	Total P (g)
Jongbloed, 1987	Mixed	Mixed	90	451
Jongbloed, 1987	Mixed	Mixed	100	500
Jongbloed, 1987	Mixed	Mixed	110	549
Hendriks and Moughan (1993)	Cross	Boar	85	412
Hendriks and Moughan (1993)	Cross	Boar	109	471
Hendriks and Moughan (1993)	Cross	Gilt	85	383
Hendriks and Moughan (1993)	Cross	Gilt	109	494

PHOSPHORUS STORES IN ANIMAL TISSUE AND LOSSES IN PRODUCT

In the body, about 80–85% of total P is stored in the bones and teeth, and non-skeletal P is mainly in red blood cells, muscle and nervous tissue (McDowell, 1992). Bone contains Ca and P in a ratio of approximately 2:1. Total body P content of growing pigs has been measured by a number of authors, and values for typical slaughter weights are shown in Table 14.2.

Cows' milk contains about 0.9 g P kg^{-1} (Gueguen *et al.*, 1989), while cattle contain about $7–9 \text{ g P kg}^{-1}$ empty body weight (ARC, 1980).

Broiler chickens are estimated to contain 3% ash, of which 18% is P (Edwards, 1992), while eggs contain about 2 g P kg^{-1}.

PHOSPHORUS REQUIREMENT OF RUMINANTS

Phosphorus nutrition of ruminants has been reviewed by ARC (1980), NRC (1988) and Gueguen *et al.* (1989). Ruminants recycle large amounts of P in saliva and the Ca:P ratio in the intestine may differ significantly from that in the diet (NRC, 1988). Rumen bacteria have a 2.0–2.5-fold higher P requirement than the host animal and impaired rumination could lead to inadequate supply of P to the bacteria (Meschy, 1994). Some of this P is lost via the faeces, so that endogenous faecal losses are higher when ruminants are fed forages than when fed concentrate diets, because of greater saliva secretion (Gueguen *et al.*, 1989). Variation in P secretion in saliva and hence faecal loss is the principal means of regulating body P in ruminants (McDowell, 1992).

Published estimates of the P requirement of cattle are shown in Table 14.3. The higher French values may be a reflection of greater use of forage in European diets and may also incorporate a 'safety margin', which is essential in feed formulation to account for variation between individual animals and in biological availability among sources of nutrients. Large

Table 14.3. Published estimates of the P requirement of growing and breeding cattle (% of diet dry matter).

Category	Source	
	NRC (1988)	Gueguen et al. (1989)
Calf milk replacer	0.6	–
Young cattle to 12 months old*	0.3–0.4	0.32–0.6
Cattle over 12 months	0.23	0.35–0.5
Mature bull	0.19	–
Cow, pregnant, non-lactating	0.24	0.23–0.28
Cow, early lactation	0.48	–
Cow, mid-lactation†	0.28–0.41	0.37–0.40

* Higher value is for younger animals (NRC, 1988) and faster-growing animals (Gueguen et al., 1989).
† Higher value for higher milk yield.

differences between individual animals in the efficiency of absorption of P have been described by Field (1984). The range in availability of P among individual sheep fed a particular diet was 63–84% (mean 70%). Other work with monozygotic twin cattle, cited by that author, suggested a strong genetic influence.

Information concerning the availability of P from various sources for ruminants is limited, but it is known that absorption declines with age (NRC, 1988). Field (1984) reported variation between ingredients in P availability for ruminants, with that from fish-meal (80%), barley and wheat (78%) being high and that from rice bran (63%) and perennial ryegrass (64%) being low.

Peeler (1972) ranked inorganic P sources in order of decreasing availability: sodium phosphate, phosphoric acid, monocalcium phosphate, dicalcium phosphate, defluorinated phosphates, bone-meal, soft phosphates. Data cited by NRC (1988) suggest that the availability of P in mixed feeds for ruminants is about 45–50% and that ruminants are well able to utilize phytate P.

Symptoms of P deficiency in cattle include reduced mineralization and increased fragility of bones, reduced appetite, slow growth, reduced milk production, stiffness of joints, anoestrus and low conception rates (NRC, 1988).

Phosphorus deficiency is common in many parts of the world among cattle grazing on soils of low P status, especially when only mature grass is available as a result of drought (Underwood, 1966). Reduced mobility due to lameness means that energy and protein deficiency frequently complicate P deficiency.

Sheep derive most of their nutrients from grazed herbage, and compound sheep feed makes little contribution to a positive P balance on farms. When sheep are reared on mainly concentrate regimes, high levels of P in the diet or a relative excess in relation to Ca can precipitate the condition of urolithiasis (urinary calculi). This is the formation of stones, or calculi, in the kidney, with resultant obstruction of urine excretion (NRC, 1980). Sheep grazing P-deficient forage are less likely to exhibit deficiency symptoms than are cattle.

The dietary requirement for P in sheep diets was estimated by Gueguen *et al.* (1989) at 0.3–0.4% for growing lambs and 0.35–0.40% for breeding ewes.

PHOSPHORUS REQUIREMENT OF PIGS

Estimates of the P requirement of pigs have been made by several authorities (Table 14.4). Older estimates were expressed in terms of total P, while those of Jongbloed and Everts (1992) are in terms of digestible P. The latter set also acknowledges the gradual fall in P requirement as the pig ages, and this forms the basis for the practice of 'phase feeding', where several diets, each lower in P level than the preceding one, are fed over the pig's life. Feeding systems that facilitate the operation of phase feeding have been developed in a number of countries, principally the Netherlands.

It is generally assumed that about one-third of plant P is available to pigs (Edwards, 1991), but recent information shows that this generalization overestimates many common ingredients, e.g. cotton-seed meal, sunflower (Cromwell and Coffey, 1993). By-products, such as maize gluten feed, which have undergone steeping or wet processing have increased P availability, while dry processing has no effect. Experiments in the USA show that pigs utilize only about 15% of the P in a maize–soybean diet and supplementation resulted in improved growth, bone strength and bone ash (Table 14.5).

Table 14.4. Published estimates of the P requirements of pigs (% of air-dry feed).

Category of pig	ARC (1981)	NRC (1988)	INRA (1984)	Jongbloed and Everts (1992)
Type of P	Total	Total	Total	Digestible
Pregnant sow	0.6	0.6	0.45–0.6	0.12–0.26
Lactating sow	0.6	0.6	0.55–0.6	0.24–0.26
Weaned pig (6–20 kg)	0.8	0.6	0.8	0.32–0.47
Finishing pig (30–90 kg)	0.55	0.4	0.5	0.16–0.23

Table 14.5. Response of growing pigs (21 to 91 kg) to P supplementation of the diet (from Cromwell, 1991).

Item	Adequate P (0.50%)	Low P (0.32%)
Daily gain (kg)	0.77	0.54
Feed : gain	3.08	3.82
Bone strength (kg)	146	77
Bone ash (%)	57.4	52.8

PHOSPHORUS REQUIREMENT OF POULTRY

Plant P is considered to be of very low availability for poultry because of their limited ability to utilize phytate (Simons, 1986). However, Cromwell and Coffey (1993) report that bioavailability of P in a number of ingredients is similar to values obtained with pigs. Edwards (1991) concluded that P in Ca phytate is not available to poultry, while the potassium and sodium salts and phytic acid are more available. Utilization of phytate P is reduced by high dietary levels of Ca and increased by higher dietary vitamin D_3. Mohammed *et al.* (1991) reported an additive effect of low dietary Ca and high dietary cholecalciferol on retention of phytate P by chickens (0–4 weeks of age). Edwards (1993) reported increased 9-day weight and bone ash and greater Ca and P retention when $1,25\text{-}(OH)_2D_3$ was added to chicken diets, but the highest P retention of phytate P was obtained when both phytase and $1,25\text{-}(OH)_2D_3$ were present in the diet. Both bone ash and incidence of rickets showed an additive effect of phytase and $1,25\text{-}(OH_2)D_3$. A less costly compound, 1-α-hydroxylated cholecalciferol, has been shown to be equally effective (Biehl *et al.*, 1995). Layers are especially sensitive to dietary Ca and P supply, as the hen will deposit in eggshells during 1 year of intensive production 30–40 times the total Ca present in its own skeleton. In broilers, body Ca and P increase more than 60 times during 6 weeks of life (Simons, 1986). A high incidence of leg deformities in growing broilers is considered a major cause of reduced output and impaired welfare. Dietary Ca:P ratio and vitamin D metabolites play significant roles in the development of the condition, which appears to have a genetic basis (Whitehead, 1994).

Dietary requirements for poultry based on available (non-phytate) P rather than total P are given by NRC (1994) and allow for lower total P in the diet while ensuring nutritional adequacy.

PHOSPHORUS BALANCE IN ANIMAL PRODUCTION

Estimates of P output from animal production may be made from manure volume and composition. However, this approach is confounded by the difficulty in obtaining a representative sample of manure. The use of a 'nutrient-balance' approach, using P levels in feed and animal product, as used by Tunney (1990), Hynds (1994) and Lenis and Jongbloed (1994), appears preferable, being subject to fewer errors.

Van Horn (1992), using a nutrient-balance approach to the feeding of dairy cows, calculated the P output at 17, 21 and 31 kg head^{-1} year^{-1} at dietary P concentrations of 0.4, 0.45 and 0.6%, respectively. This was for cows yielding 9875 l milk year^{-1} and consuming 6566 kg dry matter.

Phosphorus excretion from a grass-based system is likely to be lower, as shown in Table 14.6 using data from Moorepark for cow-feed consumption (J. Murphy, 1995, Moorepark, personal communication) and values of Gueguen *et al.* (1989) for the P content of perennial ryegrass and ryegrass silage. The P excreted in this scenario will usually be returned to the farmed area, either through defecation during grazing or by spreading of stored manure, and need have little environmental impact when correctly managed, since losses in milk and the calf at birth (ignoring P deposition in the cow carcass) exceed the P imported into the farm in concentrate feed.

Table 14.6. Estimated P excretion from dairy cows on a grass-based system.

	Quantity (kg)	P%	Total P (kg)
P input			
Grass	3400	0.4	13.6
Silage	1200	0.35	4.2
Concentrate	450	0.6	2.7
Total			20.5
P output			
Milk	5600	0.0009	5.0
Calf at birth			0.4
Excreted			15.1
Excreted (%)			74

IMPROVING PHOSPHORUS UTILIZATION AND REDUCING EXCRETION

Phosphorus excretion from animal production may be reduced by feeding lower dietary levels and by improving overall production efficiency by management, nutrition and breeding. This has been discussed by Lenis and Jongbloed (1994) and, for Irish pig production, by Lynch (1992). Improved productivity in the Moorepark pig herd from 1976 to 1992, during which time feed usage fell from 4.5 kg feed per kg meat produced to 3.5 kg feed, is estimated to have reduced P excretion from 32 to 22 kg per sow (including progeny to slaughter) per year. This is independent of any change in formulation or use of phytase.

Use of growth/performance-enhancing feed additives will also reduce P excretion by improving the efficiency of utilization of feed, while extensive systems (e.g. outdoor pig production) will increase P excretion because of less efficient feed usage.

USE OF PHYTASE IN FEEDS

Improvements in P retention from the use of exogenous phytase in pig feeds are well documented. Jongbloed et al. (1992) fed diets based on (A) maize–soya or (B) soya–cassava–hominy–sunflower, to which was added 1500 units phytase. Digestibility of P was 26 and 28% on the control diets and 43 and 55% on the supplemented diets A and B, respectively. Phytase supplementation also improved digestibility of dry matter and some amino acids (Mroz et al., 1994). Cromwell (1991) produced similar data (Table 14.7) showing improved growth rate, feed conversion efficiency and bone strength from phytase supplementation of pigs from 26 to 101 kg.

Cromwell and Coffey (1993) estimated that P excretion by finishing pigs could be reduced by 35–50% through the use of phytase and removal of inorganic P from finishing pig diets. A similar conclusion was reached by Hoppe (1992), who reported a reduction of 54% in P excretion by young pigs (up to 34 kg) when dietary P was reduced from 0.54 to 0.42% and 1000

Table 14.7. Effect of supplemental phytase on performance of growing pigs (Cromwell, 1991).

P in grower diet (%)	0.5	0.3	0.4	0.3
P in finisher diet (%)	0.4	0.3	0.3	0.3
Phytase units (gm^{-1})	–	–	500	500
Daily gain (kg)	0.88	0.72	0.89	0.87
Feed : gain	3.70	3.98	3.63	3.70
Bone strength (kg)	145	98	136	118

IU phytase included in the feed. A number of other studies in agreement with this view have been cited by Pointillart (1994). The effect of Ca level in the diet on efficacy of phytase should not be overlooked. With weanling pigs, fed a maize-soybean meal diet, Lei *et al.* (1994) found that normal levels of Ca greatly reduced the efficacy of supplemental phytase.

Edwards (1991) reported several studies confirming the effectiveness of phytase in chicken diets. It was concluded that the phytase effect occurred in the gastrointestinal tract of the chicken, and not in the feed before ingestion. However, the enzyme is liable to inactivation by the high temperatures generated during pelleting of feed. Maximal response has been from about 1000 units of phytase kg^{-1} feed.

In a later study, addition of $1,25\text{-}(OH)_2D_3$ was shown to reduce feed P in the faeces of young chickens from 55 to 35% of ingested P (Edwards, 1993). There was a significant reduction in phytate P in faeces from addition of $1,25\text{-}(OH)_2D_3$ to diets containing phytase.

Cromwell *et al.* (1995), using maize–soybean diets for finisher pigs, concluded that for every 100 phytase units kg^{-1} diet, inorganic P could be reduced by 0.0085 percentage units.

Simons *et al.* (1992) reported a reduction of 40% in P excretion from the addition of phytase to chicken feed. They reported a similar response to supplementation of layer diets, regardless of whether the diets contained high (4%) or low (3%) calcium levels.

COST IMPLICATIONS OF REDUCTION IN DIETARY PHOSPHORUS

Pig feeding has the potential to utilize industrial by-products, principally from the food and beverage industries, with economic benefit to both the pig industry and the producer of the product. Formulation of diets with lower limits on dietary total P will render some of these materials, e.g. cereal by-products, less attractive and will also limit inclusion of products such as meat and bone-meal, which are used primarily as a protein feed but contribute significant amounts of highly digestible P, and even feeds such as sunflower with very low P digestibility. The result will be more expensive feeds. While the magnitude of the cost increase will depend on prevailing ingredient prices, Lynch (1992) estimated that a reduction in the maximum allowable P in finishing pig feed from 0.8 to 0.5% would increase feed price by about 3%. All diets were formulated to contain at least 0.18% digestible P.

Knowledge of P content of feed ingredients will become more important as levels in formulation are reduced. The P content of Australian barleys is reported to have fallen from 0.41 to 0.28% between 1984 and 1990, a drop attributed to reduced usage of P fertilizer (Kellaway and Porta, 1993).

CONCLUSION

Phosphorus plays an important role in the nutrition of farm livestock. Most is found in the skeleton, but the element is involved in a wide range of functions, especially energy transfer, cell multiplication, growth and integrity. Monogastrics have a limited ability to utilize plant P and a high proportion of dietary P is excreted in manure. More efficient production, formulation of diets on the basis of digestible P rather than total P and use of phytase enzymes will all help to reduce P excretion. The potential benefits of phytase in monogastric diets is greater in maize-based diets than in those based on barley and wheat. Formulation of diets containing lower levels of total P for feeding in the later stages of growth will make an important contribution to reducing P excretion. Limitation on the levels of total P in diets may, in some cases, result in more expensive feeds.

REFERENCES

ARC (1980) *The Nutrient Requirements of Ruminant Livestock.* Commonwealth Agricultural Bureau, Slough, UK, 351 pp.

ARC (1981) *The Nutrient Requirements of Pigs.* Commonwealth Agricultural Bureau, Slough, UK, 307 pp.

Biehl, R.R., Baker, D.H. and DeLuca, H.F. (1995) 1-Alpha-hydroxylated cholecalciferol compounds act additively with microbial phytase to improve phosphorus, zinc and manganese utilization in chicks fed soy-based diets. *Journal of Nutrition* 125, 2407–2416.

Breves, G. and Schroder, B. (1991) Comparative aspects of gastrointestinal phosphorus metabolism. *Nutrition Research Reviews* 4, 125–140.

Cromwell, G.L. (1991) Feeding phytase to increase the availability of phosphorus in feeds for pigs. In: *Proceedings 52nd Minnesota Nutrition Conference*, University of Minnesota, St Paul, pp. 189–200.

Cromwell, G.L. and Coffey, R.D. (1993) An assessment of the bioavailability of phosphorus in feed ingredients for non-ruminants. In: *Proceedings Maryland Nutrition Conference for Feed Manufacturers*, University of Maryland, College Park, pp. 146–158.

Cromwell, G.L., Coffey, R.D., Parker, R.D., Monegul, H.J. and Randolph, J.H. (1995) Efficiency of a recombinant-derived phytase in improving the bioavailability of phosphorus in corn–soyabean meal diets for pigs. *Journal of Animal Science* 73, 2000–2008.

Edwards, Jr, H.M. (1991) Effect of phytase on phytate utilisation by monogastric animals. In: *Proceedings of the 1991 Georgia Nutrition Conference for the Feed Industry*, University of Georgia, Athens, USA, pp. 1–8.

Edwards, Jr, H.M. (1992) Minimising phosphorus excretion in poultry. In: *Proceedings of the 1992 Georgia Nutrition Conference for the Feed Industry*, University of Georgia, Athens, USA, pp. 124–131.

Edwards, Jr, H.M. (1993) Dietary 1,25 dihydroxycholecalciferol supplementation increases natural phytate phosphorus utilization in chickens. *Journal of Nutrition* 123, 567–577.

Field, A.C. (1984) The availability of dietary phosphorus, calcium, magnesium and salt for ruminants. In: *Proceedings of the Society of Feed Technologists*. Society of Feed Technologists, Reading, UK, pp. 12–16.

Gueguen, L. (1994) Determination of availability. *Feed Mix*, Special Issue on Phosphates, 12–15.

Gueguen, L., Lamand, M. and Meschy, F. (1989) Mineral requirements. In: Jarrige, R. (ed.) *Ruminant Nutrition: Recommended Allowances and Tables*. INRA, Paris, pp. 49–59.

Hendriks, W.H. and Moughan, P.J. (1993) Whole body mineral composition of entire male and female pigs depositing protein at maximal rates. *Livestock Production Science* 33, 161–170.

Hoppe, P.P. (1992) Review of the biological effects and the ecological importance of phytase in pigs. In: *From Research and Experience, Edition 30*. BASF, Ludwigshafen, Germany, pp. 3–15.

Hynds, S.P. (1994) A nutrient balance model for typical Irish farming systems. MSc thesis, National Council for Educational Awards, Dublin.

INRA (1984) *L'Alimentation des animaux monogastriques, porc, lapin, volailles*. Institut National de la Recherche Agronomique, Paris, 282 pp.

Jongbloed, A.W. (1987) Phosphorus in the feeding of pigs. PhD thesis, University of Wageningen, the Netherlands.

Jongbloed, A.W. and Everts, H. (1992) Apparent digestible phosphorus in the feeding of pigs in relation to availability, requirement and environment. 2. The requirement of digestible phosphorus for piglets, growing-finishing pigs and breeding sow. *Netherlands Journal of Agricultural Science* 40, 123–136.

Jongbloed, A.W. and Kemme, P.A. (1990) Apparent digestible phosphorus in the feeding of pigs in relation to availability, requirement and environment. 1. Digestible phosphorus in feedstuffs of plant and animal origin. *Netherlands Journal of Agricultural Science* 38, 567–575.

Jongbloed, A.W., Mroz, Z. and Kemme, P.A. (1992) The effect of supplementary *Aspergillus niger* phytase in diets for pigs on concentration and apparent digestibility of dry matter, total phosphorus and phytic acid in different sections of the alimentary tract. *Journal of Animal Science* 70, 1159–1168.

Kellaway, R. and Porta, S. (1993) *Feeding Concentrate Supplements for Dairy Cows*. Dairy Research and Development Corporation, Glen Iris, Victoria, Australia, 176 pp.

Khan, N. (1994) Phosphorus – the essential element. *Feed Mix*, Special Issue on Phosphates, 4–6.

Khan, N. and Jongbloed, A.W. (1994) What happens in the gut ... of the pig. *Feed Mix*, Special Issue on Phosphates, 10.

Lei, X.G., Ku, P.K., Miller, E.R., Yokoyama, M.T. and Ullrey, D.E. (1994) Calcium level affects the efficacy of supplemental microbial phytase in corn–soyabean meal diets of weanling pigs. *Journal of Animal Science* 72, 139–143.

Lenis, N.P. and Jongbloed, A.W. (1994) Modelling animal, feed and environment to estimate nitrogen and mineral excretion by pigs. In: Cole, D.J.A., Wiseman, J. and Varley, M.A. (eds) *Principles of Pig Science*. Nottingham University Press, Loughborough, UK, pp. 355–373.

Lynch, P.B. (1992) Feeding to minimise waste. In: *Proceedings Pig Health Society, 20th Annual Symposium*, Dublin, pp. 8–21.

McDowell, L.R. (1992) *Minerals in Animal and Human Nutrition*. Academic Press, San Diego, 524 pp.

Mahan, D.C. (1990) Mineral nutrition of the sow: a review. *Journal of Animal Science* 68, 573–582.

Meschy, F. (1994) What happens in the gut ... of the ruminant. *Feed Mix*, Special Issue on Phosphates, 8–9.

Mohammed, A, Gibney, M.J. and Taylor, T.G. (1991) The effects of dietary levels of inorganic phosphorus, calcium and cholecalciferol on the digestibility of phytate-P by the chick. *British Journal of Nutrition* 66, 251–259.

Moran, E.T. (1994) Low phosphorus affects broiler carcass quality. *Feed Mix*, Special Issue on Phosphates, 21–23.

Mroz, Z., Jongbloed, A.W. and Kemme, P.A. (1994) Apparent digestibility and retention of nutrients bound to phytate complexes as influenced by microbial phytase and feeding regimen in pigs. *Journal of Animal Science* 72, 126–132.

Nelson, T.S. (1967) The utilisation of phytate phosphorus by poultry – a review. *Poultry Science* 46, 862–871.

NRC (1980) Phosphorus. In: *Mineral Tolerance of Domestic Animals.* National Academy of Sciences, Washington, DC, pp. 364–377.

NRC (1988) *Nutrient Requirements of Dairy Cattle,* 6th revised edn. National Academy Press, Washington, DC, 157 pp.

NRC (1994) *Nutrient Requirements of Poultry,* 9th revised edn. National Academy Press, Washington, DC, 155 pp.

Peeler, H.T. (1972) Biological availability of nutrients in feeds: availability of major mineral ions. *Journal of Animal Science* 35, 695–711.

Peo, Jr, E.R. (1991) Calcium, phosphorus and vitamin D in swine nutrition. In: Miller, E.R., Ullrey, D.E. and Lewis, A.J. (eds) *Swine Nutrition.* Butterworth-Heinemann, Boston, pp. 165–182.

Pointillart, A. (1994) Phytates, phytases: leur importance dans l'alimentation des monogastriques. *Le Point Vétérinaire* 25, 65–72.

Pointillart, A., Fontaine, N. and Thomasset, M. (1984) Phytate phosphorus utilisation and intestinal phosphatases in pigs fed low phosphorus: wheat or corn diets. *Nutrition Reports International* 29, 473–483.

Simons, P.C.M. (1986) Major minerals in the nutrition of poultry. In: Fisher, C. and Boorman, K.N. (eds) *Nutrient Requirements of Poultry and Nutritional Research.* Butterworths, London, pp. 141–154.

Simons, P.C.M., Jongbloed, A.W., Versteegh, H.A.J. and Kemme, P.A. (1992) Improvement of phosphorus availability by microbial phytase in poultry and pigs. In: *Proceedings of the 1992 Georgia Nutrition Conference for the Feed Industry,* University of Georgia, Athens, USA, pp. 100–109.

Suttle, N.F. (1984) The availability of trace elements in animal foodstuffs. In: *Proceedings of the Society of Feed Technologists,* Reading, UK, pp. 17–31.

Thomsen, J. (1994) From rock to feed: production of phosphates. *Feed Mix*, Special Issue on Phosphates, 16–17.

Tunney, H. (1990) A note on a balance sheet approach to estimating the phosphorus fertilizer needs of agriculture. *Irish Journal of Agricultural Research* 29, 149–154.

Underwood, E.J. (1966) *The Mineral Nutrition of Livestock.* Food and Agriculture Organization of the United Nations, Rome, 237 pp.

van Horn, H.H. (1992) Environmental balance and feeding dairy cattle. In: *Proceedings 53rd Minnesota Nutrition Conference,* University of Minnesota, St Paul, pp. 67–80.

Whitehead, C. (1994) Nutritional factors of leg problems in broilers. *Feed Mix* 2(2), 24–26.

15 Nutrient-Management Planning

T.C. Daniel,[1] O.T. Carton[2] and W.L. Magette[3]

[1]*University of Arkansas, Fayetteville, AR 72701, USA;*
[2]*Teagasc, Johnstown Castle Research Centre, Wexford, Ireland;* [3]*Biological Resources Engineering Department, University of Maryland, College Park, MD 29742-5711, USA*

ABSTRACT

Nutrient management is being used in every state in the USA with especially noteworthy programmes in Wisconsin, Pennsylvania and Maryland. This best management practice for minimizing agricultural non-point pollution is popular because, if properly designed and implemented, it can protect water quality and contribute to economically and environmentally sustainable agriculture. While each programme may vary to reflect site conditions, successful programmes contain common components. The purpose of this chapter is to identify and review these common components. These components can be categorized into two important areas: the institutional and the technical.

The best technical plan will fail unless placed in the proper context and accompanied by an institutional framework that fosters implementation. Scarce resources require decision-makers to decide where the emphasis will be placed. They must identify and prioritize those watersheds that will receive attention. Clear goals and objectives for each watershed must be formulated. Provisions must be made early for a strong educational component. Each programme emphasizes the importance of cost-sharing and the voluntary nature of the programme. Innovative programmes are looking to the private sector as a means of developing and implementing nutrient-management plans. In Maryland, for example, the certification programme targets employees of the retail fertilizer industry, private consultants and government employees.

Technical components are equally important and should be developed by individuals familiar with site conditions. In the USA, state land-grant universities serve as excellent technical resources. To attain realistic yields through the use of animal manures, planners first need to know the amount of manure that will be available, as well as the nutrient content. Nutrient credits as a function of time must be identified for manure and legumes. Application rates of manure must be matched with soil-test data. Innovative methods for estimating the available nitrogen in the soil should be included where appropriate. Because soil-test phosphorus and concentrations of phosphorus in runoff are related, a threshold soil-test level of phosphorus exists and should be considered in nutrient management planning.

INTRODUCTION

Environmental concerns, coupled with the demand for cost-effective agriculture, have heightened interest in developing innovative programmes for managing nitrogen (N), phosphorus (P) and potassium (K) contained in manure and commercial fertilizer. As indicated by Shuyler (1994a), many reasons exist for adopting nutrient management as a means of cutting costs and preserving water quality without jeopardizing yields. As a result this 'common-sense' practice is being successfully implemented in the Chesapeake Bay region (Parkinson, 1994). Brodie and Powell (1995) estimate that approximately 4800 plans (146,000 ha) were produced and 2600 plans (85,000 ha) updated in Maryland at a cost of approximately \$22 ha^{-1} or \$670 per plan. For those enrolled in the programme, reductions in use of N and P have been estimated at 57 and 40 kg ha^{-1}, respectively, resulting in a \$52 ha^{-1} saving. The soil resource can also benefit from proper nutrient management. For example, when practices such as timing and proper land-application rates of manure are incorporated into the plan, the quality of both surface and groundwater can be preserved (Edwards and Daniel, 1993; Adams *et al.*, 1994).

Nutrient-management planning (NMP) is being used in every state in the USA with especially noteworthy programmes evolving in Wisconsin (Bundy *et al.*, 1994; Duffy, 1994), Pennsylvania (Beegle and Lanyon, 1994; Beegle *et al.*, 1994, 1995) and Maryland (Brodie and Powell, 1995). While the technical aspects of the programmes vary with respect to nutrient content of the manure, yield goals, soil- and manure-testing procedures, etc., each contains common fundamental components necessary for success. The purpose of this chapter is to identify and briefly review these components. As demonstrated in an article by Klausner (1995), the discussions of the respective components of NMP that follow will be, of necessity, broad. This is because the technical details of any NMP programme must be rather site-specific. For example, yield goals and the corresponding nutrient-application rates to satisfy crop needs are quite variable geographically. Thus, it would be senseless to prescribe identical nutrient-management programmes for both Florida in the USA and northern Europe. Rather, the technical details of a nutrient-management programme must be developed and adjusted by researchers and resource managers who are familiar with local conditions.

Generally, the components can be categorized into two important areas – institutional and technical – and the remainder of the chapter will briefly discuss the areas contained in each.

INSTITUTIONAL

The best NMP will fail unless placed in proper context and accompanied by an institutional framework that assists implementation. For example, even if the programme is technically very sound, if it does not contain a strong educational component the probability of success is low. The following is a discussion of important institutional components that must parallel or precede the technical aspects of NMP.

Nutrient Budget – Farm Import and Export of Nutrients

Historically, farming systems were diversified, using several cropping systems to provide feed for various livestock operations. The nutrients contained in the manure did not meet the nutrient requirements of the crops and commercial fertilizer was imported on to the farm. Most farming systems today, however, especially those that include a confined animal operation, result in an on-farm surplus of nutrients contained in manure. Large volumes of manure are produced daily and must be dealt with on the farm, land area permitting, or transported off site.

Nutrients are imported on to the farm in fertilizer, feed and replacement stock and are exported in products sold, such as meat, milk and crops. Some nutrients, of course, can enter the farm through atmospheric deposition and fixation and be lost through volatilization and/or water transport, either in overland flow or in leachate through the root zone and beyond. In a survey of dairy farms in New York State, Klausner (1995) demonstrated that, for a dairy herd of 85 cows, three times more P is imported on to the farm than is exported, resulting in an on-farm surplus of P and long-term fertility/environmental implications. Phosphorus contained in the feed accounted for 80% of the total amount imported, with commercial fertilizer constituting only a minor portion (20%). Similar on-farm nutrient budgets have been constructed in other states (Pennsylvania, Wisconsin and Maryland) with similar results. At present, land-application rates of manure are based on meeting the N needs of the crop. Should application rates of manure be based on P, as suggested as a means of minimizing eutrophication, surplus manure would be of an even greater magnitude.

These examples and experiences, especially in the Chesapeake Bay area, clearly indicate a need for an institutional structure that provides for a much improved manure-distribution system. The on-farm NMP should foster proper distribution of manure on the farm, but, more importantly, mechanisms need to be in place that also ensure proper off-farm distribution of surplus manure. On a local scale, written agreements should be required between the grower with surplus manure and the neighbour who owns the land where the manure is to be applied. On a larger scale, innovative approaches are needed to foster the transport of manure from manure-rich areas to areas where the manure is needed for crop production. For example, local communities could provide tax credits to encourage the creation of a manure-hauling industry. Low-interest loans, combined with state-supported cost-sharing, should be considered as a means of fostering such enterprises, along with other incentives used to encourage industry to perform certain tasks.

Watershed Prioritization, Problem Identification and Water-Quality Goals

Since agricultural non-point pollution is so broad in scope and costly to address, resource decision-makers must identify and prioritize those watersheds and accompanying aquatic resources that will receive attention. A procedure for accomplishing this difficult task is outlined by Sharpley *et al.* (1994); it includes a blend of technical and subjective inputs and is appropriately done by the responsible agencies and interested groups at the state/regional level. As specifics develop, local input should be sought as quickly as possible. Wisconsin's state-funded 'priority watershed' programme is an excellent example of a progressive and innovative approach for focusing scarce resources where they will produce the most results. As part of the programme, NMPs are being developed by the private sector for the priority watersheds (Sturgul *et al.*, 1994). Nutrient-management planners are reimbursed $2.43 ha^{-1} at no cost to the landowner.

An integral step in identifying priority watersheds or catchments is clearly defining the water-quality problems of the aquatic resources therein (lake, stream, river, groundwater or a combination). Numerical measures of water quality (i.e. water-quality standards) are often available as a means by which to identify such water-quality problems. Water-quality standards are established by some government authority and are typically based on a defined use (such as trout fishing, public water supply, etc.). Perhaps less available will be water-quality goals, or targets that describe some future water-quality condition. A well-known water-quality goal of US environmental legislation in the 1970s was to make most US waters 'fishable and swimmable'. Water-quality goals are often difficult to quantify scientifically, but, when articulated in non-technical language, they can help mobilize widespread public support for programmes such as NMP that are integral to achieving improved water quality. However, care must also be taken to be honest with the public and to admit that, due to the 'environmental memory' of a catchment and the hydrological travel time of nutrient movement, water-quality improvements will necessarily lag behind the widespread implementation of NMP.

In particular reference to the technical framework for NMP, achieving agreement on water-quality goals is difficult, but essential. First and foremost, it is absolutely critical that the kind of water-quality problem be clearly defined. Is the problem with surface water or groundwater, or both? If surface water is the chief problem, is the resource fresh or saline? Is eutrophication of surface water an issue, or is nitrate (NO_3) contamination of groundwater the chief concern? In answering these questions, by examining ambient water quality against water-quality standards and/or goals, the technical details of an NMP programme can be tailored to address specific resource-protection needs most effectively. For example, if the aquatic system needing protection suffers from eutrophication and P is the limiting nutrient, an NMP programme should be developed that bases manure application rates in the catchment on crop needs for P. Developing an NMP programme based on crop N requirements in this instance would be ineffective.

Targeting in a Watershed: Capability on the Horizon

Water is the transport mechanism for nutrient entry into the aquatic system, either in surface runoff or groundwater via the vadose zone. Traditionally, it was assumed that each area in a watershed contributes equally to surface runoff. Recently, however, variable-source-area (VSA) hydrology has become the accepted method for describing surface runoff in a watershed. The VSA hydrological concept is that, in a topographically defined watershed, there are sub-basins that account for the bulk of the runoff. Simply stated, some areas in the watershed account for all the surface runoff, while other areas contribute only to infiltration and recharge (Gburek and Pionke, 1995). If these can be distinguished, cost-sharing and nutrient-management programmes can target these critical areas. For example, if P in runoff is the problem (eutrophication), the programme should target P inside the VSA. If groundwater is the problem, NMPs can be N-based and targeted areas may be outside the VSAs. This approach, while relatively new, could prove to be one of the major advances in addressing agricultural non-point pollution. For the first time, this technique would allow limited resources to focus on those areas in the watershed causing the problem. Gburek and Pionke (1995) used the VSA hydrological concept in a Pennsylvania watershed to develop management strategies for land application of animal manure. They concluded that establishing different levels of management in different areas of the watershed is the most efficient approach for minimizing contamination of groundwater and/or surface water. However, the need for a strong educational programme to accompany such approaches was emphasized, because recommended practices would probably vary from farm to farm.

Education

As expressed by Gburek and Pionke (1995), strong educational programmes must accompany nutrient-management programmes, especially when decision-makers use targeting within a watershed. While growers have been the traditional clientele, there is an increasing and ongoing need for training programmes to meet the demand for certification/licensing of the private sector to develop NMPs.

Voluntary Programme

Experience has shown that legislating management to reduce agricultural non-point pollution is almost impossible. Thus, the programme should be voluntary; acceptance, implementation and success will be a natural result of a programme that is technically sound, includes appropriate incentives and has a strong educational component. Requirement of a plan, however, is another issue, especially for those growers receiving a government subsidy or within a high-priority watershed. For example, in order to meet the 40% reduction in nutrient loading to Chesapeake Bay, Pennsylvania enacted legislation requiring NMPs from concentrated animal operations (Beegle *et al.*, 1995). A confined animal operation is defined as an operation that exceeds five animal-equivalent units (455 kg live weight) per ha (two animals per acre). Also, experience reveals that not everyone is going to cooperate, even after being identified as responsible for an important water-quality problem. Where this occurs, a 'bad actor' provision must be available to resource managers to influence individuals who consistently refuse to participate in the solution to the problem.

Not everyone agrees that the Pennsylvania approach is correct. The adjacent state of Maryland, under the same constraints to reduce nutrient losses as Pennsylvania, uses a totally voluntary NMP programme. To accomplish results, Maryland uses targeted tributary-based (catchment-based) strategies to set regional NMP goals. The state relies on aggressive educational programming through county agricultural advisers to disseminate the nutrient-management message. Maryland also provides cost-sharing funds to farmers for constructing animal manure-management facilities. If a farmer chooses to accept these funds, the farmer must also develop an NMP.

Agency and Private-Sector Participation

The private sector is being looked to as a major player in the development, implementation and coordination of NMPs, because of government downsizing. One thing seems clear; too much additional work will be required to rely on already overburdened agencies. But how can we turn over a job that has been traditionally done by an unbiased public agency, so that uniformity and quality is maintained? As outlined by Shuyler (1994a, b), the private sector has a strong incentive and can potentially do a quicker and better job, given the proper direction and supervision. How does this get accomplished, because this sector needs training and supervision? Public agencies should retain a role as 'gatekeeper' to decide who can develop NMPs and which plans are sound. Training programmes are being established to certify/license professionals seeking to formulate NMPs. In one innovative approach, several states have combined to develop coordinated programmes (Parkinson, 1994). In Maryland, for example, the certification programme targets employees of the retail fertilizer industry, private consultants and government employees. Programmes in Wisconsin and Pennsylvania also encourage participation of the private sector.

Cost-Sharing

As with the educational component, cost-sharing programmes must be an inherent part of the total nutrient-management programme. In Pennsylvania, a special fund has been established to provide assistance in the form of loans and grants for implementation of NMPs. Practices are cost-shared at 80%, up to $30,000 per farm, for those having an approved NMP (Parkinson, 1994). In Virginia, a state tax credit on farm equipment relating to nutrient management is available to those growers having an NMP in place (Parkinson, 1994). This innovative programme allows a credit of 25% off the purchase price or $3750, whichever is lower, on the purchase of items necessary for implementing the NMP, such as manure spreaders and fertilizer applicators. In Wisconsin, manure-storage facilities are cost-shared up to $30,000 for those growers in the priority watersheds. In Maryland, producers can receive as much as 87.5% of the installation/construction cost of some best management practices, particularly animal-manure management facilities.

TECHNICAL

This section lists the technical components that should be incorporated in NMPs. A general discussion of the concepts is included, keeping in mind that the specific technical inputs must be derived by individuals familiar with the climate/region for which the NMP is intended. In the USA, the state land-grant universities would serve as an excellent resource in developing the required technical information for their state/region. In Ireland, this information can be obtained from the local Teagasc (national body with responsibility for agricultural research, advice and education) adviser and/or local authority.

Quantity of Manure Produced

The amount of manure produced varies with the type of operation (pigs vs. poultry) and management (pig-parlour flushing schedules). Sound determinations of these values are important for calculating manure-application rates required to meet the nutrient needs of the crop to be grown. In the USA the quantity of manure produced as a function of animal type can be roughly estimated from several published sources (Midwest Plan Service, 1985; American Society of Agricultural Engineers, 1993). More accurate values can be obtained by measuring the amount of manure produced over time. This is the preferred method.

Analysing Manure for Nutrient Content

The nutrient content of the manure should be well established over time through systematic manure analysis. Care should be taken to ensure that a representative sample is collected for analysis. Separate samples should be taken when changes in feed or rations occur or in storage areas where the manure cannot be completely mixed. Total N, P and K, moisture content and ammonium (NH_4)-N, should be determined.

Nutrient Availability and Credits from Manure and Legumes

Not all of the nutrients contained in the manure are available the first year and therefore not all can be used to replace commercial fertilizer on a kg-for-kg basis. How the manure is managed after application makes a difference in the magnitude of these values. Volatilization rates for N can be high (> 50%) if the manure is not soil-incorporated soon after application. Because of this, higher N credits are given for manure that is rapidly incorporated. The readily available nutrients in the manure can be directly credited the first year; however, mineralization is required to bring about credits the second and third year after manure application. Because of this long-term crediting, significant savings are possible. For example, in Wisconsin the dollar value the first year for N, P and K in 69 t ha^{-1} of dairy manure is estimated at $140 (Duffy, 1994), with N credits of 150 kg ha^{-1} allowed the following 2 years.

Nitrogen credits for the crop to be grown can also come from legumes gleaned in the rotation. For example, Bundy *et al.* (1992) and other workers in the Midwest area of the USA have shown that a good alfalfa stand will provide all the N needed for the following maize crop.

Soil Test

Soil testing is an excellent method for establishing a sound fertility-management programme and determining the fertility status of the soil. The level of available P and K in the soil can be easily determined from a representative sample. The test will show where the nutrients are deficient and where the manure should be applied to produce a yield response. If the test indicates that P and K are present in excess of crop needs, manure should be directed elsewhere to those fields that are deficient.

Until recently, methods were not available for estimating the amount of available N in the soil. The new approach for estimating the amount of available N in the soil measures the amount of crop-available NO_3-N in the root zone (Bundy *et al.*, 1994). The NO_3-N contained in the soil profile may result from carry-over from the previous year's fertilizer or manure application or preceding legume crop. Two soil-N tests are available in Wisconsin. One is used to assess the NO_3-N status of the soil profile (60 cm deep) prior to planting in the spring. The other is a 'pre-sidedress' soil NO_3-N test that provides an N-availability index and predicts sidedress N requirements – the amounts of N required for maize to grow to maturity from the 1 m height. Using this innovative approach to credit residual NO_3-N, the potential exists to reduce fertilizer cost and minimize NO_3-N contamination of the groundwater. The amount of residual NO_3-N in the profile is influenced by the amount of N applied and extent of leaching that occurs. Thus soil texture and precipitation patterns play a major role in determining the amount of N in the profile.

Soil-Test Phosphorus and Eutrophication

Generally, manure is land-applied and application rates are based on the N needs of the forage, with little consideration given to P requirements. Such practices often result in excessive P fertilization, because the P:N ratio found in manure is usually much higher than the P:N ratio required by plants. For instance, when manure was used to meet N needs for fescue production in north-west Arkansas, an excess of 40, 37 and 17 kg ha^{-1} of P was applied annually, using poultry, pig and dairy manure, respectively (USDA-SCS, 1992). The P excesses were greater when application rates were adjusted for N losses (i.e. volatilization). Thus, the inherent characteristics of animal manures and nutrient uptake by crops can promote P build-up in soils. For example, 65% of soils tested in Delaware were in the high to excessive P range (Sims, 1993).

Previous research indicated that the P level in the surface soil directly influences the amount of dissolved reactive P (DRP) in runoff (Sharpley *et al.*, 1994). In a recent field study, Pote *et al.* (1996) showed that distilled water extract provided the highest correlation ($r^2 = 0.82$) between the level of P in the soil and that contained in the runoff. Other extractants evaluated included Mehlich 3, Bray–Kurtz, Olsen, iron-oxide-impregnated strips and acidified NH_4 oxalate.

Because soil-test P (STP) and DRP concentrations are related, a threshold STP level exists that can result in runoff sufficiently high in DRP to cause eutrophication. The persistent nature of STP makes this source of runoff P especially problematic, and this important relationship must be considered in developing P management strategies that limit eutrophication but sustain crop production. Lacking clear eutrophic standards and research data, some states (Ohio, Michigan, Wisconsin and Arkansas) have used a subjective process, based on STP levels adequate for crop production and those 'perceived' to bring about eutrophic runoff, to identify a general threshold STP level for management purposes (Sharpley *et al.*, 1996). Use of the threshold STP-level approach is an important concept to incorporate in NMPs, because it changes the emphasis from N to P once the threshold level is exceeded. Methods of incorporating this important concept into NMPs are evolving in the USA. Some states recommend that no more manure be applied once the threshold STP level has been reached, while others allow only P additions required to meet crop uptake (Sharpley *et al.*, 1995).

Realistic Yield Goals

Yield goals that are too high result in excessive land-application rates of manure and can lead to poor nutrient efficiency and water-quality problems (Edwards and Daniel, 1993). Realistic yield goals should be established and incorporated into a rotation and manure-application schedule that produces the dry matter required of the feeding programme.

Avoiding Environmentally Sensitive Areas

No manure should be spread in some areas of the landscape, and recommendations should be incorporated into the NMP requiring the presence of a physical buffer, i.e. some type of vegetation, in proximity to environmentally sensitive areas, such as sink-holes, drinking-water wells, surface water and where surface runoff concentrates. Additional considerations should also be given to minimizing the environmental impact of manure application on flood plains and frozen ground.

Application Timing

Nutrients contained in the manure are most efficiently utilized when land application of manure coincides with nutrient uptake. For example, nutrients contained in autumn-applied manure are generally about 10% less efficient than if they are applied in the spring (Bundy *et al.* 1994). However, time constraints are usually high in the spring, and soil compaction can occur. From an environmental and agronomic standpoint, autumn application of N fertilizers is not recommended on coarse-textured soils or on shallow soils over fractured bedrock.

CONCLUSION

In the USA and increasingly around the world, NMP is recognized as an agricultural 'best management practice' for the control of pollution from production agriculture. As such, it is a practical, cost-effective technique by which farmers can remain economically viable and therefore sustainable, while protecting environmental quality. In today's parlance, NMP creates a 'win–win' situation for both farmers and the public at large, both of whom have a joint interest in a quality environment and a sustainable, economical and wholesome food supply. NMP encompasses concepts about total quality management on the farm – concepts that have a good appeal to the general public.

Despite these positive characteristics, however, widespread NMP adoption by the agricultural community will occur only if programmes are technically sound and backed by an institutional framework of support that encourages implementation. Most bureaucrats cannot develop an acceptable technical content for a successful NMP programme; likewise, most scientists cannot draft and implement the necessary institutional mechanisms for effective NMP implementation. Development of a truly effective NMP programme, therefore, must be a joint effort between agricultural scientists, various levels of government, farmers and perhaps industry and members of the public.

Involving key members of the public (such as representatives of special environmental interest groups) in developing an NMP framework can have positive benefits that lead to the wide popular appeal of, and support for, NMP. For example, a common question asked in the USA by the non-agricultural public is, 'If NMP is so sound agronomically, why don't all farmers adopt it immediately, without public incentives?' By involving key citizens in developing an overall NMP programme, the technical and sociological obstacles that inhibit NMP adoption can be presented in an non-defensive manner. Likewise, the 'lag time' that will invariably occur between NMP implementation and water quality improvement can be discussed openly and unemotionally a priori. Farmer involvement in developing an NMP programme is equally beneficial. Only by consulting with representative farmers can legitimate obstacles to NMP adoption be identified and addressed by scientists and bureaucrats.

In the USA, NMP has become both a technical tool for agricultural management and a matter of public policy for environmental management. Successful NMP programmes are ones that have been based on sound science in recognition of the practical problems associated with farming in an uncontrolled environment. As important, the successful programmes have an institutional framework that recognizes environmental quality as a public good on a par with a sustainable food supply.

REFERENCES

Adams, P.L., Daniel, T.C., Edwards, D.R., Nichols, D.J., Pote, D.H. and Scott, H.D. (1994) Poultry litter and manure contributions to nitrate leaching through the vadose zone. *Soil Science Society of American Journal* 58, 1206–1211.

American Society of Agricultural Engineers (1993) *ASAE Standards 1993: Standard Engineering Practices Data.* American Society of Agricultural Engineers, St Joseph, Michigan.

Beegle, D.G. and Lanyon, L.E. (1994) Understanding the nutrient management process. *Journal of Soil and Water Conservation* 49, 23–30.

Beegle, D.G., Lanyon, L.E. and Myers, J.C. (1994) *Manure Management for Environmental Protection: Utilization of Manure.* Pennsylvania State University Cooperative Extension Service, College of Agricultural Science, University Park, Pennsylvania, 42 pp.

Beegle, D.G., Lanyon, L.E. and Lingenfelter, D.D. (1995) *Nutrient Management Legislation in Pennsylvania.* Pennsylvania State University Cooperative Extension Service, College of Agricultural Science, University Park, Pennsylvania, 5 pp.

Brodie, H.L. and Powell, III, R.N. (1995) Agricultural non-point source water pollution control voluntary programme in Maryland. In: Steel, K. (ed.) *Animal Waste and the Land-Water Interface.* CRC Press, Boca Raton, Florida, pp. 449–457.

Bundy, L.G., Kelling, K.A. and Good, L.W. (1992) *Using Legumes as a Nitrogen Source.* Nutrient and Pest Management (NPM) programme publication No. A3517, University of Wisconsin-Extension and University of Wisconsin-Madison, College of Agriculture and Life Science Extension Publication, Madison, Wisconsin, 6 pp.

Bundy, L.G., Kelling, K.A., Schulte, E.E., Combs, S., Wolkowski, R.P., Sturgul, S.J., Binning, K. and Schmidt, R. (1994) *Nutrient Management: Practices for Wisconsin Corn Production and Water Quality Protection.* Nutrient and Pest Management (NPM) programme publication No. A3557, University of Wisconsin-Extension and University of Wisconsin-Madison, College of Agriculture and Life Science, Extension Publication, Madison, Wisconsin, 25 pp.

Duffy, K. (ed.) (1994) *The Bottom Line: an Economic Summary of Nutrient and Pest Management Practices.* Nutrient and Pest Management (NPM) programme publication No. A3566, University of Wisconsin-Extension and University of Wisconsin-Madison, College of Agriculture and Life Science, Extension Publication, Madison, Wisconsin, 25 pp.

Edwards, D.R. and Daniel, T.C. (1993) Effects of poultry litter application rate and rainfall intensity on quality of runoff from fescuegrass plots. *Journal of Environmental Quality* 22, 361–365.

Gburek, W.J. and Pionke, H.B. (1995) Management strategies for land-based disposal of animal wastes: hydrologic implications. In: Steel, K. (ed.) *Animal Waste and the Land–Water Interface.* CRC Press, Boca Raton, Florida, pp. 313–323.

Klausner, S. (1995) Nutrient management planning. In: Steel, K. (ed.) *Animal Waste and the Land-Water Interface.* CRC Press, Boca Raton, Florida, pp. 383–392.

Midwest Plan Service (1985) *Livestock Waste Facilities Handbook*, 2nd edn. Iowa State University, Ames, Iowa.

Parkinson, R. (1994) Evolution of nutrient management in the Chesapeake Bay region. *Journal of Soil and Water Conservation* 49, 87–88.

Pote, D.H., Daniel, T.C., Sharpley, A.N., Moore, Jr, P.A., Edwards, D.R. and Nichols, D.J. (1996) Relating extractable soil phosphorus to phosphorus losses in runoff. *Soil Science Society of America Journal* 60, 855–859.

Sharpley, A.N., Chapra, S.C., Wedepohl, R., Sims, J.T., Daniel, T.C. and Reddy, K.R. (1994) Managing agricultural phosphorus for protection of surface waters: issues and options. *Journal of Environmental Quality* 23, 437–451.

Sharpley, A.N., Daniel, T.C., Sims, J.T. and Pote, D.H. (1995) Determining environmentally sound soil phosphorus levels. *Journal of Soil and Water Conservation* 15, 160–166.

Shuyler, L.R. (1994a) Why nutrient management? *Journal of Soil and Water Conservation* 49, 3–5.

Shuyler, L.R. (1994b) Nutrient management, an integrated component for water quality protection. *Journal of Soil and Water Conservation* 49, 5–6.

Sims, J.T. (1993) Environmental testing for soil phosphorus. *Journal of Production Agriculture* 6, 501–507.

Sturgul, S.J., Combs, S.M. and Kelling, K.A. (1994) *Status of Wisconsin Nutrient Management Planning Programmes – 1994.* New Horizons in Soil Science No. 5, Department of Soil Science, University of Wisconsin, Madison, Wisconsin, 16 pp.

USDA-SCS (1992) *Agricultural Waste Management Handbook.* USDA, Washington, DC.

16 A European Fertilizer Industry View on Phosphorus Retention and Loss from Agricultural Soils

I. Steén

Issue Manager for Agriculture and Environment Affairs,
European Fertilizer Manufacturers' Association, Avenue E
van Nieuwenhuyse 4, 1160 Brussels, Belgium

INTRODUCTION

Phosphorus (P) is associated with environmental degradation in the form of eutrophication. Nutrient enrichment of streams, rivers and lakes can arise both from point sources, such as municipal and industrial effluent, and from diffuse sources, such as agricultural land. An obvious consequence of eutrophication is the increased occurrence of algal blooms, for which, in fresh water, P is considered to be the limiting nutrient (CAST, 1992; Danish Ministry of Environment and Energy, 1994; Foy and Withers, 1995; Stanners and Bourdeau, 1995). The European Fertilizer Manufacturers' Association (EFMA), while promoting the need to ensure an adequate supply of P to guarantee the security of agricultural production, acknowledges the need to limit losses from agricultural soils. It therefore seeks to help improve the stewardship of the land, which will incorporate sophisticated methods of soil and landscape conservation and P management and monitoring in agricultural soils. Some of these aspects are discussed here.

Phosphorus is essential to all life forms because it is a key element in many physiological and biochemical processes. It is an essential component of many plant compounds (ISMA, 1977) and is involved in nearly all energy functions. Crop production is the process of converting solar energy into chemical energy and P is involved in the trapping, transport and conversion of the sun's energy into sugars, starches and other energy-providing compounds. Phosphorus is also essential in the process of seed formation and has a marked influence on crop yield and, consequently, a significant role in agriculture – in both food and fibre production. Its functions cannot be performed by any other nutrient (ISMA, 1977; Gregory *et al.*, 1991). Phosphorus is added to soil either in manures or through inorganic fertilizers. The amounts should ensure optimum productivity with minimum adverse environmental impact.

PHOSPHORUS IN SOIL

The chemistry of soil P is complex. A typical topsoil contains around 1.3 to 3 t P ha^{-1}, but only a minute portion of this will be available to crops each season. Although between 40 and 60% of the soil P may be present as organic P – mostly within complex molecules in soil organic matter – this mineralizes slowly and may not meet the requirements of the crop. In the UK, net mineralization of soil organic P in arable soils ranged from 0.5 to 1.5 kg P ha^{-1} year^{-1} on arable soils with little organic matter, increasing to 8.5 kg P ha^{-1} year^{-1} where soils were recently ploughed out from grassland (Chater and Mattingly, 1980). There is a larger flux of P in permanent grassland soils (Brookes *et al.*, 1984).

Concentrations of P in the soil solution are very low, mostly varying from 0.03 to 0.2 mg l^{-1}. This low concentration, combined with the small volume of soil solution, means that the P in the soil solution must be renewed many times in the course of a growing season, from either inorganic or organic sources. Most soils release P into the soil solution only slowly (SCOPE/UNEP, 1988; SCOPE, 1995) and soil management must seek to favour the renewal of soil-solution P.

In natural ecosystems, P is supplied through the weathering and dissolution of rocks and minerals and mineralization of soil organic matter. The most common mineral in young soils and rocks is apatite, which has a low solubility and plant availability. Thus P is in short supply in most young landscapes and ecosystems. However, it is in even shorter supply in older soils, because, with continued weathering, iron- and aluminium-associated phosphate compounds are formed, which may be less plant-available. Therefore, in such well-weathered soils, P is usually the critical limiting element for plant and animal production (SCOPE, 1995).

Soils formed from different parent materials, even though they belong to the same soil type and may have a similar pH value, organic matter content and soluble P and potassium (K) status in the topsoil, may differ markedly in their fertility in relation to the availability of P. This may greatly change the fertilizer efficiency, even under conditions where plant-available P levels and P application rates are similar (Ciavatta *et al.*, 1990; SCOPE, 1995).

PHOSPHORUS APPLICATIONS AND PHOSPHORUS BALANCE

Phosphorus is added to soil, either in organic manures or inorganic fertilizers, to replace export in crop off-take and losses caused by leaching, erosion and runoff, and to allow for retention in soil in forms which are not immediately plant-available.

In Europe, during the last century, as in many developing countries today, soils were very deficient in P. In the 1840s and 1850s, the experiments of Lawes and Gilbert at Rothamsted showed that it was necessary to add more P than was removed by the crop to get a satisfactory yield (Johnston, 1994), and later experiments confirmed this (Potash and Phosphate Institute, 1986). Applying more P than the amount of P removed was accepted advisory policy, based on the concept of improving soil-P reserves and thus soil fertility. It has been estimated that the P content in European soils has now been increased by some 30–50% compared with the 1920s, and for many countries some 40% of soils now test as high and very high in readily soluble P (Fig. 16.1).

The official advisory policy to improve the P status of soils in the postwar period led to an increased use of P fertilizers in Europe. However, usage peaked in the early 1970s, and since 1973 the use of P fertilizers in Western Europe has gone down by 39%, or almost 2% annually. The major exceptions to this trend are Greece, Spain and Portugal, where demand continued to fluctuate, with apparent increases until the late 1980s, and

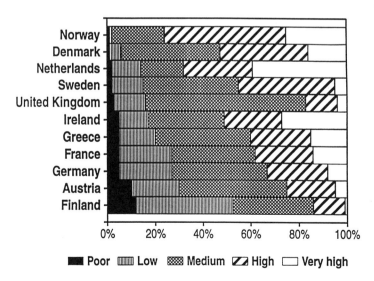

Fig. 16.1 Phosphorus status in European soils – plant-available P for each country using its standard and methodology (EFMA data).

Norway and Switzerland, where demand peaked in 1979. In Austria, Denmark, France, Germany, Norway, Sweden and Finland, P consumption has gone down by more than 50% since the individual peak year for each country. In Belgium/Luxemburg, Ireland, Italy, the Netherlands, the UK and Switzerland, the decline has ranged between 30 and 50%. Since 1992, overall consumption in Europe has apparently stabilized at about 1.6 million tons of P annually (Fig. 16.2). The industry's forecast, however, is that the annual consumption of P as mineral fertilizers will decline slowly by about 2% (12% total) until 2004/2005.

In Europe, France is the largest single market for mineral P fertilizer, followed by Italy, Spain, Germany and the UK. However, there are also differences in the amounts of P fertilizer applied to individual crops, the largest application rates to arable crops being in Ireland, Switzerland, Italy, the UK, France, Portugal and the Netherlands (Table 16.1). Table 16.2 shows the area of land under the most important crops grown in European countries in 1994/95 and the average amount of P applied to each crop. This highlights the dominance of cereals, especially wheat and barley, among arable crops. The area they occupy is about equal to that of fertilized grassland. Differences among countries in the crops grown and amounts of P applied, together with differences in soil and weather, have much to do with the P status of their soils (Lee, 1987; EFMA, internal questionnaire on phosphate status, unpublished).

The intensification of livestock production in some countries has led to the increased import of P in feedstuffs, which represents a very significant net addition to Western Europe's P balance. Disposal of animal manure in

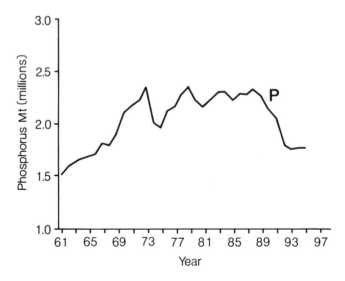

Fig. 16.2. West European P fertilizer consumption between 1961 and 1995 (FAO and EFMA data).

the high-density livestock areas has contributed to the problem of excess P in soil. The amount of P excreted by livestock in Western Europe could be some 50% higher than the amount currently applied as mineral P fertilizer. However, only a limited quantity of the P excreted by livestock is collected, i.e. while the livestock is housed, and thus can be spread on agricultural land.

The national P balance in Western European countries varies significantly (Böckman *et al.*, 1990; European Fertilizer Manufacturers' Association, 1991; Commission of the European Communities and the European Fertilizer Manufacturers' Association, 1994; EFMA, internal questionnaire on phosphate status, unpublished), because of differences in husbandry practices and hence differences in the use of mineral fertilizers and animal manure. Based on application of mineral fertilizers only and the amount of P removed by crops, only a few countries exceed a net annual accumulation of 4–5 kg P ha^{-1} on arable land and the accumulation rarely exceeds 9 kg P ha^{-1} annually. However, when the quantities of P contained in animal manure are also considered, the positive balance for some Western European countries is substantial, particularly in the Netherlands, where it exceeds 40 kg P ha^{-1} year^{-1}. For the majority of Western European countries, the positive P balance ranges between 9 and 18 kg P ha^{-1} annually, with a few exceptions where it is well below 9 kg P ha^{-1} (Table 16.3).

AVAILABILITY OF PHOSPHORUS IN SOIL

The availability of P for crop production, using water-soluble phosphates, manure and sewage sludge, has been determined in long-term experiments at Rothamsted on soils of varying texture, soil organic matter content and pH (Johnston and Poulton, 1992). When more P was applied than removed, P residues accumulated in soil in plant-available forms. Although only about 15% of the extra total P was extractable by Olsen's bicarbonate reagent (Olsen P), the remaining 85% supplied much of the P taken up by crops when this was the only P source. Soils with little plant-available P often yielded much less than P-enriched soils, even when large amounts of P fertilizer were applied. Many other field trials have also shown that plant yields on P-depleted soils were not raised to those on P-enriched soils, even with large applications of P fertilizer. Such results raise the question to what level soils should be enriched. In the absence of any other limitation to yield, it would be expected that crop yields would reach a maximum at a given level of plant-available P in soil. Johnston *et al.* (1986), in a series of experiments with four arable crops, related yield to Olsen P and showed that below a certain value of Olsen P yield declined dramatically and increased little above it. In commercial practice, this would have serious financial implications for a farmer growing crops on soils below the critical

Table 16.1. Average phosphate application rates (P kg ha^{-1}) to crops in Western Europe, 1994/95.

Arable land	Austria	Belgium	Denmark	Finland	France	Germany	Greece	Ireland	Italy	Netherlands	Norway	Portugal	Spain	Sweden	Switzerland	UK
Wheat	13	17	10	17	24	13	9	31	31	7	19	24	19	9	21	22
Barley	13	17	8	17	17	11	9	31	28	8	17	22	18	8	21	21
Rye, oats, rice	10	17	8	17	17	11	15	31	37	4	17	15	12	7	21	20
Grain, maize, incl. CCM	27	17	0	0	27	28	20	0	44	17	0	26	44	0	38	0
Total cereals	15	17	9	17	23	13	11	31	34	8	17	22	19	7	23	21
Potato	25	26	17	44	39	28	44	116	35	35	39	35	31	35	38	76
Sugar beet	25	28	17	35	35	31	28	71	39	29	0	39	48	22	38	25
Oilseed rape	24	26	11	20	22	21	0	34	22	11	20	0	17	17	35	22
Sunflower, soya	15	0	0	0	17	14	11	0	31	0	0	0	2	0	24	11
Pulses (peas, beans)	4	22	8	0	24	16	9	0	22	26	0	15	3	20	31	13
Vegetables	21	28	13	44	22	31	37	0	35	37	33	35	44	31	70	33
Fodder incl. silage maize	17	9	6	17	22	12	17	67	4	15	12	13	17	9	27	22
Tobacco	22	9	0	0	44	15	28	0	35	0	0	0	26	0	22	0
Other	13	9	8	13	17	15	28	39	4	13	17	0	7	17	31	9
Total arable land	17	18	9	21	23	15	17	43	24	21	15	21	17	10	27	23

Perm. crops (fruit, vineyard)	8	17	0	26	6	12	20	0	17	17	22	11	7	2	24	11
Grassland fertilized	10	15	8	16	10	4	0	13	0	9	8	11	9	7	10	10
Grassland non-fertilized	0	0	0	0	0	0	0	0	0	0	0	0	0	0	0	0
Total perm. grassland	1	13	7	16	8	4	0	10	0	9	8	1	0	3	3	4
Mean P application for agricultural area in use	7	15	9	18	16	11	7	13	16	14	14	8	10	9	8	9

CCM, corn cob maize.

Table 16.2. Area of land and yields of the most important crops grown in West European countries in 1994/95 and the amount of P applied to each crop.

Crop	Hectares ('000)	Yield (kg ha^{-1})	P (kg ha^{-1})
Wheat	16,431	5,415	20
Barley	11,467	4,029	16
Rye, oats, rice	4,824	3,951	15
Grain, maize, incl. CCM	3,899	7,732	32
Total cereals	36,621	5,035	20
Potato	1,517	29,903	39
Sugar beet	2,103	51,961	34
Oilseed rape	2,383	2,724	21
Sunflower, soya, linseed	2,945	1,639	12
Pulses (peas, beans)	1,617	3,252	15
Vegetables	1,625		37
Fodder incl. silage maize	12,504		14
Tobacco	189		31
Other	2,025		15
Total arable land	63,528		19
Perm. crops (fruit, vineyard)	10,294		11
Grassland fertilized	25,075		9
Grassland non-fertilized	31,813		
Total perm. grassland	56,888		4
Agricultural area in use	130,710		12

value. Similar relationships have been found in many other experiments elsewhere. Defining a critical level of soil P to give advice to farmers about applying P requires consideration of a number of factors. Two of the most important are the need not to compromise the economic viability of the farm enterprise, while at the same time soil-P levels should not be raised excessively, which would increase the risk of adverse environmental impact from P loss to water. For nitrogen (N), it is comparatively easy to arrive at an economic optimum yield and its associated N application. Currently, in the UK, for example, for wheat it would be the point at which 1 kg N ha^{-1} produces only 3 kg grain; for potatoes it would be 5 kg tubers ha^{-1} for 1 kg N. For P, the calculation is much more difficult, because soil P, rather than freshly applied P, is much more important in controlling yield in some circumstances. Johnston *et al.* (1986) determined the Olsen P value at which yield was one standard error less than the asymptotic yield and this gave some realistic levels of Olsen P which could be used for advisory purposes. More recently, the relationship between wheat yields and Olsen

Table 16.3. Phosphate balance (input − output) in European agriculture (P kg ha^{-1}) in 1994/95.

Mineral fertilizers − removal		All sources − removal		
< 5	5–10	< 10	10–20	> 20
Austria	Finland	Austria	Denmark	Netherlands
Denmark	Greece	Germany	Finland	
France	Ireland	Sweden	France	
Germany			Greece	
Netherlands			Ireland	
Norway			Norway	
Sweden				

P at four levels of N has been measured over 6 years on a sandy clay loam. The average Olsen P value to achieve 99% of the asymptotic yield was 19 mg P kg^{-1}, but 95% of the asymptotic yield was obtained at only 11 mg P kg^{-1}. A soil at 19 mg P kg^{-1} would not be considered to be excessively enriched and would be realistic both economically and agronomically (A.E. Johnston, personal communication).

In European agriculture, nearly 90% of the cereals are small-seeded – wheat, barley, etc. (Table 16.2) and usually grown in narrow rows, 10–15 cm apart. Most of the cereals are autumn-sown, so that there is a narrow window of opportunity to achieve drilling under ideal conditions. Farmers wish, therefore, to avoid unnecessary autumn work, which would include the application of P and K fertilizers. Thus, in the UK, for example, because of the preferred cropping and cultural practices, it is often advised to apply P fertilizers only to more responsive crops, such as potatoes and vegetables, and not to cereals when they are grown in rotation. For such responsive crops, the aim would be to have an Olsen P value not greater than 25 mg P kg^{-1} soil.

In other farming systems, a totally different approach might be more applicable. For example, with a preponderance of spring-sown crops grown on widely spaced rows, band placement of P either before or at drilling could be much more efficient and might lead to advice suggesting that soils be maintained at quite low levels of Olsen P.

PHOSPHORUS IN RIVERS AND LAKES

European studies on P concentrations in various surface-water bodies show that in catchments with little or no human activity P concentrations in rivers are generally less than 25 µg P l^{-1}, while those exceeding 50 µg P l^{-1} indicate an anthropogenic influence, such as sewage effluent and agricultural losses (Fig. 16.3). In rivers heavily polluted by sewage effluent, P levels may exceed 1000 µg l^{-1}. There is a positive correlation between population density and annual mean P concentration in river water. It is generally less than 20 µg P l^{-1} in catchments with fewer than 15 inhabitants km^{-2}, and higher than 200 µg P l^{-1} in those with more than 100 inhabitants km^{-2}. In Europe, P concentrations in rivers are lowest in the Nordic countries, medium or lower in the southern and eastern European countries, although slightly

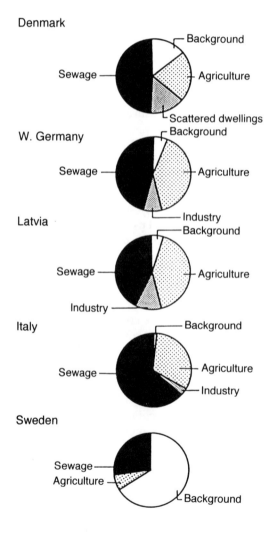

Fig. 16.3. Sources of phosphorus discharge to rivers and lakes (from Danish Ministry of Environment and Energy, 1994).

lower in southern European countries, and highest in the Western European countries. The highest concentrations are found in rivers in a band stretching from Northern Ireland, across southern England and the central parts of Europe through Romania and Moldavia to the Ukraine. In these countries more than 80% of the rivers usually have a P concentration exceeding 125 µg P l^{-1} (Stanners and Bourdeau, 1995). In general, concentrations exceed 100 µg P l^{-1} in the downstream reaches in all of the largest rivers in Europe (Danish Ministry of Environment and Energy, 1994; Stanners and Bourdeau, 1995). There is little evidence to indicate the proportion of P in rivers which comes from point and diffuse sources. There is much evidence that decreasing the release of P from point sources has not had a dramatic effect on the P concentration in water bodies (Foy and Withers, 1995). This could be because diffuse sources are still a major source of P or P is being released from sediments on lake bottoms.

The Organization for Economic Cooperation and Development (OECD) has presented a series of relationships showing the probability of a freshwater body being in a particular trophic category, e.g. oligotrophic, mesotrophic, eutrophic, hypertrophic, and the P concentration in the water body (OECD, 1982). These relationships suggest that there is an increasing probability of a water body being eutrophic as the P concentration increases from 20 to 90 µg P l^{-1}. Above this upper concentration, the probability of the water body being hypertrophic increases.

These very small concentrations must be related to the amounts of P which need to be lost from soil to achieve such a concentration. For example, the amount of P needed to increase the concentration to 90 µg l^{-1} in leachate is extremely small and depends on the volume of through drainage. Assuming that the leaching water resides in soil long enough, 90 µg P l^{-1} equates to a loss of only 0.09 kg P ha^{-1} with 100 mm through drainage and 0.45 kg P ha^{-1} with 500 mm through drainage.

PHOSPHORUS LOSSES FROM SOIL

Losses by Leaching

When large amounts of phosphate fertilizers are applied to light-textured sandy soils just before heavy rainfall, P may be lost from the surface soil by leaching. However, P leached from the upper horizon may be retained in lower horizons if these are enriched in clay (Mattingly, 1970). Dissolved P can also be transported through cracks and worm-holes into the drainage system (Voss and Preuße, 1976).

Direct measurement of P losses by leaching from mineral soils under field conditions is costly and the widely held view that P losses via this pathway are negligible is mainly derived from indirect evidence. Experiments on a silty loam at Rothamsted showed that the top 23 cm of soil on plots given 33 kg P ha^{-1} annually as fertilizer for more than 100 years contained almost twice as much total P as unfertilized land (Johnston and Poulton, 1992). There has been some enrichment of the 23–30 cm depth but only slight enrichment at 30–46 cm. The most likely explanation is that P had leached from the topsoil and some had been retained in the subsoil. Where farmyard manure supplying 40 kg P ha^{-1} was applied each year in the same experiments, the 23–30 and 30–46 cm layers were appreciably more enriched with P compared with the fertilizer-treated soil and there was some slight enrichment below 46 cm. Where farmyard manure and superphosphate were applied together, even more P had been leached. There had also been appreciable leaching, probably below 46 cm, where superphosphate had been applied to permanent grassland. There might be several possible explanations for these observations. One would be leaching of simple organic compounds, which are then mineralized to inorganic P, which, in turn, is adsorbed in the subsoil. Another could be that soil organic matter provides many low-energy bonding sites for P, from which it can be readily removed. However, the retention of P in the subsoil does not preclude the transport of some P out of the soil in drainage water.

The data above suggest that the leaching of P from agricultural soils was affected by the form in which P was added and the depth and nature of the subsoil. Of particular importance is the observation of increased leaching where organic manures were added or there was more soil organic matter derived from grassland. This could explain the greater P enrichment of surface waters in situations where much manure is applied to soils. This is especially so where a soil is already enriched with P and has few adsorption sites in the subsoil and only a small water-holding capacity.

High soil-P status often occurs on farms with intensive livestock production. Some of these farms also apply P mineral fertilizers to achieve additional yields. Too high livestock density can cause P surpluses which may exceed the P adsorption capacity of soils and thus increase the probability for P to reach groundwater. There is concern about P leaching in the Netherlands, Belgium and Denmark, and this may be due to large applications of manure to soil with little ability to retain P.

Recent studies at Rothamsted suggest that soils may have a P threshold value for Olsen P above which the risk of P leaching increases rapidly. This threshold value is likely to vary according to soil mineralogical, chemical and hydrological properties. Lysimeter studies at various centres in Europe support the findings at Rothamsted about P leaching (Amberger and Schweiger, 1978; Jung and Jürgens-Gschwind, 1977; Wiechmann, 1972). In the Mediterranean area, the potential risk of nutrient losses from soils by leaching is very small, because of the lack of rain (Sequi *et al.*, 1991). In some areas of southern Europe, there is so little rainfall that the leaching of any nutrient from soil through drainage does not occur unless soils are excessively irrigated.

Concern about P losses in leaching can be important on soils with artificial drainage systems, especially when water passes quickly to the drains via cracks and fissures in the overlying soil, and on soils with raised water-tables. As shown previously, very small amounts of P can raise the concentration of P in drainage to the critical level at which the biological balance within the receiving water is disturbed. The possibility of increased P losses from artificially drained fields affecting the P enrichment of water bodies within a catchment will depend on the proportion of drained fields within the catchment. Where fields are not drained, the longer residence time of water in the soil will give opportunity for P sorption on to sites within the soil matrix. However, this may be offset by increased losses in surface runoff.

Phosphorus losses via erosion and surface runoff

Phosphorus losses through surface runoff and erosion may be substantial. For instance, in the central Apennines, on impoverished soils, losses ranged from 2.6 to 30.6 tons soil ha^{-1} year^{-1}, containing 2–4 kg P ha^{-1} year^{-1} (Sequi *et al.*, 1991). In the greater part of Europe, erosion is considered to be the main route of P losses from agricultural land (Sharpley *et al.*, 1994b). Because erosion of phosphate-enriched soil can be a major hazard, this again suggests that soils should not be fertilized beyond the optimum for crop production (Bechtle, 1974). Erosion is a feature of bare soils not protected by vegetation. It becomes important on thawing soils overlying frozen subsoils and when heavy rain falls on dry soils. The Mediterranean regions are particularly vulnerable, with 34% of soil classified as high-erosion-risk and 50% as medium-erosion-risk. Water and wind erosion also occurs in northern Europe. In some areas, farmers have been encouraged to increase the number of grazing animals, and overgrazing and treading have so damaged the vegetation that erosion is becoming a serious problem in high-rainfall areas. Particularly affected are parts of western Ireland. Severe erosion, removing 1 cm soil ha^{-1}, could remove up to 150 kg P in the soil. Furthermore, there may be considerable erosion of fine particles from minimal slopes (2–3%), which may contain relatively large amounts of adsorbed P (Isermann, 1990, 1991).

There is also the possibility of mechanical erosion on heavy-textured soils, which are often artificially drained and also develop deep, wide cracks in summer. Autumn cultivations may cause surface soil enriched with P to fall into the cracks, increasing the movement of entrained P down through the profile. This decreases the distance that soluble P in drainage has to travel to reach drains and minimizes the opportunity for P sorption on sites within the subsoil. Enrichment of subsoils with P as a result of this phenomenon may erroneously be explained by leaching of P from surface soil.

Surface runoff is a feature of undulating, hilly terrain, especially in high-rainfall areas. In these situations in Europe, livestock husbandry dominates, and with livestock husbandry goes the need to dispose of large quantities of manure, usually in the form of slurry. The opportunities for spreading slurry under ideal conditions are limited, and heavy rain following application will invariably result in surface runoff containing water-soluble phosphate. These losses are often exacerbated by the fact that permanent grassland usually has low infiltration rates for water.

In drier areas, suitable for arable cropping, patterns of rainfall may give intense rainfall events. In such regions no-till husbandry systems may be practised and fertilizer applied to the surface will be at risk of loss if surface runoff occurs soon after fertilizer application.

PHOSPHORUS MANAGEMENT ON THE FARM

The need to balance an adequate supply of P to guarantee agricultural production while minimizing P losses from agricultural soils requires improved stewardship of the land, including improved, sophisticated methods of soil and landscape conservation, as well as nutrient management and monitoring (IFA/IMPHOS, 1989; Böckman *et al.*, 1990; Sharpley and Withers, 1994).

On any farm, the policy adopted for P applications, using either mineral fertilizer or manure or both, and the frequency and size of application must be part of an overall management strategy to maintain soil fertility. The effectiveness of such a nutrient management policy should be tested by soil analysis. Monitoring changes will help ensure that the soil-P status is being maintained at a level suitable for the soil type and yield potential.

Losses of P can be decreased by improved management practices, such as better crop rotations, including maintaining vegetative cover as long as possible, mulch sowing, reduced tillage, border strips, riparian buffer zones and timing of P application for manures and fertilizers (Commission of the European Communities and the European Fertilizer Manufacturers Association 1994). Application rates for P should be adjusted to soil P status and related to crop off-take. On farms where organic manures are used, their P content should be allowed for, and incorporation into soil rather than surface spreading should be adopted. On livestock farms with soils of a high P status, manure should be applied to meet the crops' P need and not according to their N requirement.

Recommendations for P applications should supplement any P available in manure with a precise amount of fertilizer P to ensure P sufficiency to obtain the economically optimum yield while minimizing P losses (EFMA and IFA, 1990; IFA, 1992).

FERTILIZER INDUSTRY INITIATIVES

The fertilizer industry has taken initiatives to improve the overall efficiency of nutrient utilization within agriculture. This work has been carried out in close association with universities, governments, national authorities and advisory services and research institutes. National industry initiatives have strongly supported improvements in fertilizer recommendations, soil-sampling and analytical services, computer modelling for fertilizer recommendations, extension programmes and the development of Codes of Good Agricultural Practice, all aiming at the best farm nutrient management and site-specific use of nutrients, regardless of the source (Commission of the European Communities and the European Fertilizer Manufacturers' Association, 1994).

Given its agronomic research and product-based knowledge, the fertilizer industry wishes to play an active role in developing farming systems which meet new requirements and which make sensible use of the benefits offered by its products.

CONCLUSIONS

Phosphatic fertilizers contribute to improved soil fertility and yields and, in general, they are applied according to recommendations. As many soils have been enriched with P to appropriate levels, the overall use of P fertilizers within Europe has declined since the early 1970s. However, excessive application of P tends to occur in areas with high livestock densities.

There is a significant contribution of P from agriculture to water bodies. The major P contributors are eroded soil and surface runoff. The magnitude of the problem varies among regions and therefore there is no single solution. At farm level, a field-by-field nutrient-management plan is required to ensure an adequate supply of P to meet crop requirements, while minimizing excessive build-up of soil P, which, whether by soil erosion, surface runoff or leaching, can increase the risk of P enrichment of surface-water bodies.

The fertilizer industry suggests that on-farm P management practices should include: P balance calculations, regular soil testing and P application according to P balance and soil test, adjusted to the crops' need; improved P management, especially in the use of animal manure, to ensure that there is no excessive build-up of soil P; implementation of reduced-tillage techniques where and when appropriate, and maintenance of a vegetative cover as long as possible; care in the timing of fertilizer and manure applications and placement of fertilizers and incorporation of manure into the soil, where these are appropriate; and improved water/

irrigation management and the implementation of Codes of Good Agricultural Practice.

ABSTRACT

The essential need for phosphorus (P) in crop production, together with the risk of P losses from agriculture to fresh water bodies, is discussed. Also the relationship between crop yield and readily soluble P in soils is presented, together with data on the concentration of P in drainage from agricultural land. The various routes by which P can be lost from agricultural soils and their relative importance are considered. Finally, methods by which P losses from agricultural land might be reduced and the fertilizer industry's view are presented.

ACKNOWLEDGEMENT

The author is grateful to A.E. Johnston for his contribution to this chapter and, in particular, a draft document 'P, its retention in and loss from agricultural soils', currently being prepared for EFMA.

REFERENCES

Amberger, A. and Schweiger, P. (1978) P-Bilanz eines langjährigen Lysimeterversuches. *Bodenkultur* 29(4), 325–332.

Bechtle, W. (1974) *Oberflächenabfluß, Bodenabtrag und Nährstoffauswaschung in steilen Hanglagen – Erosionsversuch Eßlingen-Mettingen.* Wasser und Boden, Stuttgart.

Böckman, O., Kaarstad, O., Lie, O.H. and Richards, I. (1990) *Agriculture and Fertilizers. Fertilizers in Perspective: Their Role in Feeding the World, Environmental Challenges, Are There Alternatives?* Report from Agricultural Group, Norsk Hydro, Oslo, Norway.

Brookes, P.C., Powlson, D.S. and Jenkinson, D.S. (1984) Phosphorus in the soil microbial biomass. *Soil Biology and Biochemistry* 16, 169–175.

Chater, M. and Mattingly, G.E.G. (1980). Changes in organic phosphorus contents of soils from long continued experiments at Rothamsted and Saxmundham. In: *Rothamsted Experimental Station Report for 1979*, Rothamsted Experimental Station, Harpenden, Part 2, pp. 41–61.

Ciavatta, C., Antisari, L.V. and Sequi, P. (1990) Interference of soluble silica in the determination of orthoP. *Journal of Environmental Quality* 19(4), 761–764.

Commission of the European Communities and the European Fertilizer Manufacturers' Association (European Commission/EFMA) (1994) *The Fertilizer Industry of the European Union: The Issues of Today, the Outlook for Tomorrow.* Brussels, Belgium.

Council for Agricultural Science and Technology (CAST) (1992) *Water Quality: Agriculture's Role.* Task Force Report, CAST, Iowa, USA.

Danish Ministry of Environment and Energy (1994) *European Rivers and Lakes – Assessment of Their Environmental State.* European Environment Agency, Copenhagen.

European Fertilizer Manufacturers' Association (EFMA) (1991) Animal manure in Europe. A study paper. EFMA, Paris.

European Fertilizer Manufacturers' Association (EFMA) and the International Fertilizer Industry Association (IFA) (1990) *A Code of Best Agricultural Practices to Optimize Fertilizer Use.* EFMA and IFA, Paris and Brussels.

Foy, R.H. and Withers, P.J.A. (1995) *The Contributions of Agricultural P to Eutrophication.* Proceedings No. 365 of the Fertiliser Society, The Fertiliser Society, Peterborough.

Gregory, D.I., Schultz, J.J. and Engelstad, O.P. (1991) Phosphate fertilizers and the environment. Discussion paper for United Nations Industrial Development Programme (UNIDO) International Conference on Ecologically Sustainable Industrial Development, 14–18 October 1991, Copenhagen, Denmark.

IFA (1992) *Sustainable Agricultural Systems for the Twenty-first Century: the Role of Mineral Fertilizers.* IFA, Paris.

IFA/IMPHOS (1989) *Role of Phosphates in Balanced Fertilization Seminar,* Vols 1 and 2. IFA/IMPHOS, Marrakech, Morocco.

Isermann, K. (1990) Share of agriculture in nitrogen and phosphorus emissions into the surface waters of Western Europe against the background of their eutrophication. *Fertilizer Research* 26(1–3), 253–269.

Isermann, K. (1991) Territorial, continental and global aspects of C, N, P and S emissions from agricultural ecosystems. In: *Proceedings of NATO Advanced Research Workshop (ARW) on Interations of C, N, P, and S Biochemical Cycles,* 4–8 March 1991, Melreux, Belgium.

ISMA (1977) *Importance of P in Agriculture.* ISMA, Paris.

ISMA (1980) *Handbook on Phosphate Fertilization.* ISMA, Paris.

Johnston, A.E. (1994) The Rothamsted classical experiments. In: Leigh, P.R. and Johnston, A.E. (eds) *Long-Term Experiments in Agricultural and Ecological Sciences.* CAB International, Wallingford, pp. 9–37.

Johnston, A.E. and Poulton, P.R. (1992) The role of phosphorus in crop production and soil fertility: 150 years of field experiments at Rothamsted, United Kingdom. In: Schultz, J.J. (ed.) *Phosphate Fertilizers and the Environment.* International Fertilizer Development Center, Muscle Shoals, USA, pp. 45–64.

Johnston, A.E., Lane, P.W., Mattingly, G.E.G., Poulton, P.R. and Hewitt, M. (1986) Effects of soil and fertilizer P on yields of potatoes, sugarbeet, barley and winter wheat on a sandy loam soil at Saxmundham, Suffolk. *Journal of Agricultural Science, Cambridge* 106, 155–167.

Jung, J. and Jürgens-Gschwind, S. (1977) *Ergebnisse von Lysimeteruntersuchungen in de Großanlage Limburgerhof. (Die wichtigsten Befunde aus 50 Jahren Versuchstätigkeit, 1927 bis 1977).* BASF, Limburgerhof.

Lee, J. (1987) Land resources and their use in the European Community. In: Barth, H. and 'Hermiler, P. (eds) *Scientific Basis for Soil Protection in the European Community.* Elsevier, Barking, pp. 29–63.

Mattingly, G.E.G. (1970) Residual value of basic slag, Gafsa rock phosphate and superphosphate in a sandy podzol. *Journal of Agricultural Science, Cambridge* 75, 413–418.

OECD (1982) *Eutrophication of Waters: Monitoring, Assessment and Control.* Organization for Economic Cooperation and Development, Paris.

Potash and Phosphate Institute (1986) *P for Agriculture: a Situation Analysis.* Potash and Phosphate Institute (PPI), Atlanta.

Scientific Committee on Problems of the Environment (SCOPE) (1995) *P in the Global Environment: Transfers, Cycles and Management.* John Wiley and Sons, Chichester.

SCOPE/UNEP (1988) *P Cycles in Terrestrial and Aquatic Ecosystems.* Proceedings of Workshop arranged by the Scientific Committee on Problems of the Environment (SCOPE) and the United Nations Environmental Programme (UNEP) and organized by the Department of Agrobiology and Forestry of the Polish Academy of Sciences, 1–6 May 1989, Czerniejewo, Poland.

Sequi, P., Ciavatta, C. and Antisari, L. (1991) Phosphate fertilizers and P loadings to river and seawater. *Agrochimica* 35, 1–3; *Gennaio-Giugno* 35, 200–211.

Sharpley, A.N. and Withers, P.J.A. (1994) The environmentally-sound management of agricultural P. *Fertilizer Research* 39, 133–146.

Sharpley, A.N., Chapra, S.C., Wedepohl, R., Sims, J.T., Daniel, T.C. and Reddy, K.R. (1994) Managing agricultural Phosphorus for protection of surface waters: issues and options. *Journal of Environmental Quality* 23(3), 437–451.

Stanners, D. and Bourdeau, P. (eds) (1995) *European Environment: The Dobrís Assessment.* European Environment Agency, Copenhagen.

Voss, W. and Preuße, H.-U. (1976) *Die Gewässerbelastung durch den Oberflächenaustrag gelöster and fester Substanzen, Forschungs Benat* 30, 229–238.

Wiechmann, H. (1972) *Überlegungen zum Phosphataustrag aus Böden.* Stuttgart-Hohenheim pp. 253–269.

17 European Perspective on Phosphorus and Agriculture

F. Mariën
Commission of the European Communities, Rue de la Loi 200, B-1049 Brussels, Belgium

INTRODUCTION

Phosphorus (P) is, or was, very frequently a limiting factor in the growth of natural vegetation. Undeveloped or very-low-intensity agriculture areas and natural pristine ecosystems have shown us that P deficiency is the normal condition.

No one will disagree with the good intentions of scientists of the nineteenth century who proposed the use of P to improve agricultural yields, both in quantity and quality.

Such a statement is still true today, and P can still be a useful input to agriculture. But, fortunately, with over a century of experience, we have benefited from a considerable amount of scientific and technical progress. This has shown us some of the disadvantages and negative side-effects this element may have on our environment.

This is a very opportune occasion to present the latest scientific knowledge and discuss possible management issues of P or, more appropriately, of nutrients. Phosphorus can hardly be separated from nitrogen and oligoelements, either in the uptake of nutrients to ensure the optimum metabolism in plants, humans or animals, when present in adequate quantities or in the eutrophication of surface waters. Problems of eutrophication exist because we tend to forget that we live in an ecosystem with an equilibrium of its own. Any outside interference will result in a disruption of this equilibrium. Although eutrophication can also sometimes be a naturally occurring process, its increasing incidence in recent years is most of the time directly attributable to elevated levels in waters of nutrients from anthropogenic sources. The two main sources of anthropogenic nutrients are urban waste water and agriculture.

1997 CAB INTERNATIONAL. *Phosphorus Loss from Soil to Water*
(eds H. Tunney, O.T. Carton, P.C. Brookes and A.E. Johnston)
The views expressed in this contribution are those of the author and do not necessarily reflect the views or policies of the European Commission.

THE URBAN WASTE-WATER SOURCE

Although it can be difficult to ascertain figures, we can estimate that P emanating from urban waste water may account for some 50% of the total loss to water. It is no wonder that the urban waste-water treatment directive introduced quality requirements to be met for treating waste waters to be discharged into sensitive areas.

Phosphorus in urban waste water has two main sources: the human metabolism and detergents. In some cases, specific industrial effluent may also contribute. Where countries have taken steps to phase out the use of P in household detergents (prevention at source), it has resulted in more or less a halving of the P load in urban waste water.

The remaining P load, coming from metabolic sources, cannot, however, be approached by prevention-at-source measures, but has to be dealt with in the course of waste-water treatment. To that aim, the urban waste-water treatment directive (91/271/EEC) requires treatment plants above 10,000 population equivalents (p.e.) to provide for P removal when effluents are discharged into sensitive areas, before 31 December 1998. The following standards are laid down for the discharges:

- 2 mg l^{-1} total P for treatment plants between 10,000 and 100,000 p.e.;
- 1 mg l^{-1} total P for treatment plants above 100,000 p.e.;
- minimum percentage reduction of 80%.

These can be reached by standard techniques, such as simultaneous precipitation in activated sludge plants. Compared with nitrate removal, P removal requires less costly investment. Furthermore, by providing waste-water treatment plants with P removal systems, a far larger reduction in the nutrient input to waters can be achieved than in the nitrogen sector, where the input to waters from diffuse sources has a distinctly larger impact, even if all point sources are being dealt with.

THE AGRICULTURE SOURCE

Some say that the main source of P loss comes from a diffuse source, namely agriculture. This is, at present, probably true in most regions of the Community. For a long time, scientific studies have shown that P leaching is negligible and that the main cause of P loss from agriculture is through soil erosion (Hanotiaux, 1978). The up-to-date information presented in this book throws new light on P loss to water from agriculture. To reduce P loss to water it may be necessary to control P inputs on to soil but it may also be, in some situations, more effective to adopt soil-erosion counter-measures.

The volume and concentration of loss depends on the levels of P present in the soil. For a long time we have imported tonnes and tonnes of P for European agriculture, while only negligible amounts have been exported. If we calculate an estimate of P balance for European agriculture, we shall see that the inputs can still be several times the outputs. This element has been building up and the concentration of P in soils has been increasing to such an extent for the past 35 years that it is now difficult to find soils in Europe which are deficient in P (Tunney, 1992).

Intensive livestock farming has been and still is contributing very greatly to the increase in the concentration of P in soils.

Unfortunately, while the P concentration was increasing in the soil, so was the P loss to water, and we can therefore expect that in some regions of the Community there may be problems of eutrophication due to P. This is also one of the reasons for the present discussion on P loss to water. It should be noted that before P is leached or transported to water it is being spread on to soil. With the preventive principle in mind, we should probably allow ourselves to give some thought to whether it is necessary to apply more quantities of P in order to increase plant growth efficiency but ignoring environmental consequences and sometimes even costs.

It is sensible to calculate the adequate P supply to plants in order to guarantee optimum yields in agriculture, but this may not be sufficient to control P emissions to water. If we want to defend a coherent water policy dealing with eutrophication, we should be concerned with nutrient management, that is, essentially with nitrates and P from all anthropogenic sources.

THE EUROPEAN WATER POLICY

In 1991 the Council adopted Directive 91/676/EEC 'concerning the protection of waters against pollution caused by nitrates from agricultural sources'. Although this directive is primarily concerned with preventing and reducing water pollution caused or induced by nitrates from agricultural sources, it also demands measures to be taken against eutrophication:

- in the criteria to be used to designate vulnerable zones, member states should examine whether natural freshwater lakes, other freshwater bodies, estuaries, coastal waters and marine waters are found to be eutrophic or may become so in the near future;
- the guidelines for the codes of good agricultural practice and the measures to be included in the action programmes should contain requirements for the spreading of all nitrogen fertilizers and this would also probably have consequential effects for P.

Directives 91/271/EEC and 91/676/EEC are not the only measures which have been taken at a European level. Community water policy started in 1973 and has since adopted a wide range of water legislation (approximately 30 legal texts; a list is appended to this chapter). Some of these texts also contain some measures with regard to P.

- In Council Decision 75/437/EEC, the contracting parties should undertake to limit pollution of the maritime area from land-based sources by organic and elemental P. Recommendations under the Paris Convention have required 50% reduction in emissions of nutrients to the North Sea.
- In Council Directive 75/440/EEC, member states are to set limit values for P, expressed as phosphate (P_2O_5), in water being abstracted for drinking purposes.
- In Council Directive 76/464/EEC, member states are to take the appropriate steps to reduce pollution of waters by inorganic compounds of P and elemental P.
- In Council Decision 77/586/EEC, the contracting parties are to take the appropriate measures to reduce the pollution of the Rhine by inorganic compounds of P and elemental P.
- In Council Directive 78/659/EEC, no limit value for P is set, but values of 0.2 mg l^{-1} for salmonid and of 0.4 mg l^{-1} for cyprinid waters, expressed as orthophosphate (PO_4), may be regarded as indicative in order to reduce eutrophication.
- In Council Directive 80/68/EEC, member states are to take the necessary steps to limit the introduction into groundwater of inorganic compounds of P and elemental P so as to avoid pollution of this water.
- In Council Directive 80/778/EEC, member states are to set a maximum admissible concentration for total P in drinking water not exceeding 5 mg l^{-1} and a guide level of 0.4 mg l^{-1}, expressed as P_2O_5. It is to be noted here that in its recent proposal amending Directive 80/778/EEC, the Commission did not put forward a value for P.
- In Council Decision 83/101/EEC, the contracting parties are to take all appropriate measures to prevent, abate, combat and control pollution of the Mediterranean Sea caused by discharges of:
  - inorganic compounds of P and elemental P;
  - substances which have, directly or indirectly, an adverse effect on the oxygen content of the marine environment, especially those which may cause eutrophication, from rivers, coastal establishments or outfalls, or emanating from any other land-based sources within their territories.

ADDITIONAL MEASURES

Financial Aids

In the framework of Council Regulation 2078/92/EEC 'on agricultural production methods compatible with the requirements of the protection of the environment and the maintenance of the countryside', a Community aid scheme is instituted to contribute to the achievement of the Community's policy objectives regarding agriculture and the environment. In 1995, 779 million European currency units (ECU) was available for this scheme, which may include aid for farmers who undertake, *inter alia*:

- to reduce substantially their use of fertilizers;
- to reduce the proportion of sheep and cattle per forage area;
- to ensure the upkeep of abandoned farmland or woodlands;
- to improve the training of farmers with regard to farming or forestry practices compatible with the environment;
- to take some agricultural land out of productive use, ensuring long-term 'set-aside' for purposes connected with the environment.

It is to be noted that, of the hundreds of proposals received in the framework of Regulation 2078/92/EEC, only very few included measures for P and eutrophication of waters.

Objectives 1, 5a and 5b regions eligible for financial aid for programmes of rural development may also include measures (farmers' training and education, investment at the farm) which help agriculture to be more friendly to the environment. This may allow member states to define additional and more specific action not foreseen at a Community level to combat eutrophication and P loss to water.

Community Research Programmes

The Community Research Programme on Agriculture and Fisheries, a part of the Fourth Framework Programme on Research, Technological Development and Demonstration, contains among its research tasks a section on agriculture–environment interactions, which supports, *inter alia*, mechanical equipment and techniques promoting practices that are more environmentally friendly and giving greater protection to the environment (water, soil, landscape, etc.) to reduce contamination and erosion.

Companies, universities and research centres are invited to submit transnational project proposals meeting the objectives of the work programme. The best projects are to be selected for Community funding.

THE NEED FOR FURTHER COMMUNITY LEGISLATION ON PHOSPHORUS/PHOSPHATES

The process of obtaining Community legislation is long and time-consuming. It can take many years from the moment the Commission starts work on a proposal until it is adopted by the Council. Before working on a draft proposal, the Commission has to provide sufficient arguments in order to be able to justify its initiative to the Parliament and Council.

If we want a directive, we shall have to find a good reason for it. For P, the Commission invited members of the committee foreseen under article 9 of Directive 91/676/EEC to give their opinion on the need for a specific action at Community level. The general consensus was that no further action was needed for the moment.

If we examine the framework for the objectives for agriculture and forestry and for the management of water resources in the Fifth Action Programme for the Environment (COM(92)23 final), the Commission has set out two commitments for 1995.

1. The examination of the need for a directive on phosphate reduction.
2. A reduction programme for phosphate use.

Therefore, the possibility of having Community legislation dealing with P still exists, but considering the measures already existing at Community level, and the variety of possibilities offered to member states to define specific instruments or to ask for financial aid, such legislation is not envisaged in the immediate future.

The Commission is well aware of the close relationship between P and other nutrients, in particular nitrates, in the process of eutrophication and the degradation of the quality of Community waters. This is one of the reasons why, in July 1994, the Commission adopted a proposal for a Council Directive on the ecological quality of water. This legislation, when adopted, will play a major role in the management and control of the quality of Community waters.

The proposal for a Council Directive on the ecological quality of water is a new departure for European Community (EC) water legislation. It considers surface waters as a potential resource, not just from an economic point of view, but also from the point of view of nature conservation, although it recognizes that measures taken to protect the ecological quality of the water body will also enhance its resource value.

The proposal was presented by the Commission on 15 June 1994. It requires member states:

- to monitor the overall ecological quality of their surface waters (Article 3);

- to make an assessment of the sources of pollution and adverse anthropogenic influences (Article 4);
- to define operational targets to maintain or improve ecological quality (Article 5);
- to establish integrated programmes in order to meet those targets (Article 6).

Most of the decisions regarding the timetable for reaching good ecological quality and the contents of the integrated programmes are left to member states, but they must be made in a transparent manner with full consultation of the public and of interested parties (Article 7).

In theory, the proposal will provide a framework for any measures required in order to tackle any remaining P problems. However, we shall have to see how the proposal develops and how it is implemented. It may be that there is still a need for specific measures to counter specific problems.

CONCLUSION

So far, there is not sufficient evidence to justify more Community action in the form of legislation on P. The Commission believes that the measures available to member states – financial aids, research programmes and the existing legal Community framework – provide for sufficient cover of the problem.

Some believe that there may be a lack of clarity and coherence in the water legislation, but this would have very little impact on the P issue, and in any case the Commission is envisaging proposing a new common base for the Community water legislation. This could take account of some of the criticisms of water legislation, such as the need for more flexibility and more subsidiarity. However, such an approach would require more actions to be defined and enforced at national or regional levels within member states.

The control of P emission (or loss) to water might be such an example of subsidiarity for those regions in the Community suffering from eutrophication of waters resulting from excess P.

REFERENCES

COM(92)23 final – vol. II. *Towards Sustainability: a European Community Programme of Policy and Action in Relation to the Environment and Sustainable Development.* Commission of the European Communities, Brussels.

Hanotiaux, G. (1978) Entraînements d'éléments nutritifs suite au phénomène d'érosion en région limoneuse. *Pédologie* 28 192–204.

Tunney, H. (1992) Some environmental implications of phosphorus use in the

European Community. In: *Phosphorus, Life and Environment: from Research to Application*, Proceedings of the 4th International IMPHOS Conference, Ghent, Belgium. World Phosphate Institute, Casablanca, Morocco, pp. 347–359.

LIST OF EXISTING COMMUNITY WATER LEGISLATION

Council Directive 73/404/EEC of 22 November 1973 on the approximation of the laws of the Member States relating to detergents.

Council Decision 75/437/EEC of 3 March 1975 concluding the Convention for the prevention of marine pollution from land-based sources.

Council Directive 75/440/EEC of 16 June 1975 concerning the quality required of surface water intended for the abstraction of drinking water in the Member States.

Council Directive 76/160/EEC of 8 December 1975 concerning the quality of bathing water.

Council Directive 76/464/EEC of 4 May 1976 on pollution caused by certain dangerous substances discharged into the aquatic environment of the Community.

Council Decision 77/585/EEC of 25 July 1977 concluding the Convention for the protection of the Mediterranean Sea against pollution and the Protocol for the prevention of the pollution of the Mediterranean Sea by dumping from ships and aircraft.

Council Decision 77/586/EEC of 25 July 1977 concluding the Convention for the protection of the Rhine against chemical pollution and an Additional Agreement, signed in Berne on 29 April 1963, concerning the International Commission for the protection of the Rhine against pollution.

Council Decision 77/795/EEC of 12 December 1977 establishing a common procedure for the exchange of information on the quality of surface fresh water in the Community.

Council Directive 78/659/EEC of 18 July 1978 on the quality of fresh waters needing protection or improvement in order to support fish life.

Council Directive 79/869/EEC of 9 October 1979 concerning the methods of measurement and frequencies of sampling and analysis of surface water intended for the abstraction of drinking water in the Member States.

Council Directive 79/923/EEC of 30 October 1979 on the quality required of shellfish waters.

Council Directive 80/68/EEC of 17 December 1979 on the protection of groundwater against pollution caused by certain dangerous substances.

Council Directive 80/778/EEC of 15 July 1980 relating to the quality of water intended for human consumption.

Council Decision 81/420/EEC of 19 May 1981 on the conclusion of the Protocol concerning cooperation in combating pollution of the Mediterranean Sea by oil and other harmful substances in cases of emergency.

Council Directive 82/176/EEC of 22 March 1982 on limit values and quality objectives for mercury discharges by the chlor-alkali electrolysis industry.

Council Directive 82/242/EEC of 31 March 1982 on the approximation of the laws of the Member States relating to methods of testing the biodegradability of non-ionic surfactants and amending Directive 73/404/EEC.

Council Decision 83/101/EEC of 28 February 1983 concluding the Protocol for the protection of the Mediterranean Sea against pollution from land-based sources.

Council Directive 83/513/EEC of 26 September 1983 on limit values and quality objectives for cadmium discharges.

Council Decision 84/132/EEC of 1 March 1984 on the conclusion of the Protocol concerning Mediterranean specially protected areas.

Council Directive 84/156/EEC of 8 March 1984 on limit values and quality objectives for mercury discharges by sectors other than the chlor-alkali electrolysis industry.

Council Decision 84/358/EEC of 28 June 1984 on the Agreement for cooperation in dealing with pollution of the North Sea by oil and other harmful substances (Bonn Agreement).

Council Directive 84/491/EEC of 9 October 1984 on limit values and quality objectives for discharges of hexachlorocyclohexane.

Council Decision 85/613/EEC of 20 December 1985 concerning the adoption, on behalf of the Community, of programmes and measures relating to mercury and cadmium discharges under the convention for the prevention of marine pollution from land-based sources.

Council Decision 86/85/EEC of 6 March 1986 establishing a Community information system for the control and reduction of pollution caused by the spillage of hydrocarbons and other harmful substances at sea.

Council Directive 86/280/EEC of 12 June 1986 on limit values and quality objectives for discharges of certain dangerous substances included in List I of the Annex to Directive 76/464/EEC.

Council Directive 90/656/EEC of 4 December 1990 on the transitional measures applicable in Germany with regard to certain Community provisions relating to the protection of the environment.

Council Directive 91/271/EEC of 21 May 1991 concerning urban waste water treatment.

Council Directive 91/676/EEC of 12 December 1991 concerning the protection of waters against pollution caused by nitrates from agricultural sources.

18 Views on Phosphorus and Agriculture – Paris Commission

S. Sadowski
*Oslo and Paris Commissions, New Court,
48 Carey Street, London WC2A 2JQ, UK*

INTRODUCTION

The Paris Convention and the Commission which was created to oversee it are an important part of an institutional framework established by governments to work for the prevention of marine pollution in the north-east Atlantic. The present arrangements began to develop in the early 1970s with the adoption in 1972 of the Convention for the Prevention of Marine Pollution by Dumping from Ships and Aircraft (the 'Oslo Convention'), followed by the adoption in 1974 of the Convention for the Prevention of Marine Pollution from Land-Based Sources (the 'Paris Convention'). Each of these Conventions established a Commission, which should meet at regular intervals and whose duties include overall supervision over the implementation of the Conventions. These Commissions are known as the Oslo Commission and the Paris Commission.

By the end of the 1980s, these two Conventions were beginning to show their age. Environmental concerns and policies had developed significantly in this period and the countries involved agreed that the time was ripe to update and modernize this framework. Consequently, a new Convention was adopted at the joint ministerial meeting of the Oslo and Paris Commissions in Paris in September 1992, which merges and modernizes the existing Oslo and Paris Conventions. When it enters into force this Convention – the Convention for the Protection of the Marine Environment of the North-East Atlantic ('OSPAR Convention') – will replace the present Conventions.

Nutrients are indirectly addressed by the Oslo Convention – for example, as regards exercising control over the dumping at sea of sewage sludge, an activity which will be terminated completely by the end of 1998. However, this organization does not concern itself specifically with nutrient issues and will not be referred to again in this chapter. Similarly, no further reference will be made to the OSPAR Convention of 1992, because this treaty has not yet entered into force. It should, however, be borne in mind that the OSPAR Convention will become the relevant international legal instrument when it does come into force, which may be towards the end of 1997, depending on the progress of the ratification process by the signatories.

The maritime area covered by the three Conventions is the same, i.e. the north-east Atlantic, extending westwards to the east coast of Greenland, eastwards to the continental North Sea coast, south to the Straits of Gibraltar and northwards to the North Pole. This area, of course, includes the North Sea.

The Contracting Parties to the Paris Commission are Belgium, Denmark, the European Economic Community (EEC), France, Germany, Iceland, Ireland, the Netherlands, Norway, Portugal, Spain, Sweden and the UK.

The broad objective of the Paris Commission is to supervise the implementation of the Paris Convention – that is, to take all possible steps to prevent pollution of the sea by adopting, individually and jointly, measures to combat marine pollution from land-based sources.

The Paris Commission's interest in nutrient-related issues began more than 10 years ago, when the incidence of unusual algal blooms was first notified to the Commission in 1983. In 1985 the Paris Commission reviewed information about eutrophication and algal blooms which had been submitted by several contracting parties and discussed a possible cause–effect relationship between these blooms and increased concentrations of nutrients in coastal waters.

Although nutrients are not specifically mentioned in the Paris Convention, Article 9 of the Convention does provide a mechanism whereby, when pollution from land-based sources originating from the territory of one contracting party is likely to prejudice the interests of other states, the contracting parties concerned undertake to consult each other with a view to negotiating a cooperation agreement.

Consequently, a consultation meeting was held in Denmark in the autumn of 1985, at which most contracting parties provided detailed information about the situation in their coastal waters at that time. Following this consultation meeting, the meeting of the Paris Commission in 1986 agreed to establish a working group on Nutrients (NUT). The first meeting of the NUT Working Group was held later in 1986, and NUT has met annually since then.

THE NORTH SEA CONFERENCE PROCESS

Another vital part of the institutional framework as regards nutrients in the marine environment in this region is the North Sea Conference process. There have been a series of North Sea Conferences held at ministerial level, as follows:

- 1984   Bremen
- 1987   London
- 1990   The Hague
- 1993   Copenhagen (Intermediate Ministerial Meeting, where nutrients were a topic)
- June 1995   Esbjerg

The key product of these conferences has been a series of ministerial declarations. Although these declarations do not constitute international law as such, they are powerful political statements of intent, which are fully considered in the more legalistic framework of the Paris Commission. Although not all of the parties to the Paris Commission are also North Sea countries, the Paris Commission has nevertheless taken on board many of the policies which have been developed in the North Sea Conference framework and has developed programmes and measures with a view to fulfilling objectives of the North Sea Conferences which relate to nutrients. Thus, the political impetus and public profile which is provided by the North Sea Conference process can be described as being complementary to the implementation of these policies by the Paris Commission.

MEASURES ADOPTED

The NUT working group has developed a number of measures, which have then been formally adopted by the Paris Commission.

- *PARCOM Recommendation 88/2* on the reduction in inputs of nutrients to the Paris Convention area. This recommendation established a 50% reduction target for inputs of nitrogen (N) and phosphorus (P) into problem areas between 1985 and 1995. This measure was adopted following the Second North Sea Conference in 1987, at which ministers agreed on such reduction targets in the North Sea area.
- *PARCOM Recommendation 89/4* for a coordinated programme for the reduction of nutrients. In this recommendation, certain sectors, including agriculture, were identified for action.
- *PARCOM Recommendation 92/7* on the reduction of nutrient inputs from agriculture into areas where these inputs are likely, directly or indirectly, to cause pollution and to this end to reduce:

- ammonia volatilization;
- leaching of N, mainly nitrate;
- leaching, runoff and erosion losses of P
- farm-waste discharges.

Fifteen measures were identified for the purpose of reducing leaching, runoff and erosion losses of P. These are described below.

A particularly important feature of the Paris Commission's work in recent years has been the recognition that not only is it important to prepare and adopt suitable measures to address pollution problems, but it is also essential to ensure that they are implemented in practice. In 1995, the Paris Commission published the last implementation report relating to PARCOM Recommendation 88/2. The contracting parties concerned reported as follows with respect to progress towards meeting the reduction target for P inputs.

- Belgium, Denmark, Germany, the Netherlands, Norway, Sweden and Switzerland reported that they expected to reach a reduction in P inputs of the order of 50% by the end of 1995 into areas where these inputs were likely, directly or indirectly, to cause pollution.
- France expected to reach a reduction in P inputs into such areas of the order of 25% by the end of 1995.

In the work to establish measures to reach the 50% reduction target, it soon became clear that the agriculture sector would be not only the most important sector to handle but also the most difficult. In 1992, as a first step to reducing nutrient losses from the agriculture sector, the Paris Commission adopted PARCOM Recommendation 92/7, in which contracting parties agreed that, in applying all, or some, of these measures, preference should be given to those which involve reduction of emissions at source and also that the measures could include regulatory and/or advisory measures and financial instruments. This recommendation also makes the point that this list of measures is not exhaustive.

The 15 measures specified in this recommendation as being relevant to reducing leaching, runoff and erosion losses of P are as follows.

1. *Timing of non-application of manure.* The utilization efficiency of the nutrients contained in animal manures is dependent upon the time of application, i.e. the manure should be applied shortly before or during the early growing season when crops are able to assimilate the nutrients. The spring application may be promoted by prohibition of manure application at certain times, e.g. during autumn or winter. Alternatively, certain months during spring or early summer may be set for the more appropriate time of application.

2. *Improved application techniques of manure.* The gaseous loss of ammonia from field-applied manure has direct implications for the efficient use of N in animal manures. The use of specialized equipment, for example trailing hoses, mainly for the application of slurry or urine in crops having a dense canopy, reduces the volatile losses substantially. Direct injection into soil or immediate ploughing down of the manure into bare soil are efficient means of reducing ammonia volatilization. Application techniques which ensure an even distribution of manure on the field allow crops to take up a larger fraction of the manurial nutrients, which reduces losses.

3. *Ground conditions.* Manure should not be applied on frozen or snow-covered soil, on steep slopes or under conditions which do not rule out the risk for surface runoff of the manure (such as close to watercourses). Ground conditions, such as infiltration capacity of the soil, also greatly influence ammonia volatilization and should be considered before application of manure. Soil texture is an important parameter to take into account. Sandy soils are prone to higher losses than clay soils and may be candidates for set-aside programmes, protection zones, etc.

4. *Fertilizer advice programmes.* Advice and information to farmers, along with the setting up of fertilizer planning, including animal manure, are important means of promoting efficient use of fertilizer. Prognosis tools for nutrient application may be an important addition to fertilizer planning. Prognosis tools for nutrient application include assessment of soil-nutrient contents, soil-nutrient mineralization potentials and/or the use of calculation methods.

5. *Winter crop cover.* Land without plant cover during the autumn and winter will lose more N by leaching and be at greater risk of nutrient loss by runoff and erosion compared with land with plant cover.

6. *Restrictions on tillage in autumn.* Tillage operations, such as ploughing or disc-harrowing, in autumn loosens the soil and liberates soil particles, which in turn are exposed to runoff. The migration of particles and soil organic matter down the soil profile and possibly into drainage pipes may also be increased due to tillage practices in autumn. Autumn tillage also increases the mineralization of N in the soil and therefore also the potential losses if, for instance, no autumn green cover is established.

7. *Green fallowing.* The leaching and erosion losses from land left in green fallow are much lower than losses from arable land or grazed grassland. Conversion of current arable land into green fallow will, in the long term, reduce losses.

8. *Afforestation programmes.* Leaching and erosion losses from forests are generally much lower than losses from arable land.

9. *Vegetation zones along watercourses.* If sufficiently wide, vegetation strips along watercourses may prevent nutrients in surface runoff from entering the watercourses. Stabilization of banks is an additional effect of vegetation strips.

10. *Correct irrigation practices.* Excess irrigation will result in increased runoff and/or increased leaching losses.

11. *Measures to reduce erosion.* Erosion losses can be reduced by sowing alternate strips of different crops, by establishment of plant cover to protect soils against the impact of raindrops, by planting strips of woodland, by leaving areas in edges of fields under green cover, by appropriate cultivation practices, etc. In certain cases, drainage reduces erosion losses. However, it may carry the risk of increased leaching.

12. (a) *Animals per hectare.* High application rates of animal manure are crucial to the losses of N from agricultural land. Limitation of the livestock density on a farm basis is one indirect means of setting standards for application rates. The parameter limiting the livestock density may be chosen to be the number of livestock units on average per hectare. If the livestock density on individual farms exceeds a certain number of animals (e.g., two livestock units per hectare), agreement should be made on the reduction of the number of animals. Alternatively, the surplus manure may be exported to neighbouring farmers or processing plants. Conversion factors from the number of animals to livestock units (and maximum limits to livestock density per hectare) are likely to differ among countries due to local soil properties, to the rearing of different species and to differing feeding practices.

(b) *Nutrients per hectare.* The amount of nutrients in animal manure that is allowed to be applied to land each year, taking into account crop needs, can be used to regulate applications of animal manure and sewage sludge. This is usually based on the total content of N and/or P in organic manures allowed per hectare per year on a farm basis.

13. *Assessment of nutrients in manure.* It is necessary to know as precisely as possible the content of nutrients in the animal manures in order to take full account of the manure as fertilizers and to be able to derive accurate fertilizer planning for the individual fields. Information on the nutrient content of a variety of manures may be made available to farmers by normative values. Alternatively, the total amount of N and P produced in manure by animals on the farm can be calculated. Assessment of nutrient losses in animal housing and storages allows calculation of the amount of N and P in manure in storage. Individual farmers may be advised to measure the nutrient concentration in the slurry before field application.

14. *Manure distribution.* The allocation of a surplus of manure among farms is important to the efficient use of the nutrients in manure. The distribution of manure among farms may be promoted by use of large lorries transporting the manure from, e.g. a dairy farm to the field of a crop farmer. Contracts between manure producers and receivers will support the distribution among farms and thus reduce pollution. Control of farmers with contracts by the authorities is one way to ensure good agricultural practice in managing surplus manure. The use of central storages or slurry banks will further allow crop farmers to make proper use of surplus manure produced within animal husbandry.

15. *Requirements for manure storage capacity.* Adequate storage capacity for animal manures enables farmers to apply the manure to land at times when the risk is minimized of both direct pollution and nutrient losses by runoff, leaching and volatilization.

CONCLUSION

Discussions about future work on nutrients in the framework of the Paris Commission took place at the 1995 meeting of the NUT working group, which was held in London from 19 to 22 September. The NUT established its draft work programme for the coming year, which was forwarded to the 1996 meeting of the Commissions in June 1996 for adoption. This draft work programme which will be updated is summarized in Table 18.1. Work related to nutrients from agriculture is as follows.

1. *Balanced fertilization.* Efforts will continue towards the establishment of an operational definition of balanced fertilization. This involves finding a form of words that can reconcile the tension between the needs of agriculture on the one hand and the protection of the environment on the other. Whether it will be possible to find a form of words which is not so vague as to have little practical meaning and which can be accepted by all parties as being of general application remains to be seen. Some parties are more pessimistic than others about the prospects of success and consider that a mechanism is needed which can be applied flexibly enough to enable the numerous relevant variations in circumstances to be taken into account – e.g. soil and weather conditions, type of crop, nature and use of surrounding waters.
2. *Mineral surpluses.* Contracting parties will be providing more national data and surplus calculations, including comparative calculations for different catchment areas.
3. *Calculation methodologies.* Work on methods of calculating nutrient discharges, emissions and inputs will be continued, with a view to ensuring that data collected by the countries concerned are soundly based and are comparable. The NUT is giving priority to work towards the development of a common methodological framework for assessing nutrient inputs. Better validation will be sought by combining source assessment and riverine measurement approaches. Data gathering, modelling and catchment mapping will also be developed.

More generally, but still relevant to agriculture and to P, NUT will also continue work on a strategy to combat eutrophication, including consideration of how the Common Procedure for the Identification of the Eutrophication Status of the Maritime Area of the Convention – which is still under development – should be incorporated in this strategy. In this connection, NUT has proposed that an *ad hoc* working group should be convened in

Table 18.1. Draft work programme for the Working Group on Nutrients, 1996/1997.*

Subject (lead country in parentheses)	Activity (priority to be finalized by NUT, 1996)

1. Agriculture
a. BEP, balanced fertilization (Norway)
 	
b. Surplus
 	
Calculation/methodologies (UK) (see also point 4)
	a. Finalize operational definition of balanced fertilization; further work on measures
	b. Examine reports on
	  (i) national data for application rates (fertilizer/manure), surplus calculations
	  (ii) calculations of mineral surplus for two catchment areas/administrative regions
	c. Further work on calculation methodologies

2. Recommended measures in the Report on Nutrients in the Convention Area (§57 of Part B) (Germany)	Examine draft PARCOM measure/s
3. Establish emission inventories for nutrient sources a. Municipal and industrial sewage, and direct and indirect industrial discharges (France) b. Forestry (Finland) c. Aquaculture (Germany) d. Traffic (Germany) e. Atmospheric emissions from sources other than traffic (no lead country)	a., b., and c. Examine reports prepared by the lead countries d. Examine an evaluation of nutrient emissions from traffic e. Secretariat to provide an overview of information available in other forums
4. Development of common methodologies for the quantification of inputs, discharges, losses and emissions of nutrients (UK)	Review the need for harmonization of existing methodologies or preparation of new methods
5. Definition of nutrient-related terminology	To consider further work in the light of the conclusions of PRAM 1996† and OSPAR 1996†

6. Making the strategy to combat eutrophication operational	Consider further work in the light of the outcome of the *ad hoc* working group on a conceptual framework to combat eutrophication and case-studies on balanced fertilization
7. Research and development programmes to promote development of techniques for reducing discharges of nutrients from all sources	Consider and make proposals
8. To consider the justification for further reduction targets for the different sources of ammonia emissions	If appropriate, make proposals for new measures, based on justifications submitted by contracting parties
9. Establish methodologies for assessing the effectiveness of the measures taken to reduce nutrient inputs to the aquatic environment	Discuss common methodologies in the light of information received
10. Preparations for the 1997 Ministerial Meeting of the Oslo and Paris Commissions	

* Oslo and Paris Conventions for the Prevention of Marine Pollution Working Group on Nutrients (NUT), London, 19–22 September 1995.
† 1996 meeting of the Programmes and Measures Committee.

early 1996 to discuss the framework for these various concepts and how they might be put into practice, taking the North Sea as the first case-study.

Finally, in view of the limited success in achieving the reduction targets specified in PARCOM Recommendation 88/2 and in the light of the reaffirmation by ministers at the North Sea Conference in Denmark in June 1995 of their continued commitment to substantial reductions in inputs of nutrients to the North Sea, NUT also began consideration of further measures with a view to securing further reductions in nutrient inputs. So far there has only been an introductory debate in the PARCOM framework of this issue, including the possibility of new reduction targets. This debate will continue.

19 Phosphorus Loss in Runoff, Leaching and Erosion

SURFACE RUNOFF

19.1 Phosphorus Fractionation in Grassland Hill-Slope Hydrological Pathways

R.M. Dils and A.L. Heathwaite
Department of Geography, Sheffield University, Sheffield S10 2TN, UK

Introduction

The form of phosphorus (P) in runoff from agricultural land is determined by the P source, flow routeing and chemical transformations during transport. As the hydrological pathway determines the rate of water movement, and, in turn, the contact time between 'new' water and the soil, P forms present in infiltrating water are modified depending on the route taken. Routes that transport rainwater from hill slopes to streams with relatively little modification of the initial P form include: infiltration–excess overland flow, saturated surface runoff from variable and partial source areas and preferential flow through soil macropores and artificial subsurface drains. Water percolating through the unsaturated zone takes slightly longer to reach the stream and saturated through flow is considerably delayed. Consequently, different hydrological pathways might be characterized by different P forms and the amount of P transported. Because P fractions vary in their bioavailability, the relative proportion of each fraction in different pathways will influence the overall runoff composition and its impact on the receiving stream. Some of these aspects were studied at increasing spatial scales: the experimental plot, the hill-slope transect and the small catchment.

Site and methods

A small (120 ha) mixed agricultural catchment near Ashby de la Zouch, Leicestershire, was selected and soil-water samples were obtained from the plot and hill-slope transect during base- and storm-flow conditions. The soils are freely draining, fine loams overlying impermeable clays which are underlain by tile drains at 0.7 m. At the catchment outlet, discharge was monitored continually and an automatic sampler sampled runoff that occurred at stream-flow levels exceeding base-flow. To monitor P transported in hill-slope hydrological pathways, a series of overland flow troughs, macropore water samplers, dip wells and piezometers were installed in banks at top, mid- and bottom slope positions along two hill-slope transects (150 m in length) and three hydrologically isolated experimental plots (10 m × 1.5 m). Regular sampling was combined with more intense sampling during storm events. Results are presented for the period September 1994 to March 1995.

Results

Catchment scale. Stream base-flow P concentrations remained fairly constant; total P (TP) values ranged from 0.12 to 0.36 mg P l^{-1}, with an equal distribution between the soluble and particulate phases. These concentrations exceed the recommended value of 0.1 m P l^{-1} for rivers set to prevent the eutrophication of receiving lakes (Department of the Environment, 1993). The importance of storm events to P export was observed consistently, peak loadings exceeded pre-storm values by a factor of 20. Intensive sampling of three storm events in autumn demonstrated an 'exhaustion' effect; both the magnitude and the P fractions in runoff decreased with consecutive storms. The first storm (55.5 mm) had a peak TP concentration of 2.0 mg P l^{-1} and the soluble inorganic fraction dominated (76–86%). The large soluble P losses probably occurred as a result of the 'flushing out' of P mineralized during the long, dry summer, when there was no aqueous transporting medium (Jordan and Smith, 1985). In subsequent runoff, TP concentrations decreased to about 1.2 mg P l^{-1}, dominated by the particulate fraction (50–77%). The lower TP concentrations may have resulted from a depleted soil P-store, which could be leached, or from a lower rainfall intensity.

Experimental-plot and hill-slope-transect scale. Preliminary results indicate that subsurface flow, particularly through soil macropores and tile drains, may be important pathways for the transport of P. In macropore water samples taken over the winter period, molybdate-reactive P (MRP) concentrations ranged from ~0.001 to 0.48 mg P l^{-1} and, in a third of all samples, TP exceeded 1 mg P l^{-1}. Runoff through a tile drain had a peak concentration of 0.95 mg TP l^{-1} during a 10.0 mm winter-storm event, compared with the respective stream concentration of 0.64 mg TP l^{-1}. As expected, overland flow concentrations were large during winter-storm events; MRP averaged 1.8 mg P l^{-1} and TP 2.48 mg P l^{-1}. In all samples, the soluble fraction was dominated by the inorganic component (61–98%).

Acknowledgement

This research was funded by the Ministry of Agriculture, Fisheries and Food, Award No. AE 8750.

References

Department of the Environment (1993) *Methodology for Identifying Sensitive Areas (Urban Waste Water Treatment Directive) and Designating Vulnerable Zones (Nitrates Directive) in Northern Ireland.* Consultative Document, April 1993, Environment Service, Department of the Environment in Northern Ireland, Belfast, 59 pp.

Jordan, C. and Smith, R.V. (1985) Factors affecting leaching of nutrients from an intensively managed grassland in County Antrim, Northern Ireland. *Journal of Environmental Management* 20, 1–15.

19.2 Soil-derived Phosphorus in Surface Runoff from Grazed Grassland
P.M. Haygarth and S.C. Jarvis
Soil Science Group, Institute of Grassland and Environmental Research, North Wyke, Okehampton, Devon EX20 2SB, UK

Introduction

Inputs of phosphorus (P) to grassland soils as inorganic fertilizer can exceed 40 kg P ha^{-1} year^{-1} and P is also recycled in manures and livestock excreta. Grasslands, therefore, can contain a large reservoir of P, some of which is vulnerable to loss in surface and subsurface pathways. Because 70% of the UK land area is in grassland that occupies sloping ground in high-rainfall areas, the potential for P loss and the effects of such losses on the eutrophication of surface waters may be significant on a national scale. Surface and estuarine water concentrations of P in excess of 10 µg P l^{-1} are all that are required for algal growth (Sharpley and Smith, 1989). Results on the frequency, magnitude and forms of P determined in water lost from 1 ha, hydrologically isolated, grazed plots are presented.

Methods

One-hectare, hydrologically isolated, field plots were established in 1982 on old, unimproved pasture on land sloping between 5 and 10%, in Devon, south-west England (SX 650995). Gravel-filled ditches to 30 cm at the boundaries collect water, which represents surface runoff and surface lateral flow to 30 cm combined. This 'surface runoff' water is channelled through a continuous level-recording reservoir before being exported through V-notch weirs (Talman, 1983). Annual inputs of fertilizer and slurry both supply 16 kg P ha^{-1}. Samples were collected from the drainage weirs during spring and autumn 1994, and were stored and filtered (Haygarth *et al.*, 1995) before molybdate-reactive P (MRP) was determined using a Tecator 5020 flow injection analyser with an autosampler (Tecator Ltd., Method Application ASN 60–03/83). Total P (TP) was determined on unfiltered waters, using a persulphate method, based on Eisenreich *et al.* (1975).

Results and discussion

Losses of TP and MRP in runoff were determined in up to ten events on seven plots during 1994 (Table 19.1). On 10 May, 16 kg P ha^{-1}, as triple superphosphate, was applied in dry conditions after a soil moisture deficit of 38 mm had accumulated. Then, in the period 14–18 May, over 50 mm of rain fell, which resulted in drainage equivalent to 20 mm. This event resulted in the largest P loss recorded during the year; the maximum concentration exceeded 1700 µg P l^{-1}, equivalent to a maximum loss of over 18 g P h^{-1}. The resulting loss was estimated to represent over 0.5 kg P ha^{-1}, a not inconsiderable fraction of the triple superphosphate added 6 days earlier. The timing of P additives, both as fertilizer and slurry, in combination with climatological occurrences, is crucial in influencing the loss of P in runoff from grassland. During such low-frequency but high-intensity events, excretal returns from cattle grazing the plots may have influenced P losses also. In addition, poaching effects, soil-moisture status and rainfall intensity may also be important and represent an interaction between local geomorphology, climate and management. Therefore, modelling such

Table 19.1. Summary of data for molybdate-reactive phosphorus (MRP), total phosphorus (TP) and TP loading determined in surface runoff from seven 1 ha grazed grassland plots during eight (TP) and ten (MRP) rainfall events throughout 1994.

	MRP (µg P l^{-1})	TP (µg P l^{-1})	TP load (mg P ha^{-1} h^{-1})
Mean	40	122	680
Standard error	19	32	339
Minimum	< 10	26	3.7
Maximum	1,296	1,773	18,510
Number observations	70	56	56

runoff losses is difficult, because of the factors listed here, especially low-frequency, high-intensity rainfall events. If the extreme events are excluded and data from the higher-frequency, lower-magnitude events are used, 14% of the relationship between TP concentration and surface discharge was explained by the following empirical model, which describes the straight line in Fig. 19.1:

TP (µg l^{-1}) = 58 + 42 × surface discharge (l s^{-1})

The mechanisms that release P to surface runoff during low-intensity events are poorly understood and further mechanistic studies are required before improved water- and agricultural-management strategies can be implemented with confidence to reduce P transfer from soil to water (Haygarth and Jarvis, 1996).

Surface runoff P was monitored during a single storm event using a high-

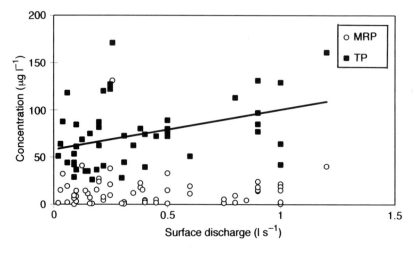

Fig. 19.1. Relationship between TP and MRP and surface discharge, determined in surface lateral flow to 30 cm from grassland plots, excluding the low-frequency, high-magnitude events.

resolution, replicated, sampling programme. In summary, TP concentration changes resemble the shape of the discharge hydrography but displaced by *c.* 5 h (Fig. 19.2). There was a closer correlation of TP load with temporal changes in drainage flow. In contrast, MRP did not appear to follow the discharge curve and remained relatively unaffected by discharge, perhaps illustrating the dominance of particulate and organic P forms in storm runoff. Colloidal and particulate MRP has been found to be an important constituent of total reactive P in soil runoff water, and these forms may be readily transported through surface waters to be bioavailable to algae (Haygarth *et al.*, Chapter 20, this volume).

Conclusions

Losses of P in runoff are affected by the timing of management practices, such as the application of fertilizer and slurry, and grazing in relation to rainfall. In the extreme, 0.5 kg P ha^{-1} was estimated to have been lost during one event, but further research is required to confirm this estimate and the mechanisms that lead to P loss. However, on a national scale, the transfer of P from grassland to surface waters is a major source of diffuse P pollution in the UK, especially because of the large area of grassland and the geomorphological and climatological conditions in which grasslands occur.

Acknowledgements

The authors are grateful to the Ministry of Agriculture, Fisheries and Food, London, who fund this work.

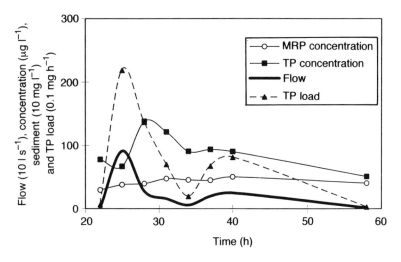

Fig. 19.2. TP concentration, MRP concentration and TP load in relation to flow, determined in surface lateral flow to 30 cm from grassland plots during a single storm, December 1994.

References
Eisenreich, S.J., Bannerman, R.T. and Armstrong, D.E. (1975) A simplified phosphorus analytical technique. *Environmental Letters* 9, 45–53.
Haygarth, P.M. and Jarvis, S.C. (1997) Surface runoff of soil derived phosphorus from grassland lysimeters. *Water Research* 31, 140–148.
Haygarth, P.M., Ashby, C.D. and Jarvis, S.C. (1995) Short-term changes in the molybdate reactive phosphorus of stored soil waters. *Journal of Environmental Quality* 24(6), 1133–1140.
Sharpley, A.N. and Smith, S.J. (1989) Prediction of soluble phosphorus transport in agricultural runoff. *Journal of Environmental Quality* 18, 313–316.
Talman, A.J. (1983) A device for recording fluctuating water tables. *Journal of Agricultural Engineering Research* 28, 273–277.

19.3 Agricultural Phosphorus Load and Phosphorus as a Limiting Factor for Algal Growth in Finnish Lakes and Rivers

O.P. Pietiläinen
Finnish Environment Agency, PO Box 140, FIN-00251 Helsinki, Finland

Introduction
Agriculture comprises the biggest source of phosphorus (P) and nitrogen (N) into Finnish watercourses. About three-quarters of the total P (TP) load from agriculture is bound to mineral soil particles and only 5% of this P is potentially available for algal growth (Ekholm, 1994). However, dissolved reactive P (DRP) appears to be totally bioavailable. Consequently, the amount of DRP from agriculture is of crucial importance in controlling eutrophication of surface waters in Finland. This study estimated the amount of DRP and TP load from agricultural areas to estimate whether primary production in Finnish lakes and rivers is limited by P or N or both.

Material and methods
The amounts of P lost were measured in four small agricultural drainage basins (agricultural fields 39–100%) and three afforested basins (fields 0–2%). The detailed characteristics of the basins and the methods of sampling and chemical analyses used have been described by Pietiläinen and Rekolainen (1991). To evaluate whether P or N was the limiting factor for algal growth, upper-layer samples (0–2 m) were taken in summer 1982–1991, when there was little water flow, from 1090 pelagial lake stations and from 554 stations in rivers and straits (Pietiläinen and Kauppi, 1993).

Results and discussion
The DRP load from the mainly agricultural basins was 0.15–0.18 kg P ha^{-1} year^{-1} and TP load 0.42–0.86 kg P ha^{-1} year^{-1}, which included loads from the cultivated land of 0.15–0.41 kg DRP ha^{-1} and 0.77–1.7 kg TP ha^{-1}. These were about one order of magnitude higher than the loads from forested basins (DRP 0.01–0.04 kg DRP ha^{-1} year^{-1} and TP 0.05–0.11 kg TP ha^{-1} year^{-1}). Of the TP, DRP was, on average, 25% from the cultivated fields and 19% in the forested basins.

The following criteria were used when deciding whether the primary production was limited by P or N in lakes and rivers: P was considered the limiting nutrient when the ratio of dissolved inorganic N (DIN) to dissolved inorganic P (DIP) was > 12, and N when the ratio was < 5 (Forsberg *et al.*, 1978).

The number of P-limited lakes was 488 (45% of all the lakes studied) and N-limited lakes 291 (27%), while 311 (28%) of the lakes were limited by both nutrients. The number of P-limited rivers was 227 (41%), N-limited rivers 137 (25%) and rivers limited by both nutrients 190 (34%). Phosphorus limitation in lakes was often strong, i.e. the DIN:DIP ratio was much higher than 12, while in most cases N limitation was weak, i.e. the ratio was only slightly smaller than 5. Therefore, according to this approach, the importance of N as a limiting nutrient in Finnish surface waters is probably somewhat overestimated.

Finnish lakes and rivers can be classified into six groups according to their inorganic nutrient concentrations and nutrient limitation. Group 1 consists typically of large lakes which are strongly P-limited (Fig. 19.3). These lakes have a surplus of nitrate (NO_3)-N throughout the year, as a result of a high ratio of N to P in the incoming water and a low denitrification capacity. In group 2, P limitation is somewhat weaker than in group 1 and this group includes very oligotrophic and clear-water lakes. Most small humic lakes in Finland belong to group 3, in which both nutrients can often simultaneously limit algal growth. Group 4 includes some very humic lakes in the north. Group 5 consists of lakes heavily loaded with nutrients from agriculture, industry or domestic sewage waters. Most of these lakes are shallow and the internal P load can be significantly

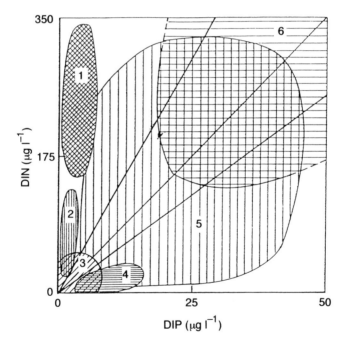

Fig. 19.3. A schematic representation of the classification of Finnish fresh waters into six groups, according to their inorganic nutrient concentrations and nutrient limitation during the growing season (May–September). The shaded areas indicate typical variations of DIN and DIP concentrations for each group. The middle line indicates the DIN:DIP ratio of 7:1, the upper line the ratio of 12:1 and the lower line the ratio of 5:1. The more the ratio exceeds 12, the stronger is the P limitation, and the more the ratio fall below 5, the stronger is the N limitation. More detailed information is in the text.

higher than the external P load (Knuuttila *et al.*, 1994), which results in fast changes in DIN:DIP ratios. Consequently, in these lakes, both P and N often exceed the limits for algal growth simultaneously, and pollution-control measures should be focused equally on both nutrients. Several agriculturally loaded rivers belong to group 6. They are loaded so heavily that low inorganic nutrient concentrations occur only occasionally, during low-flow periods. In these rivers, if algal growth is limited, it is because of other factors, like the lack of light.

References
Ekholm, P. (1994) Bioavailability of phosphorus in agriculturally loaded rivers in southern Finland. *Hydrobiologia* 287, 179–194.
Forsberg, C., Ryding, S.-O., Claesson, A. and Forsberg, A. (1978) Water chemical analyses and/or algal assay: sewage effluent and polluted lake water studies. *Proceedings of the International Association of Theoretical and Applied Limnology* 21, 352–363.
Knuuttila, S., Pietiläinen, O.-P. and Kauppi, L. (1994) Nutrient balances and phytoplankton dynamics in two agriculturally loaded shallow lakes. *Hydrobiologia* 275/276, 359–369.
Pietiläinen, O.-P. and Kauppi, L. (1993) DIN/DIP ratios in Finnish lakes and rivers – useful information for water pollution control? *Vesitalous* 6, 1–7 (in Finnish).
Pietiläinen, O.-P. and Rekolainen, S. (1991) Dissolved reactive phosphorus and total phosphorus load from agricultural and forested basins to surface waters in Finland. *Aqua Fennica* 21, 127–136.

19.4 Increase in Soluble Reactive Phosphorus Transport in Grassland Catchments in Response to Soil Phosphorus Accumulation
R.V. Smith,[1] R.H. Foy[1] and S.D. Lennox[2]
[1]*Agricultural and Environmental Science Division,*
[2]*Biometrics Division, Department of Agriculture for Northern Ireland, Agriculture and Food Science Centre, Newforge Lane, Belfast BT9 5PX, UK*

Introduction
Enrichment by phosphorus (P) is the primary cause of eutrophication in fresh waters. With the control of P point sources, the contribution of agricultural P is receiving increasing attention (Sharpley *et al.*, 1994). Recent studies in the Lough Neagh catchment area (4400 km^2) in Northern Ireland have revealed an upward trend in soluble reactive P (SRP) loadings over a 17-year period within the range 13.0 to 14.5 g P ha^{-1} year^{-1} (Foy *et al.*, 1995). This increase was postulated to originate from P accumulation in soils caused by an imbalance between agricultural inputs and outputs. The present study attempted to identify whether there had been a comparable temporal increase in SRP concentration in drain flow from a small grassland catchment which had no domestic or industrial sources of P and an absence of farm buildings.

Methods
The land drain used drained 6 ha of very gently sloping, gleyed soil, developed on a basaltic till belonging to Greenmount Agricultural and Horticultural College, County Antrim. Soluble reactive P concentrations in drain flow measured for a two-year period in January 1981–December 1982 were compared with those for January 1990–December 1991. Because of the observed variability

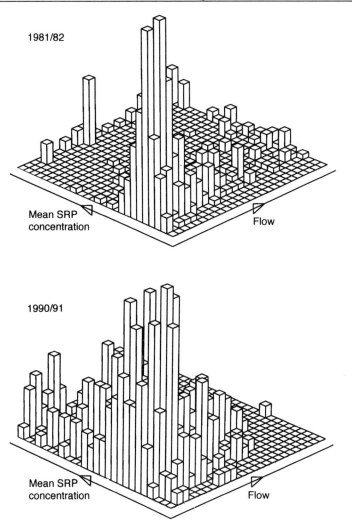

Fig. 19.4. Three-dimensional histograms of all daily soluble reactive P (SRP) concentrations and flows from the Greenmount Agricultural and Horticultural College catchment during 1981–1982 and 1990–1991. Mean SRP concentration increases are shown in intervals of 4.0 µg P l^{-1} and flows at intervals of 2 m^3 ha^{-1} day^{-1}.

in Olsen P values, input/output P budgets were used to predict changes in soil-P status.

Results and discussion

The P accumulation rate for the Greenmount catchment of approximately 24 kg ha^{-1} year^{-1} suggested an increase in Olsen P of 1.0 mg P kg^{-1} year^{-1}. Soluble reactive P concentrations in drain flow from the catchment measured on a daily basis for the 2-year periods (Fig. 19.4) show that the median concentration had increased by 10 µg P l^{-1} in 1990–1991 compared with 1981–1982. This difference was only apparent in mean concentrations for the two periods after data associated with high-flow events, which were more frequent in 1981–1982, were excluded from the comparison. This rate of increase of 1.1 µg P l^{-1} year^{-1} was interpreted as reflecting an increase in SRP concentration in soil solution and is comparable to the increase in background SRP concentration of 1.5 µg ± 0.54 µg P l^{-1} year^{-1} which was reported recently for a 17-year period from diffuse sources in the much larger catchment of Lough Neagh (Foy *et al.*, 1995).

Expressed on a catchment-area basis, the annual rate of background increase of SRP in the surface waters, i.e. streams and rivers of the Lough Neagh catchment, corresponds to a loss of only about 0.01 kg P ha^{-1} year^{-1}. This compares with the annual accumulation rate of total P in the soil of 10 kg P ha^{-1} year^{-1}. Although the resulting SRP increase in Lough Neagh appears to be small, it is very significant in terms of the ecology of the lake. It has been sufficient to reverse the decrease in Lough Neagh algal populations which followed decreases in P loadings from sewage-treatment works since 1981.

Bailey (1994) reported that in 1993 almost 50% of soils tested for farmers in Northern Ireland had P contents higher than the recommended index 2 range and many were seriously overloaded. He suggested that the key to managing soil fertility is to balance the amounts of nutrients applied in fertilizers and manures against the actual needs of crops and soils. In Holland, they have approached this problem of overuse of P by defining their objective as a 'loss standard' for farms of 5 kg P ha^{-1} year^{-1} by the year 2000. This 'loss standard' is the P surplus on holdings (i.e. inputs – outputs of P). In Northern Ireland the present 'loss standard' averages 10 kg P ha^{-1} year^{-1}. Achieving a 'loss standard' of 5.0 kg P ha^{-1} year^{-1} in Northern Ireland would reduce fertilizer use by 85%, with savings of £7,650,000 per annum, together with considerable environmental benefits. However, in the future, an environmentally acceptable 'loss standard' may well need to be set below 4.0 kg P ha^{-1} year^{-1}.

References

Bailey, J.S. (1994) Nutrient balance: the key to solving the phosphate problem. *Topics: The Journal of the Milk Board for Northern Ireland* November, 16–17.

Foy, R.H., Smith, R.V., Jordan, C.J. and Lennox, S.D. (1995) Upward trend in soluble phosphorus loadings to Lough Neagh despite phosphorus reduction at sewage treatment works. *Water Research* 29, 1051–1063.

Sharpley, A.N., Chapra, S.C., Wedepohl, R., Sims, J.T., Daniel, T.C. and Reddy, K.R. (1994) Managing agricultural phosphorus for protection of surface waters: issues and options. *Journal of Environmental Quality* 23, 437–451.

19.5 Phosphorus Loss to Water from a Small Low-Intensity Grassland Catchment

H. Tunney, T. O'Donnell and A. Fanning
Teagasc, Johnstown Castle Research Centre, Wexford, Ireland

Introduction

In New Zealand, total phosphorus (P) losses from pasture land of about 1.67 kg ha^{-1} year^{-1} were more than ten times larger than losses from pine forest and unimproved land (Cooper and Thomsen 1988). In Ireland, there is increasing evidence that losses of water-soluble P of 1 kg ha^{-1} year^{-1} occur on fertilized grassland, which is equivalent to approximately 0.2 mg P l^{-1} in drainage water (Hayward et al., 1993). This loss can cause serious eutrophication in the water but is of little consequence in agronomic terms. This study aimed to measure P losses in overland flow from a small catchment not receiving fertilizer to estimate background P losses from low-intensity-managed grassland.

Materials and methods

Drains were reopened and cleared in 1994 in a small, 1.4 ha catchment, located on Warren field at Johnstown Castle. The site, on a gley soil with poor infiltration, was reclaimed originally from woodland in the 1970s and sown to grass, which has received no fertilizers and has been cut and the herbage removed. The catchment has a gentle slope of about 4% to an open drain, lined with butyl rubber, at one side of the site; the other three sides of the catchment are isolated by a drain 1 m deep. The overland flow is channelled through this lined drain to a tank with a V-notch weir and automatic flow recorder and sampler. The water flow from tile drains, 1 m deep, which includes a small continuous background flow from a spring in the field, is also measured. Water samples are taken once per week, both from the tank measuring the drain flow and from the automatic sampler on the overland flow tank, and analysed for molybdenum-reactive P (MRP). The results are based on these flow measurements, plus evapotranspiration and rainfall measurements, recorded 2 km away at Johnstown Castle.

The measurements of overland flow reported here were recorded between November 1994 and April 1995 inclusive, which was the only overland flow in the period July 1994 to June 1995.

Results

Water monitoring for the 6 months (Fig. 19.5) and the water balance for the catchment (Fig. 19.6) show that, of 572 mm total rainfall, actual evapotranspiration (AE) was 123 mm, overland flow 267 mm and 25 mm was collected as drain flow; this left 157 mm unaccounted for. It appears likely that water unaccounted for flowed through the soil but was not captured by the drainage system. The average MRP in the water (sampled once each week) was 0.021 (range 0.005–0.149) and 0.030 mg P l^{-1} (range 0.005–0.054) for drainage and overland flow, respectively, which equates to losses of 0.087 and 0.007 kg P ha^{-1} in overland flow and drain water, respectively. When the figure for drainage water was increased to include the water unaccounted for, the total MRP loss to ground and drainage was estimated at 0.055 kg P ha^{-1} for the 6-month period and would be approximately the same for 12 months. There was little loss of MRP in overland flow on this low-fertility site and the losses would probably be lower still on a site with lower overland flow.

Conclusion

The loss of MRP in this low-intensity grassland is of the order of 0.15 kg P ha^{-1} $year^{-1}$ when the soil had a P-test value (Morgan's) of 2 mg P l^{-1}.

References

Cooper, A.B. and Thomsen, C.E. (1988) Nitrogen and phosphorus in streamwater from adjacent pasture, pine and native forest catchments. *New Zealand Journal of Marine and Fresh Water Research* 22, 279–291.

Hayward, J., Foy, R.H. and Gibson, C.E. (1993) Nitrogen and phosphorus budgets in the Erne System, 1974–89. *Biology and the Environment. Proceedings of the Royal Irish Academy* 93B, 33–44.

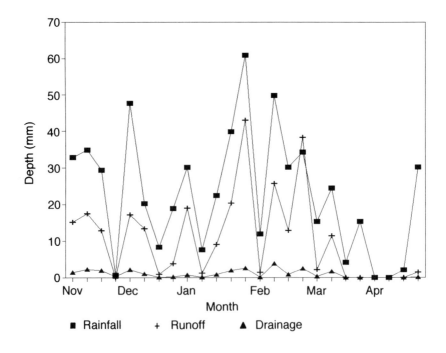

Fig. 19.5. Rainfall, ■; surface runoff, +; and drainage, ▲; all in mm (10 day averages), between November 1994 and April 1995 at Warren field, Johnstown Castle. Total rainfall in the period was 572 mm.

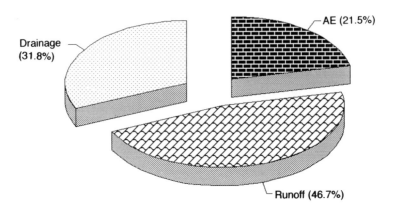

Fig. 19.6. The water balance for the period between November 1994 and April 1995 at Warren field, Johnstown Castle.(AE, evapotranspiration).

LEACHING

19.6 The Relation Between Accumulation and Leaching of Phosphorus: Laboratory, Field and Modelling Results

O.F. Schoumans and A. Breeuwsma
DLO – The Winand Staring Centre for Integrated Land, Soil and Water Research, Wageningen, The Netherlands

Introduction

Phosphorus (P) leaching from agricultural soils to surface water is becoming a considerable problem where there are heavily fertilized sandy soils in areas with shallow groundwater tables (Breeuwsma *et al.*, 1995). In order to simulate the relation between accumulation and leaching of P to groundwater and surface water, a process based on an agricultural nitrogen model (ANIMO) has been adapted to include a new description of P sorption. Process description and parameter assessment of inorganic P reactions with sandy soils were based on results of batch experiments. Calibration has been described by Schoumans (1995); validation was by column experiments. The effect of manure applications on P leaching from plots with different soil-P levels has been followed to obtain field information on P leaching. Finally, model simulations were made.

Process description

Following Van der Zee *et al.* (1989), the reaction of orthophosphate with sandy soil is described by two processes: a fast and reversible adsorption reaction on the surface of aluminium and iron hydroxides and a slow, diffusion-controlled, process. The slow reaction can be interpreted as: (i) diffusion through the solid phase of aluminium and iron hydroxides, where P precipitates with aluminium and iron; or (ii) diffusion into the micropores, where P adsorbs at the inner surface. The adsorption and precipitation reactions can be described as:

$$Q = \frac{KcQ_m}{1 + Kc} \qquad \frac{\partial S}{\partial t} = \sum_{i=1}^{n} k_i (C - CE_i)^{b_i}$$

Three empirical, parallel, precipitation terms properly describe the mean results of laboratory studies on 60 samples (Schoumans, 1995).

Methods

A column experiment of Van der Zee *et al.* (1989) was used to validate the process description, with sorption and desorption parameters determined separately in batch experiments. In the column experiment, three P pulses were given sequentially: first 3 mol P m^{-3} and then two pulses of 0.2 mol P m^{-3} (Fig. 19.7).

In a field study, three P applications (0, 48, 87 kg P ha^{-1}) were given in triplicate on plots on an acid sandy soil. The degree of P saturation (DPS) of the top (40 cm) of each plot was different. Concentrations of P in soil water were measured before and 5 weeks after manure application (Fig. 19.8).

Model simulations, made with the validated P model, estimated the mean annual P leaching to deeper layers and to surface water (Fig. 19.9) of a wet soil under pasture with water-tables at 0.3 m (high) and 1.0 m (low) below the surface. Model simulations were made for different DPS of the top (40 cm) of soil.

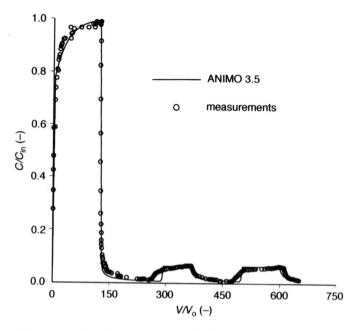

Fig. 19.7. Column experiment for one sample (validation).

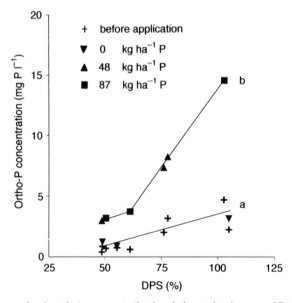

Fig. 19.8. Mean orthophosphate concentration in relation to the degrees of P saturation of the topsoil (0–40 cm) before (a) and after (b) manure application.

Results
The results of the laboratory column experiments agree well with the prediction of the model (Fig. 19.7). Field data (Fig. 19.8) show large concentrations of P in the topsoil (3–15 mg P l^{-1}) after the application of manure during the first 5 weeks after manure application. This caused additional leaching losses from the topsoil of 1–7 kg P ha^{-1} at a DPS of 50–100%. Modelling results indicate an increase in annual losses of 2–22 kg P ha^{-1} year^{-1} under these conditions (Fig. 19.9). At zero application these losses reduce to 1–14 kg P ha^{-1} year^{-1} (Fig. 19.9). Leaching rates to surface waters are significantly lower (0.5–4 kg P ha^{-1} year^{-1}) (Fig. 19.9).

Conclusions
Phosphorus losses by leaching are well described by the model; soluble P is related to the DPS and model results indicate high P losses from P-saturated soils.

References
Breeuwsma, A., Reijerink, J.G.A. and Schoumans, O.F. (1995) Impact of manure on accumulation and leaching of phosphate in areas of intensive livestock farming. In: Steele, K. (ed.) *Animal Waste and the Land–Water Interface.* Lewis Publishers, Boca Raton, Florida, pp. 239–249.
Schoumans, O.F. (1995) *Validation of the Process Description of Abiotic Phosphate Reactions In Acid Sandy Soils.* Rapport 381 (in Dutch). The Winand Staring Centre, Wageningen, the Netherlands.
Van der Zee, S.E.A.T.M., Leus, F. and Louer, M. (1989) Prediction of phosphate transport in small columns with an approximate sorption kinetics model. *Water Resources Research* 25, 1352–1365.

19.7 Organically Combined Phosphorus in Soil Solutions and Leachates
W.J. Chardon
Research Institute for Agrobiology and Soil Fertility (AB-DLO), PO Box 129, 9750 AC Haren, Netherlands

Introduction
A substantial part of the total phosphorus (P) in the soil solution can be dissolved organic P (DOP). The DOP is generally more mobile than inorganic orthophosphate and is potentially an important P source for surface-water eutrophication. Results relating to soils receiving animal manure indicate the importance of the DOP fraction, which constituted the largest part of the total P in soil solutions below 50 cm. In a manured sandy soil column, more than 90% of P leached was in organic forms, and in leachates from lysimeters growing maize, DOP was 77% of total P. Thus characterization of the DOP fractions in soil solutions will improve our understanding of P leaching from soils.

Methods and results
Time series of soil solutions. The soil solution from plots that received large amounts of animal manure between 1971 and 1982 was sampled in 1988 and 1989, together with soil samples at seven depth intervals (0–20, 20–30, ..., 70–80 cm) (Del Castilho *et al.*, 1993). The soil solution was extracted from field-moist samples by centrifugation at 2039 g and then filtered through a 6 μm paper

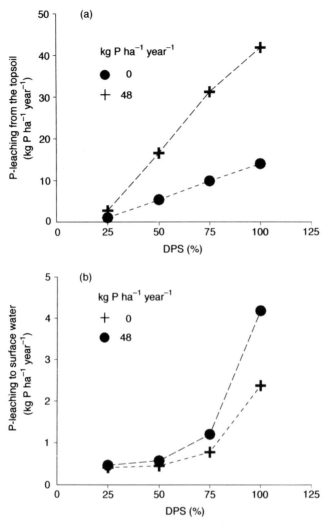

Fig. 19.9. Mean phosphorus leaching (model simulations for 20 years) from the topsoil and to surface water (ditch) at two application rates and four degrees of P saturation (DPS) of the topsoil (0.4 m).

filter. Organic P was calculated as total-P minus orthophosphate ('molybdate-reactive P'). Results from the plot that had received a total of 160 t ha^{-1} year^{-1} of pig manure slurry show that total P in the soil solution decreased with depth below the plough layer (30 cm), and that % DOP increased with depth to a value of 70% of total P (Fig. 19.10). The same trends were found for plots that received less manure or fertilizer. For grassland soil, Ron Vaz *et al.* (1993) also found that, in centrifuged soil solutions, the contribution of DOP to total P increased with sampling depth.

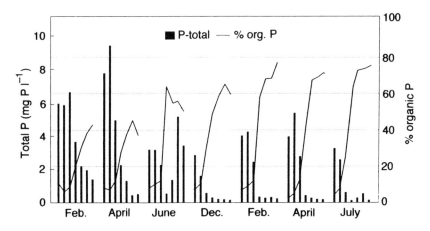

Fig. 19.10. Total P content and DOP as % of total P in soil-solution samples; samples were taken on seven occasions. Vertical bars are the values at seven depth intervals, from left to right: 0–20, 20–30, ... 70–80 cm.

Manure leaching experiment. Leaching of P from a sandy loam enriched with cattle slurry was studied in the laboratory. A polyvinyl chloride (PVC) column (inner surface 1.85 dm^2) was filled with a layer of 70 cm of pure sand, with an initial P content of 6.5 mg kg^{-1}. On top of this was placed a layer of 16 cm of a sandy loam containing 530 mg P kg^{-1}, which was mixed with cattle slurry, which supplied the equivalent of 300 mg P kg^{-1} dry soil. The manured column was compared with a blank, where no slurry was mixed with the sandy loam soil. The columns were percolated twice weekly with 100 ml water (0.54 cm) for 4 months. The percolate was collected, but not filtered, and analysed weekly for orthophosphate and total P. Leaching of P was negligible in the blank column; leaching from the manured column was significant, but corresponded to only 0.5% of P applied with the manure. Of the P that was leached, 96% was in organic forms. This result is comparable with those of Gerritse (1981), who applied dried pig slurry to sandy soil columns; he found that, after 1 year, the cumulative amount of P leached corresponded to 0.8% of the P applied, with 88% of the amount of P leached in the organic form. Thus it appears that only a small fraction (< 1%) of the P in animal manure has a high mobility, and almost all this P is in organic compounds.

Lysimeter experiment. Lysimeter pots were filled successively with 5 cm gravel, 15 cm white sand, 15 cm grey sand (1% organic matter) and 25 cm field soil (4% organic matter). The pots were dug into the soil so that the top of the lysimeter corresponded with the soil surface. The site was outside and maize (*Zea mays* L.) was the test crop. Beside normal rainfall, additional water was added, if required by the crop. Nitrogen (N) was added each year as follows: (1) no fertilization, (2) 250 kg mineral N ha^{-1}, (3) 200 t animal manure ha^{-1} and (4) 200 t manure and 250 kg mineral N ha^{-1}. All pots received 71 kg P ha^{-1} as superphosphate annually. The amount of manure added in (3) and (4) supplied the equivalent of 300 kg P and 1000 kg N ha^{-1} year^{-1}. The leachate was sampled at intervals during 7 years and analysed for orthophosphate and total P. On average, DOP

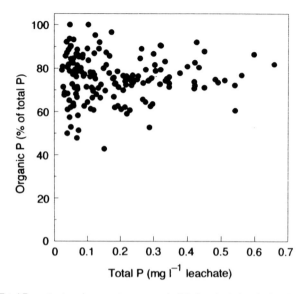

Fig. 19.11. Total P content and percentage organic P in leachate in a lysimeter experiment.

accounted for 75% of total P leached from treatment (3) (Fig. 19.11); for other treatments, this fraction was comparable. Thus, neither total P added nor the forms of P influenced this percentage. Plots receiving animal manure had a higher total P in the leachate than those receiving none.

Conclusions

The DOP fraction constitutes the largest part of total P in soil solutions below a depth of 50 cm where animal manures have been applied. In manured sandy soil columns, more than 90% of P leached was in organic forms. In leachates from lysimeters where maize was grown, DOP was 77% of total P. Thus, characterization of the DOP fractions in soil solutions is essential to improve our understanding of P leaching from soils.

Acknowledgements

Part of this work was carried out with the support of the Environment Research Programme 1990–1994 of the European Community, contract EV5V-0469.

References

Del Castilho, P., Chardon, W.J. and Salomons, W. (1993) Influence of cattle-manure slurry application on the solubility of cadmium, copper and zinc in a manured acidic, loamy-sand soil. *Journal of Environmental Quality* 22, 689–697.

Gerritse, R.G. (1981) Mobility of phosphorus from pig slurry in soils. In: Hucker, T.W.G. and Catroux, G. (eds) *Phosphorus in Sewage Sludge and Animal Waste Slurries*. Reidel, Dordrecht, pp. 347–369.

Ron Vaz, M.D., Edwards, A.C., Shand, C.A. and Cresser, M.S. (1993) Phosphorus fractions in soil solution: influence of soil acidity and fertilizer additions. *Plant and Soil* 148, 175–183.

19.8 Phosphorus Losses in Drainage Water from an Arable Silty Clay Loam Soil

G. Heckrath, P.C. Brookes, P.R. Poulton and K.W.T. Goulding
IACR-Rothamsted, Harpenden, Hertfordshire AL5 2JQ, UK

Introduction

Phosphorus (P) inputs are a key factor in controlling the trophic state in fresh water. The movement of P from a terrestrial to an aquatic environment can therefore pose a threat to water quality. The accumulation of P in agricultural soils resulting from fertilizer, organic manure and sewage sludge applications, as observed in several countries in the European Union (EU), might enhance the potential for P losses (Sibbesen, 1989). Because of the high P-retention capacity of most soils, vertical movement of P through the soil profile into drainage water was generally considered of little importance (Sharpley and Menzel, 1987). Recent data from the Broadbalk experiment at Rothamsted raise questions about this.

Methods

Details of the experimental site (Johnston, 1969) and the methods used (Heckrath *et al.*, 1995) are given elsewhere. The soil is silty clay loam of the Batcombe series developed in clay-with-flints. The experiment consists of 20 plots, which have pipe drains installed along the middle of each plot to a depth of c. 65 cm. For about 150 years, plots have received either no P, superphosphate to supply 35 or 17 kg P ha^{-1} year^{-1} or farmyard manure supplying 40 kg P ha^{-1} year^{-1}. Drainage water was collected five times between October 1992 and February 1994 and analysed for total P (TP), total dissolved P (TDP), total particulate P (TPP), dissolved organic P (DOP) and dissolved molybdate-reactive P (MRP). Bicarbonate-extractable P (Olsen P) was determined on air-dried soil from the plough layer (0–23 cm).

Results

Olsen P ranged from 7 to 90 mg P kg^{-1} soil, as a result of different P applications and different off-takes of P in grain and straw, whose yields varied in response to different rates of nitrogen (N) fertilizer. Phosphorus in drainage water and its relative proportions in different P fractions varied substantially among the five drainage events and among the different plots (Heckrath *et al.*, 1995). Molybdate-reactive P was the largest P fraction in the drainage water, range 65–85% of the TP. Dissolved organic P accounted, on average, for less than 5% of TP. Compared with other data from heavy soils (e.g. Sharpley and Menzel, 1987), TP in drainage water was large, exceeding 1 mg P l^{-1} on several plots during most drainage events.

The concentrations of TP were small in drainage water from plots with less than 60 mg Olsen P kg^{-1} soil. There was then a rapid increase of TP in drainage water up to the maximum concentration of Olsen P (Fig. 19.12). A simple split-line model described this relationship well for all five drainage events. The slopes were not significantly different from zero below 60 mg Olsen P kg^{-1}. Above this, the slopes differed significantly ($P < 0.01$), reflecting the variation in soil and weather conditions among the five events. This implies that, up to 60 mg Olsen P kg^{-1} soil, P was strongly retained in the plough layer. However, beyond this point, P losses through drainage water were much more closely related to the P concentrations in the plough layer than commonly suggested. The sharp increase in TP beyond 60 mg Olsen P kg^{-1} soil could imply a shift towards more

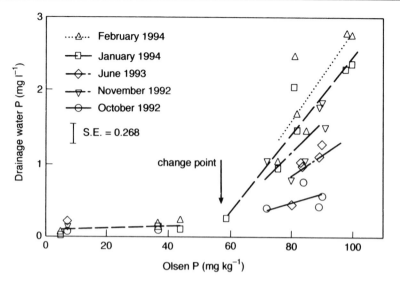

Fig. 19.12. Relationship between total P (TP) in drainage water and Olsen P from different Broadbalk plots. The split-line model shown fitted a common initial line and change point (at 57 mg Olsen P kg^{-1} soil), but separate changes in slope in the steep section for the five drainage events; 90% of the variance was accounted for.

P being held on low-energy sites (Holford and Mattingly, 1976). Lookman (1995) found a similar split-line pattern relating water-extractable P from the plough layer of a range of acid sandy soils to their degree of P saturation. A build-up of calcium-bound P in highly P-saturated soils was believed to represent the more mobile P fraction in those soils.

Phosphorus leaving the plough layer of Broadbalk was prevented from resorption in the P-deficient subsoil. The mechanisms involved could either be preferential flow or transport of P in forms which make it less susceptible to sorption (e.g. colloidal or organically associated P). Whatever the mechanism, these results indicate a potential for an enhanced contribution of P losses through drain flow on clay loam soils, once a certain concentration of Olsen P is exceeded in the plough layer.

References

Heckrath, G., Brookes, P.C., Poulton, P.R. and Goulding, K.W.T. (1995) Phosphorus leaching from soils containing different P concentrations in the Broadbalk experiment. *Journal of Environmental Quality* 24, 904–910.

Holford, I.C.R. and Mattingly, G.E.G. (1976) A model for the behaviour of labile phosphate in soil. *Plant and Soil* 44, 219–229.

Johnston, A.E. (1969) Plant nutrients in Broadbalk soils. In: *Rothamsted Experimental Station Report for 1968*, Rothamsted Experimental Station, Harpenden, Part 2, 93–115.

Lookman, R. (1995) Phosphorus chemistry in excessively fertilized soils. PhD thesis, Katholieke Universiteit Leuven, Louvain, Belgium.

Sharpley, A.N. and Menzel, R.G. (1987) The impact of soil and fertilizer phosphorus on the environment. *Advances in Agronomy* 41, 297–324.

Sibbesen, E. (1989) Phosphorus cycling in intensive agriculture with special reference to countries in the temperate zone of Western Europe. In: Tiessen, H. (ed.) *Phosphorus Cycles in Terrestrial and Aquatic Ecosystems.* University of Saskatchewan, Saskatoon, Canada, pp. 112–122.

19.9 Evidence of Phosphorus Movement from Broadbalk Soils by Preferential Flow

D. Thomas, G. Heckrath and P.C. Brookes
IACR-Rothamsted, Harpenden, Hertfordshire AL5 2JQ, UK

Introduction

Phosphorus (P) leaching was measured in drainage water from the Broadbalk experiment at Rothamsted (Heckrath *et al.*, 1995; Heckrath *et al.*, Chapter 19, this volume). The results did not agree with the previous understanding of the vertical movement of P down the soil profile (Sharpley and Menzel, 1987), which assumes that most soils have a high P adsorption capacity and therefore lose negligible amounts of P to water. This study aimed to find out why P was not adsorbed in the subsoil but was leached in drainage water passing out of the plots via drains at about 0.65 m depth.

Methods

Details of the experimental site are given elsewhere (Johnston, 1969). Soil was sampled to a depth of 1 m on five Broadbalk plots: unmanured (nil), farmyard manure (FYM) and three inorganically fertilized plots, the latter receiving 35 kg P ha^{-1} year^{-1} but differing amounts of nitrogen (N). Adsorption isotherms were obtained by equilibrating 2.5 g of soil with 25 ml aliquots of solutions containing P in the range 0–60 mg P l^{-1} at room temperature and determining the concentration of P in the equilibrium solution. Langmuir two-surface equations were fitted to the data.

Four Prenart Super Quartz porous cups with a nominal pore size of 2–4 µm were installed at a depth of 60 cm either side of the drains on several plots. Samples of soil solution were collected at a vacuum of 0.25 bar over a 24 h period. Unfiltered solutions were analysed for molybdate-reactive P (MRP) and total P (TP). Soil solution was also extracted, using a modification of the method of Linehan *et al.* (1989). Field-moist soil was taken from the plough layer in 15 cm cores and sieved through 6.25 mm. About 200 g of soil were centrifuged on a swing out rotor at 2700 rpm for 1.5 h to extract the soil solution, which was analysed for MRP without further treatment.

Results

Adsorption isotherms showed that subsoil below plough depth had a greater number of sites available for adsorbing P than surface soil from the plough layer. The 'maximum monolayer adsorption capacity' (MMAC) was in the order of 750 mg P kg^{-1} for subsoils from the 75–100 cm and 50–75 cm depths, while soil from the plough layer (0–23 cm) had an MMAC of about 200 mg P kg^{-1}. In the absence of P in the equilibrating solution, desorption of P was observed for plough-layer soils, except that from the nil plot. Desorption was negligible or not detected in surface soil from the nil plot and subsoils below 50 cm. Where desorption was measured, the amount and the MMAC were both related to Olsen P in the plough-layer soil.

There was a large variation between the soil solution P concentrations sampled by the porous cups in each plot and there appeared to be no relationship between this P concentration and Olsen P in the plough layer. Also MRP in the solutions in the porous cups was substantially lower than that in drainage water. However, P in the soil solution extracted from the plough layer by the centrifugation technique showed a close correlation with MRP in drainage water.

The results show that P was being leached in drainage water despite the large adsorption potential of these subsoils. Also, very little P was present in the soil solution at depth. One possible explanation for the loss of P in the drainage water is that it was flowing down large, possibly permanent, cracks in the soil (preferential flow), thus reducing the effective sorption capacity of the Broadbalk subsoil.

References
Heckrath, G., Brookes, P., Poulton, P. and Goulding, K.W.T. (1995) Phosphorus leaching from soils containing different P concentrations in the Broadbalk experiment. *Journal of Environmental Quality* 24, 904–910.
Johnston, A.E. (1969) Plant nutrients in Broadbalk soils. In: *Rothamsted Experimental Station Report for 1969*, Rothamsted Experimental Station, Harpenden, Part 2, pp. 93–115.
Linehan, D.J., Sinclair, A.H. and Mitchell, M.C. (1989) Seasonal changes in Cu, Mn, Zn and Co concentrations in the root zone of barley (*Hordeum vulgare* L.). *Journal of Soil Science* 40, 103–115.
Sharpley, A.N. and Menzel, R.G. (1987) The impact of soil and fertilizer phosphorus on the environment. *Advances in Agronomy* 41, 297–324.

19.10 Phosphorus Leaching in the Brimstone Farm Experiment, Oxfordshire

K.R. Howse,[1] J.A. Catt,[1] D. Brockie,[1] R.A.C. Nicol,[1] R. Farina,[1] G.L. Harris[2] and T.J. Pepper[2]
[1]*Soil Science Department, IACR-Rothamsted, Harpenden, Hertfordshire AL5 2JQ, UK;*
[2]*ADAS Anstey Hall, Maris Lane, Trumpington, Cambridge CB2 2LF, UK*

Introduction
The Brimstone Farm experiment, sited near Faringdon, Oxfordshire, on a clay soil, was started in 1978 to study the effects of drainage and cultivation on nitrate leaching. It has 20 hydrologically isolated 0.2 ha plots, most of which have mole drains at 0.6 m depth and 2 m lateral spacing; Cannell *et al.* (1984) gave details of the site facilities. Recently, the leaching of phosphorus (P) has been studied. Some measurements of P losses in selected events were begun in February 1990, and a full programme was started in October 1993. Leaching losses of soluble P and total P (TP) are now measured under winter cereals established after ploughing, on plots given four main treatments.

1. A lower P fertilizer rate (16 kg P ha^{-1} year^{-1}) compensating for previous crop uptake.
2. A higher P fertilizer rate (33 kg P ha^{-1} year^{-1}).
3. Drain flow restricted by U-bends in the exit drains from the plots.
4. Drain flow unrestricted.

In addition to losses through the mole drains, TP and soluble P are measured in combined surface runoff and interflow in the cultivated topsoil (Ap horizon), which is collected in a plough furrow at the lower margins of each plot. Phosphorus losses in this plough furrow are also measured on a single, minimally cultivated plot with no mole drains.

Methods

Soluble (molybdate-reactive) P (MRP) was measured in water samples using a SHENA segmented flow-colorimetric analyser, and TP by inductively coupled plasma-arc spectrometry following digestion in aqua regia. Samples were taken on a flow-related basis during major flow events and at frequent time intervals during other periods of lower flow. Annual loadings were calculated by integrating concentrations with measured flow rates over time, using Simpson's rule. To compensate for differences in total annual flows between plots, the loadings were also normalized to a unit volume of drain flow, equivalent to a water depth of 100 mm.

Results

Results for 1990–1993. In single drain-flow events sampled in this period, MRP was fairly constant, at 0.03–0.15 mg P l^{-1}; often there was a slight dilution with increasing flow rate. In contrast, TP was more variable, at 0.06–1.31 mg P l^{-1}; the concentration often increased with flow rate, probably because much P was in particulate form and the more turbulent flow in storm conditions caused more disaggregation of the soil. In the combined surface runoff and Ap interflow, MRP ranged from 0.05 to 0.72 mg P l^{-1}, and TP from 0.07 to 3.13 mg P l^{-1}. Concentrations of MRP and TP in both drain flow and combined surface runoff plus Ap interflow showed few differences between runoff events at different times after P fertilizer application. This suggests that, for most of the time, losses of P are controlled by processes in the soil unrelated to fertilizer applications, although they may be related to soil-P reserves built up by P fertilizer applications in the past.

Results for 1993–1995. In the winter of 1993/94, TP in drain flow ranged from 0.09 to 0.50 mg P l^{-1} and increased with increasing drain-flow rate. Over the whole drainage season, approximately 0.7 kg total P ha^{-1} was lost in drain flow. In the same winter, on four plots with restricted drain flow, MRP ranged from 0.02 to 0.43 mg P l^{-1}. There was a dilution effect with increasing flow rate, and the mean MRP loading for the whole winter drain-flow period was 0.135 kg P ha^{-1}. On eight plots with unrestricted drain flow, MRP ranged from 0.02 to 1.15 mg P l^{-1} and the mean loading was 0.240 kg P ha^{-1}. Normalized to unit volume of drain flow, the mean annual loss on the unrestricted plots was 70.6 g P ha^{-1} per 100 mm drain flow, but on the restricted plots it was only 49.9 g P ha^{-1} per 100 mm drain flow, a decrease of 29%.

In the winter of 1994/95, MRP ranged from 0.02 to 0.45 mg P l^{-1} on plots with restricted drain flow and from 0.02 to 1.67 mg P l^{-1} on plots with unrestricted drain flow. The respective mean annual loadings in drain flow were 0.047 and 0.097 kg P ha^{-1}, and the normalized mean losses were 23.8 and 49.4 g P ha^{-1} per 100 mm drain flow, indicating a decrease of 52% by drain-flow restriction. Total P was not measured in 1993/94, but in 1994/95 drain-flow restriction decreased the mean annual concentration of TP by 19%, annual TP loading by 26% and loss per 100 mm drain flow by 29%.

Because of wet autumn weather, no P fertilizer was applied in the 1993/94 season and the following year it was delayed until 2 December 1994. Over the succeeding 3–4 weeks, the drain flow contained considerably higher concentrations of soluble P than either before or after. The December drain flow was a mean of 23% of the total for the whole of the 1994/95 winter, but it contained a mean of 51% of the winter's MRP loading. The two P application rates resulted in different soluble P losses in 1994/95; normalised to 100 mm drain flow, the plots given the lower rate lost 38.3 and 21.7 g P ha^{-1} with unrestricted and restricted flow, respectively, whereas those given the higher rate lost 60.5 and 25.8 g P ha^{-1} with unrestricted and restricted flow, respectively. The percentage increase in MRP loss in drain flow resulting from the higher fertilizer application rate was therefore 37% with unrestricted drain flow but only 16% with restricted drain flow. Expressed on a similar basis, the increase in TP in drain flow resulting from the higher application rate was a mean of only 3%.

The total amount of soluble P lost in the combined surface runoff and Ap interflow of the undrained, minimally cultivated plot was 1.195 kg P ha^{-1} in 1993/94 and 0.979 kg P ha^{-1} in 1994/95, when 16 kg P ha^{-1} as fertilizer was applied in December 1994. However, there was much less total flow in 1994/95 – 270 mm compared with 449 mm in 1993/94.

Conclusions

The concentrations of soluble P in most drain water and surface runoff plus Ap interflow samples from the Brimstone Farm experiment are greater than the minimum (20 µg P l^{-1}) thought to cause eutrophication (OECD, 1982). For a period of 3–4 weeks following application of P fertilizer to wet soil, the concentrations were increased, suggesting that fertilizer application should be restricted to times when the soil is dry, so that rain in the succeeding few weeks does not lead to drain flow or surface runoff to streams. Fertilizer application in excess of the amount required to compensate for previous crop uptake led to a 37% increase in annual P loading. Surface runoff and Ap interflow also contained more soluble P than drain flow, and on undrained, minimally cultivated land resulted in a ninefold increase in annual loading compared with the loss through the mole drains of ploughed land. However, restricting the drain flow so that water was held for longer periods in contact with the soil decreased soluble P losses by 29–52%.

References

Cannell, R.Q., Goss, M.J., Harris, G.L., Jarvis, M.G., Douglas, J.T., Howse, K.R. and Le Grice, S. (1984) A study of mole drainage with simplified cultivation for autumn-sown cereals on a clay soil. 1. Background, experiment and site details, drainage systems, measurement of drainflow and summary of results 1978–80. *Journal of Agricultural Science, Cambridge* 102, 583–594.

OECD (1982) *Eutrophication of Waters: Monitoring, Assessment and Control.* OECD, Paris.

19.11 Phosphorus Input to a Brook through Tile Drains under Grassland

C. Stamm,[1] R. Gächter,[2] H. Flühler,[1] J. Leuenberger[1] and H. Wunderli[1]

[1]*Swiss Federal Institute of Technology Zürich, Institute of Terrestrial Ecology, Soil Physics, Grabenstr. 3, CH-8952 Schlieren, Switzerland;* [2]*Swiss Federal Institute for Environmental Science and Technology, CH-6047 Kastanienbaum, Switzerland*

Introduction

Lake Sempach is one of the eutrophic lakes of the Swiss Plateau (Gächter and Stadelmann, 1993). It is in a catchment characterized by intensive pig production, with an enormous phosphorus (P) surplus in the balance between P applied and removed. Today, eutrophication in this catchment is mainly caused by P input from agricultural land where grassland is dominant and from which surface runoff can lead to large P loads in brooks (Von Albertini et al., 1993). However, it is uncertain whether all P input stems from surface runoff or if the widespread drainage systems also contribute to P losses from the land. Amounts of P discharged and P concentrations of drainage waters under intensively managed grassland are reported here.

Methods

The catchment in which this study was made is characterized by soils of low permeability and, on average, 1200 mm precipitation per year. The soils at the two sites used are representative of gley and brown earth soils. The meadows growing on them are cut five to seven times between April and October and manure is spread after each cut. This regime is common practice in the region.

Between April and October 1994, drain discharge was measured continuously, using a V-notch weir and an inductive flowmeter. Data were stored on data loggers, which controlled automatic sampling devices such that, if discharge exceeded a given limit, water samples were taken every 15 min. Samples were collected and filtered (0.45 µm) within 24 h. The samples were analysed for molybdate-reactive phosphate (MRP). For some events, total P (TP) was measured in the unfiltered and filtered samples, to estimate total dissolved phosphate (TDP). Particulate P (PP) was determined as the difference between TP and TDP, soluble organic P (SOP) as the difference between TDP and MRP. In May 1995, samples of two events were also filtered through 0.05 µm filters to estimate colloidal P.

Results and discussion

The dominant P fraction was MRP (Fig. 19.13), with PP being of minor importance; SOP was not detected except for one event. For most samples, MRP concentrations were very similar for filtrates through 0.45 µm and 0.05 µm filters, suggesting that colloidal transport was not very relevant or that the colloids involved were smaller than 0.05 µm.

The P dynamics were characterized by a simultaneous increase and decrease in MRP with discharge rate (Fig. 19.13). During base flow, MRP ranged between 50 and 150 µg P l^{-1}, whereas maximum concentrations reached values of 600–1500 µg P l^{-1} during peak flow. Extreme values up to 4800 µg P l^{-1} were measured during events shortly after manure application.

The MRP fluxes equate to losses which varied between a few grams up to 500 g ha^{-1}. Based on discharge measurements and a statistical relationship between concentration and discharge data, the total losses per ha were estimated to range between 600 and 1900 g MRP (site I) and 1100 and 1600 g (site II) for the period of April to October. Even the lower estimates are above the accepted critical losses of 400 g ha^{-1} $year^{-1}$.

Phosphorus is strongly adsorbed to the soils of the catchment and a large concentration of soluble P can be found only in the uppermost centimetres of the profile. Experiments with artificially coloured water show the occurrence of preferential flow in these soils, which suggests that this transport mechanism would explain the increased P concentration in drain flow during storms.

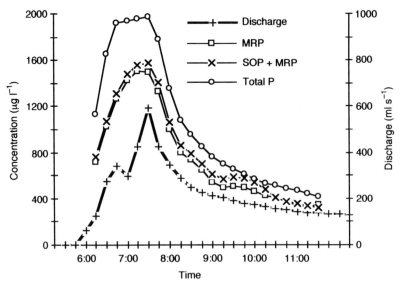

Fig. 19.13. Discharge and concentration of different P fractions during one event (18.5.1994) at site II. MRP, molybdate-reactive phosphate; SOP, soluble organic P.

References

Gächter, R. and Stadelmann, P. (1993) Gewässerschutz und Seenforschung. In: Sempachersee. *Mitteilungen Naturforschende Gesellschaft Luzern* 33, 343–378.

Von Albertini, N., Braun, M. and Hurni, P. (1993) Oberflächenabfluss und Phosphorabschwemmung von Grassland. *Landwirtschaft Schweiz* 6, 575–582.

EROSION

19.12 Phosphate Losses in the Woburn Erosion Reference Experiment

J.A. Catt,[1] A.E. Johnston[1] and J.N. Quinton[2]

[1]*Soil Science Department, IACR-Rothamsted, Harpenden, Hertfordshire. AL5 2JQ, UK;*
[2]*Silsoe College, Cranfield University, Silsoe, Bedfordshire. MK45 4DT, UK*

Introduction

The Erosion Reference experiment at Woburn Experimental Farm, Bedfordshire, was started in 1988, jointly by Rothamsted Experimental Station and Silsoe College, to measure the effects of different cultivation treatments on surface runoff and soil loss. From the start, soluble phosphorus (P) has been measured in the runoff and total P (TP) in the eroded soil. The eight 0.08 ha plots are sited on a 5° slope at 100–107 m above sea level on a loamy sand soil (Cottenham series) derived from the Lower Greensand (Cretaceous). The plots are arranged in two blocks, each of which has four plots with different pairs of the following cultivation treatments.

1. Cultivation and drilling across the slope, parallel to the contour.

2. Cultivation and drilling up and down the slope, perpendicular to the contour.
3. Standard ploughing and seed-bed preparation after removal of crop residues.
4. Minimum tillage with retention of crop residues.

All plots grow the same crop each year in the following rotation: potatoes, winter wheat, winter barley, sugar beet, winter wheat, winter barley. This rotation and the fertilizer treatment for each crop are typical of those used by farmers working sandy soils in the Woburn area. This study summarizes the effects of the treatments on losses of soluble P and TP and proposes an optimum cultivation treatment for minimizing P loss by soil erosion.

Methods

The collector tanks at the foot of each plot were emptied after each erosion event and the volumes of runoff and weights of sediment measured. Samples of both were taken for chemical analysis. Molybdate-reactive P (MRP) was determined in unfiltered runoff using a segmented flow colorimetric analyser and TP in the sediment by aqua regia extraction and inductively coupled plasma-arc spectrometry.

Results

The total loads of MRP in runoff and TP in eroded soil lost from the plots in the first three crop periods were very variable (Table 19.2) and were approximately proportional to the very variable amounts of total runoff and soil loss, respectively. The mean concentrations of MRP in runoff for each crop-treatment combination over these 3 years were within the narrow range 1.175–2.117 mg P l^{-1}; those for TP in eroded soil were slightly more variable (0.054–0.457 g P kg^{-1}).

Under potatoes in 1989 and winter barley in 1990/91, the treatment yielding the most runoff and soil loss, and consequently the largest amounts of MRP and TP, was cultivation up and down the slope; the treatment yielding the smallest amounts was cultivation across the slope. However, under winter wheat in 1989/90, when there were much larger amounts of runoff and MRP in runoff and of eroded soil and TP in eroded soil from all plots, the treatment yielding the largest amounts of all of these was standard cultivation after removal of previous crop residues, and that yielding the least was minimal cultivation with retention of residues.

In the 1989/91 period, phosphate fertilizer was applied only to the potatoes planted in spring 1989; the rate was 102 kg P ha^{-1}. Expressed as percentages of this application, the losses of MRP in runoff from the single event that occurred under potatoes ranged from 0.005% (across-slope cultivation) to 0.040% (up and down). If TP in the eroded soil is added, these percentages increase to 0.13% and 3.68%, respectively. Adding the subsequent losses under winter wheat (1989/90) and winter barley (1990/91) increases the MRP percentages for the same two treatments to 0.268 and 0.332 and the MRP plus TP to 10.05% and 12.91%. However, because of the larger difference under wheat (1989/90) between the P losses from plots given minimal and standard cultivations, the full ranges over the three crops were 0.26% MRP and 8.87% TP under minimal cultivation to 0.34% MRP and 14.09% TP under standard cultivation.

In February 1991, the mean concentrations of TP in the topsoils of the eight plots ranged from 857 to 1071 g P kg^{-1}. Total P concentration factors (eroded soil/plot soil) were calculated from these values and from the calculated mean TP contents of all eroded soil samples taken from each plot over the 1989/91 period. Under potatoes in spring 1989 and winter wheat in 1989/90, the mean

Table 19.2. Mean values (g P ha^{-1}) of MRP in runoff and TP in eroded soil lost from plots under four cultivation treatments for crops grown in 1989–1991 on a sandy loam soil.

		Crop and cropping year		
Cultivation treatment	Form of P	Potatoes 1989	Winter wheat 1989/90	Winter barley 1990/91
Across slope	MRP	5	267	2
	TP	123	9834	6
Up and down slope	MRP	41	291	7
	TP	3713	9039	75
Standard, residues removed	MRP	18	324	6
	TP	955	13004	63
Minimal, residues retained	MRP	28	234	3
	TP	2881	5869	18
SED	MRP	31	62	–
	TP	2220	6550	–

SED, standard error of the difference between means, not given for 1990/91 because of zero values for some plots.

concentration factors exceeded 1.0 for all treatments except across-slope cultivation for potatoes (Table 19.3). However, under winter barley in 1990/91, they were < 1.0 for all treatments. Mean concentration factors for clay, based on particle-size analyses of the same eroded soil and plot soil samples, showed almost the same pattern (Table 19.3), but those for other size fractions were quite different. Also clay percentages and TP contents for individual plots and events were strongly correlated ($r^2 = 0.92$). These relationships suggest that most of the transported TP resides in the clay fraction, where it is probably sorbed on layer silicate and iron oxide particles.

Conclusions

Losses of MRP in runoff from the Woburn plots have generally been small, with a maximum of 0.324 kg P ha^{-1} year^{-1} from 14 events in an unusually wet winter, although they are significant with respect to water quality. Losses of TP in eroded soil have been much larger, with a maximum of 13 kg P ha^{-1} year^{-1} for the same cultivation treatment as the MRP maximum, namely standard cultivation with crop residues removed. In drier periods with less runoff and erosion, the treatment resulting in the largest MRP and TP losses was cultivation up and down the slope. The best treatment for minimizing P losses in all rainfall conditions is therefore minimal cultivation with crop residues retained and crop rows planted across the slope. The largest cumulative TP + MRP loss over 3 years was equivalent to 14.4% of the single P fertilizer application made in that period; however, almost all of this was in particulate form, probably associated with the clay fraction, and was not necessarily derived directly from the fertilizer.

Table 19.3. Mean concentration factors (eroded soil/plot soil) for TP and clay content for the four cultivation treatments under three crops in 1989–1991.

Crop and year	Cultivation treatment			
	Across slope	Up and down slope	Standard, residues removed	Minimal, residues retained
Potatoes 1989				
TP	0.887	2.162	1.303	1.746
Clay	0.826	2.833	1.265	2.393
Winter wheat 1989/90				
TP	2.118	1.453	1.820	1.751
Clay	2.843	1.652	2.335	1.857
Winter barley 1990/91				
TP	0.253	0.893	0.750	0.397
Clay	0.301	1.019	0.910	0.410

19.13 Storm-Event Transport of Phosphorus in the Absence of Surface Runoff Generation

O.S. Hodun and T.P. Burt
School of Geography, University of Oxford, Mansfield Road, Oxford OX1 3TB, UK

Introduction

In catchments where land-use practices directed towards the reduction of surface (especially linear) erosion and associated phosphorus (P) transport are successful, other diffuse sources, such as channel erosion and subsurface leaching, may control the level of P pollution in surface waters. The input of P from such sources, although negligible in the base flow, may increase significantly after rainfall and thus deserves special attention. The absolute magnitude of P loss, as well as chemical speciation and potential bioavailability, was assessed in a catchment with low erosion rates during a series of runoff events in December 1994 which lasted for approximately 120 h. Channel erosion was assumed to be the only contributor to particulate-matter loss, therefore amounts and forms of P in suspended sediment and in stream-channel deposits were compared and enrichment ratios (ER) for different species calculated.

Materials and methods

The study area, in the headwaters of the River Stour, a typical central England agricultural watershed, surface area 6.2 km^2, is characterized by gentle slopes and well-drained brown calcareous soils. The stream in which P was measured had background P levels of about 0.02 and 0.04 mg P l^{-1} for dissolved molybdate-reactive P (MRP) and total P (TP), respectively. Stream-water samples (0.5 l) were collected automatically at intervals ranging from 15 min to 4 h. Samples of the stream-channel material (upper 1–2 cm, 11 samples) were collected manually at equal intervals along the stream, each sample consisting of five to six randomly selected subsamples. The 130 water samples were analysed for MRP

(Murphy and Riley, 1962). In 60 samples dissolved P (DP), dissolved unreactive P (DUP), particulate P (PP) and TP were determined also (Johnes and Heathwaite, 1992). Suspended matter retained by 0.45 μm filters was air dried and weighed. Channel material was air-dried and 0.5-mm-sieved. Inorganic P in 0.05 g of air-dried channel material and suspended matter, only for 40 samples with more than $0.1\ g\ l^{-1}$ suspended matter, was fractionated by sequential extraction with ammonium chloride (NH_4Cl) ('loosely bound P' or 'labile P'), sodium hydroxide (NaOH) ('iron (Fe)- and aluminium (Al)-bound P'), and hydrochloric acid (HCl) (calcium (Ca)-bound P) (Hieltjes and Lijklema, 1980). Phosphorus in the first two extractions, as well as MRP, was considered to be bioavailable (BAP).

Results and discussion

There were five consecutive rainfall events, with a total of 37 mm precipitation. Significant levels of P pollution were detected; TP concentrations in the water ranged from 0.1 to $3.48\ mg\ l^{-1}$, average $0.76\ mg\ l^{-1}$, whereas MRP averaged $0.07\ mg\ l^{-1}$ with a maximum of $0.18\ mg\ l^{-1}$ (Table 19.4). As the amount of sediment decreased, its P content increased dramatically (Fig. 19.14), but the

Table 19.4. Characteristics of P losses during the observation period.

Species	MRP	DUP	NH_4Cl-P*	NaOH-P*	HCl-P*	PP	TP	BAP
Average conc. (mg l^{-1})	0.07	0.014	0.07	0.23	0.08	0.67	0.76	0.37
Max. conc. (mg l^{-1})	0.18	0.16	0.24	1.67	0.85	3.40	3.48	1.86
Absolute amount (kg)	0.60	0.12	0.66	2.04	0.69	5.90	6.63	3.30
% of total loss	9.0	1.8	10.0	30.8	10.4	89.0	100	49.8

* Calculated from the weighted average content of the fractionated samples.

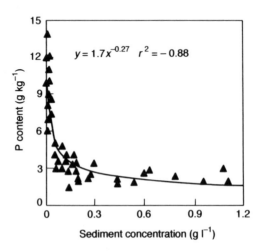

Fig. 19.14. Relationship between sediment P content and sediment concentration of runoff.

Table 19.5. Amounts of phosphorus (g P kg^{-1}), extracted by the method of Hieltjes and Lijklema (1980).

	Average	Max.	Min.	ER
Suspended sediment				
NH$_4$Cl-P	0.19	0.43	0.03	2.1
NaOH-P	0.59	1.06	0.43	2.8
HCl-P	0.20	0.38	0.10	2.2
EIP	0.98	1.69	0.68	2.6
TP content (average of all samples analysed)	2.24	14.00	0.89	
TP content (average for fractionated samples only)	1.98	4.80	0.89	
Channel material				
NH$_4$Cl-P	0.09	0.16	0.02	
NaOH-P	0.21	0.35	0.13	
HCl-P	0.09	0.12	0.04	
EIP	0.38	0.56	0.23	

minute total amounts involved mean that the overall contribution of such sediment is small. The average TP content calculated for all the digests, 2.24 g P kg^{-1}, was only about 12% higher than the sum of the fractionated samples, 1.98 g P kg^{-1}. An average of 0.98 g P kg^{-1} of inorganic P was extracted from suspended sediment, almost 80% of which was found to be labile or bound to Fe and Al compounds (Table 19.5). More than 6 kg of P was lost from the catchment, about half of it (3.3 kg) in bioavailable form. The total extracted inorganic P (EIP) in channel material gradually increased downstream (from 0.23 g P kg^{-1} to 0.56 g P kg^{-1}), with, on average, 0.38 g P kg^{-1}. The highest ER of 2.8 was found for the NaOH-extracted fraction, whereas for NH$_4$Cl-P and HCl-P it was 2.1 and 2.2, respectively (Table 19.5). The average ER for EIP was 2.6, which reflected the significance of the NaOH-extractable ('Fe- and Al-bound P') fraction.

References

Hieltjes, A.H.M. and Lijklema, L. (1980) Fractionation of inorganic phosphates in calcareous sediments. *Journal of Environmental Quality* 9, 405–407.

Johnes, P.J. and Heathwaite, A.L. (1992) A procedure for the simultaneous determination of total nitrogen and total phosphorus in freshwater samples using persulphate microwave digestion. *Water Resources* 26, 1281–1287.

Murphy, J. and Riley, J.P. (1962) A modified single solution method for the determination of phosphate in natural waters. *Analitica Chimica Acta* 27, 31–36.

19.14 Impact of Different Tillage Practices on Phosphorus Losses from Agricultural Fields

A. Klik[1] and J. Rosner[2]

[1]*University for Renewable Resources Vienna, A-1190 Vienna, Austria;* [2]*Federal Government of Lower Austria, A-3430 Tulln, Austria*

Introduction

Soil erosion and the subsequent transport of sediment from the land surface to a receiving water body is a function of complex soil, hydrological and climatological processes. Erosion and sediment transport within a field (rill and inter-rill erosion) are a function of raindrop impact, which causes soil-particle detachment, and the energy of overland flow, which contributes to detachment and carries suspended sediment down-slope. These processes are highly dependent on the nature of the soil surface at the time of rainfall and, specifically, on the extent of surface cover by plant residues or growing plants. Soil surface cover and soil erosion are inversely proportional, so that husbandry practices promoting the maintenance of surface cover throughout the year will decrease erosion.

Materials and methods

In 1994, the effect of different tillage practices on soil erosion and the transport of particulate and dissolved nutrients was investigated at three different locations in Austria (Table 19.6).

All plots were sown with maize in 1994. Small grains, as a cover crop, are rotated with maize or sunflower. The three tillage systems: (i) conventional; (ii) conservation; and (iii) no tillage with mulching, are compared on plots 4 by 15 m. Each plot is isolated by vertical boards set 15 cm into the ground to prevent runoff entering the plots. Runoff water and sediment from the plots were collected and measured after every storm event and samples retained for physical and chemical analyses. A rain-gauge at each site measured precipitation at 5 min intervals and the data were stored using an automatic data-logging system.

Results

In the 1994 growing season, four heavy rainstorm events in Mistelbach and one in Pyhra led to surface runoff and soil erosion; the total amounts of both are shown in Table 19.7.

In Mistelbach total phosphorus (P) losses with conventional tillage were about 300 kg ha^{-1}. Conservation tillage and no tillage plus mulching reduced the amount of eroded soil significantly (Table 19.7) and therefore the loss of total P. Between 0.3 and 2% of the total P was dissolved P (Table 19.8).

Much smaller amounts of soil and P were lost at Pyhra. Compared with conventional tillage, conservation tillage reduced P losses by 40% and no tillage by 70% (Table 19.8). At the third site, in Gießhübl, no soil erosion and no surface runoff occurred during the 1994 growing season.

Table 19.6. Mean annual precipitation and soil characteristics at three sites in Austria at which soil erosion and nutrient loss were measured.

	Mistelbach	Pyhra	Gießhübl
Mean annual precip. (mm)	539	734	763
Slope (%)	12	16	5
Soil texture	Silt loam	Sandy loam	Clay loam
Clay (%)	18	17	17
Organic matter (%)	2.0	2.1	2.0

Table 19.7. Surface runoff and weight of soil eroded from the plots in Mistelbach and Pyhra during the 1994 growing season.

	Conventional till		Conservation till		No till	
	Runoff (mm)	Eroded soil (t ha^{-1})	Runoff (mm)	Eroded soil (t ha^{-1})	Runoff (mm)	Eroded soil (t ha^{-1})
Mistelbach	80.6	317.1	73.5	43.8	79.9	26.0
Pyhra	13	67.5	19	42.8	20	20.8

Conclusions

On a silt loam and a sandy loam, but not on a clay loam, large rainfall events caused much soil to be eroded and therefore large losses of total P with conventional tillage systems. Conservation tillage and no tillage with surface mulching decreased these losses. Losses of dissolved P ranged between 0.3 and 2% of the total P lost.

Table 19.8. Phosphorus losses (kg P ha^{-1}) due to soil erosion and runoff in Mistelbach and Pyhra during the 1994 growing season.

		Conventional till		Conservation till		No till	
	Precip. (mm)	Total P	Dissolved P	Total P	Dissolved P	Total P	Dissolved P
Mistelbach							
26.5.94	51.7	183.2	0.48	23.3	0.36	24.1	0.21
3.6.94	44.3	8.8	0.05	0.8	0.03	0.1	0.09
30.6.94	115.6	100.3	0.27	18.9	0.05	3.9	0.28
19.7.94	52.1	5.3	0.05	0.4	0.01	0.2	0.01
Total	263.7	297.6	0.85	43.4	0.45	28.3	0.59
Pyhra							
1.6.94	87.5	54.0	0.10	31.2	0.06	17.1	0.10

20 Catchment Studies, Modelling and Management

CATCHMENT STUDIES

20.15 Catchment Studies of the Loss of Phosphorus from Agriculture to Surface Water

R. Grant, B. Kronvang and A. Laubel
National Environmental Research Institute, Department of Freshwater Ecology, Vejlsøvej 25, DK-8600 Silkeborg, Denmark

Introduction

In many shallow surface waters, eutrophication may be controlled by diffuse loss of phosphorus (P) from arable land. Danish shallow lakes are eutrophic above a threshold P concentration of 0.080–0.120 mg P l^{-1}, and this P concentration is exceeded in most Danish non-point-polluted streams and lakes. Increased effort is needed to abate diffuse P loading (Kronvang *et al.*, 1995), and this requires further knowledge of the diffuse sources.

This project aimed to identify and quantify the sources of P loss at a catchment level under varying delivery conditions. Results are presented for the year May 1993–April 1994 from an intensively farmed stream catchment, Gelbæk (Grant *et al.*, 1995).

Methods

The soil type is mainly sandy loam and about 50% of the arable land is tile-drained. Drainage water from two representative tile-drained catchment areas and stream water were measured and sampled, both hourly (intensive) and weekly (discrete), with more frequent sampling during storm events. Phosphorus in rill erosion within the catchment was estimated.

Results and discussion

Importance of sampling strategy. Compared with intensive sampling, the discrete sampling strategy, normally used for monitoring purposes, underestimated annual total P loss by 120% for drainage water and 27% for stream water; this was attributed to the event nature of the P flux. Sampling during runoff events is essential if reliable estimates of P loads and other sediment-bound contaminants are required.

Loss of different phosphorus forms. Determined by intensive sampling, the annual loss of total P from one tile-drained catchment was 0.627 kg P ha^{-1}, of which dissolved P was 71%. This catchment received large inputs of manure and parts of the catchment were low-lying riparian areas with a permanently high groundwater-table. For the other tile-drained catchment, annual total P loss amounted to 0.098 kg P ha^{-1}, of which dissolved P was 45%. From the Gelbæk stream catchment, annual total P loss was 0.61 kg P ha^{-1}, of which 48% was dissolved P.

Phosphorus losses were associated mainly with peak flows; 40% of the annual total P lost from tile-drained catchments and 37% of that from the stream catchment occurred during 1 month, March 1994.

Transport processes in the catchment. For the stream, as well as for the individual tile-drained catchments, there was a positive correlation between the concentration of particulate P and particulate matter. Hence the transport mechanisms of particulate P from soils to drains have been interpreted by studying the trends in particulate-matter concentration during single storm events and during the first autumn storms. Differences were explained by a decrease in the quantity of fine particles in soil macropores or in drains.

Budget for the phosphorus delivery sources. Phosphorus loss in drainage water and rill erosion can be calculated by the difference between P concentrations in stream and drain water. This assumes that the studied drained catchments were representative of the entire artificially drained area within the stream catchment. Drainage water contributed about 40% of the annual dissolved P loss and 18% of the annual particulate P loss from the Gelbæk catchment. During single storm events, however, drainage water contributed between 5 and 51% of the particulate P loss. Particulate P from rill erosion contributed only about 7% of the annual particulate P loss, thus stream-bed and bank erosion was the dominant source of particulate P.

Quantification of the phosphorus sources by 'fingerprinting'. During storm events, the physical and chemical composition of suspended matter was investigated to assess possible 'fingerprinting' techniques to quantify the sources of suspended matter and associated P within the catchment area. Sediment from different sources was analysed for grain-size composition, organic matter, organic nitrogen, trace elements, P fractions and caesium-137 (^{137}Cs). The results suggested various difficulties with such 'fingerprinting' for catchments with a heterogeneous geology and intensive farming. However, using different sets of 'fingerprinting' may be more promising.

Conclusion

On average, P lost from tiled drains contributed about 30% of the total P lost from the loamy soil in the Gelbæk catchment. About one-third of Danish arable land lies on loamy soils and more than half of this is artificially drained. Water from tile drains on loamy soils is, therefore, along with erosion, an important diffuse source of P to the aquatic environment. Internal stream erosion appears to be the main source of particulate P.

References

Grant, R., Laubel, A., Kronvang, B., Andersen, H.E., Svendsen, L.M. and Fuglsang, A. (1996) Loss of dissolved and particulate phosphorus from arable catchments by subsurface drainage. *Water Research* 30, 2633–2642.

Kronvang, B., Grant, R.R., Larsen, S.E., Svendsen, L.M. and Kristensen, P. (1995) Non point nutrient losses to the aquatic environment in Denmark: impact of agriculture. *Marine and Freshwater Research* 46, 1–11.

20.16 Sources of Soluble and Particulate Phosphorus in Surface Water in Eastern England: Relative Importance of Agricultural Versus Small Point Sources

D. Harper and G. Evans
Ecology Unit, Department of Zoology, University of Leicester, Leicester LE1 7RH, UK

Introduction

Water-management authorities (at present the National Rivers Authority (NRA) in the UK) require three types of information about phosphorus (P) and eutrophication.

First, they need an accurate knowledge of the P status of watercourses in order to implement legislation. This comes both from the European Community, particularly the Urban Waste-Waters Directive (UWWD), and from the UK government, particularly the Special Ecosystem Use Class of the Statutory Water Quality Objectives of the 1991 Water Resources Act. Thus rivers receiving excessive amounts of P can be targeted for remedial action and, equally important, rivers in a seminatural state, receiving little P, can be protected from further deterioration.

Second, they need to know the nature of the P and its sources. Most of eastern England is farmed intensively and some areas are densely populated. Few rivers are affected by agriculture or sewage effluent alone, and both sources and amounts of P must be known.

Lastly, and arguably most importantly, must come a clear demonstration of the ecological value of low-P watercourses in their own right, as opposed to the value of oligotrophy in lakes or reservoirs they may drain into. Up to now much effort has been directed at understanding and protecting standing waters, so very little is known about how stream ecology is effected by P enrichment. Eastern England has few standing waters (although the Norfolk Broads and several reservoirs are ecologically important), so it could be argued that P standards for the whole area should logically be based upon river ecology. Most knowledge of river eutrophication relates to macrophytes (Mainstone *et al.*, 1994), and the classification of river plants throughout the UK shows trophic differences among rivers (Holmes, 1995). In lowland, nutrient-enriched rivers, particularly small ones, small differences in trophy are harder to detect and the use of algae and macroinvertebrates may be necessary.

The NRA (Anglian Region) commenced routine analyses of soluble P in its rivers, most at monthly intervals, in 1990. Biological investigations under the UWWD on rivers below large sewage-treatment works where P removal had been implemented had started before 1990. In 1993 a project was begun to identify low-nutrient rivers within the region, to investigate the nature and sources of their nutrients, particularly P, and to demonstrate the ecological value of such rivers using macrophytes, algae and macroinvertebrates.

Preliminary results
The majority of rivers within the Anglian Region are rich in P. The majority of sites sampled, including nearly all sites on large rivers and those close to the London conurbation, had mean soluble P concentrations exceeding 0.1 mg l^{-1}. All rivers were affected by sewage-effluent discharges, the majority from small village sewage works. The NRA monitoring data show that all rivers receiving sewage effluent from large works (dry-weather discharge exceeding $20,000$ m^3 day^{-1}) have median soluble P values above 1 mg P l^{-1}, with lower values for streams without, or with very small, effluent discharges. There is much noise in the data, caused by such factors as infrequent sampling, catchment geology and river-channel character, but, nevertheless, preliminary conclusions are that large rivers all have large P concentrations, with the exception of the Bure, where P-stripping has been implemented to protect the Norfolk Broads. The P status of small rivers reflects the effluent loading from their catchments. First- and second-order streams without sewage effluents are often low in P, but a substantial number of these, without effluent discharges or with very small ones, have high P levels.

References
Holmes, N.T.H. (1995) *Macrophytes for Water and Other Quality Assessments.* Contract Report to the National Rivers Authority, Anglian Region, Peterborough.
Mainstone, C., Gulson, J. and Parr, W. (1994) *Phosphorus in Freshwater: Standards for Nature Conservation.* English Nature Research Reports No. 73, English Nature, Peterborough.

20.17 Biogeochemical Significance of Membrane and Ultrafilter Separation of Low-Molecular-Weight Molybdate-Reactive Phosphorus in Soil and River Waters

P.M. Haygarth,[1] M.S. Warwick[2] and W.A. House[2]
[1]*Institute of Grassland and Environmental Research, North Wyke, Okehampton, Devon EX20 2SB, UK;* [2]*Institute of Freshwater Ecology, River Laboratory, East Stoke, Wareham, Dorset BH20 6BB, UK*

Introduction
Eutrophication of inland and marine waters, associated with excessive loadings of phosphorus (P), nitrogen (N), carbon (C) and iron (Fe), is a contemporary European water-quality issue (PARCOM, 1993). Understanding the environmental mobility and bioavailability of P for algal uptake is essential if eutrophication is to be minimized (Bosdrom *et al.*, 1988). Molybdate-reactive P (MRP), usually determined on < 0.45 µm filtrate, is assumed to represent soluble, mobile and hence biologically available P. Other species, such as organic and particulate forms of P, are potentially biologically less significant, because P is retained within or attached to an organic/inorganic precipitate, itself possibly attached to

larger particles. This study investigated how MRP was associated with different particle-size fractions below 0.45 µm and assessed the variability in MRP in these fractions for a range of natural waters.

Methods

Two river waters, one from a lowland chalk stream, River Frome, and the other from an upland river containing humic materials, River Swale, and two soil waters, one a freely drained soil leachate collected at 135 cm depth and the other a storm-induced surface runoff, both from grassland soils, were sampled and handled according to the protocol of Haygarth *et al.* (1995). The samples were fractionated into < 0.45 µm, < 0.22 µm, < 0.1 µm and < 0.025 µm fractions, using composite cellulose-nitrate/acetate membranes under vacuum at 60 cmHg, while fractions less than 10,000 and 1000 molecular weight (MW) size were separated by ultrafiltration at 207 cmHg. Molybdate-reactive P was measured on each size range.

Results and discussion

In three of the four waters, the concentration of MRP in the < 1000 MW fraction was significantly smaller ($P < 0.05$) than that determined in the fraction < 0.45 µm (Table 20.1 and Fig. 20.1). In the soil-surface runoff water, dominated by eroded particles of clay and colloidal organic material, MRP in the < 1000 MW fraction was only 14% of the total; the majority, 74%, was associated with the unfiltered < 0.45 µm fraction, with the residual 12% in the < 0.45 to < 1000 MW range. In contrast, for the soil leachate and chalk-stream water, 96 and 67%, respectively, of the MRP was in the < 1000 MW fraction, with significantly different, but smaller, portions in the < 0.45 µm to < 1000 MW range. These results illustrate the variability in P associated with different particle-size fractions, including the colloidal components, in different natural waters. The apparent importance of particulate material in transport of MRP in the soil

Table 20.1. Molybdate-reactive phosphorus determined after membrane and ultrafiltration (µmol P l^{-1}).

Filter	River Frome		River Swale*		Soil leachate		Soil runoff	
	Mean	s(n−1)	Mean	s(n−1)	Mean†	s(n−1)	Mean†	s(n−1)
Unfiltered	5.80‡	0.10	0.54	0.01	2.38	0.15	5.14‡	0.40
< 0.45 µm	4.21	0.04	0.43	0.05	2.41	0.07	1.48	0.11
< 0.22 µm	4.16	0.26	0.58	0.05	2.41	0.06	1.11	0.04
< 0.1 µm	4.01§	0.09	0.61	0.04	2.42	0.05	1.26	0.16
< 0.025 µm	4.10§	0.02	0.61	0.15	2.54	0.09	1.46	0.33
< 10,000 MW	4.04§	0.03	0.72	0.05	2.55	0.05	0.97§	0.08
< 1000 MW	3.87§	0.23	0.65	0.15	2.26§	0.10	0.72§	0.09

* No statistical analyses conducted on the River Swale water, because the analytical values were close to the limit of detection.
†Significant difference between all < 0.45 µm, single-factor anova (soil leachate: $F = 14.3$, $P = 1.54 \times 10^{-6}$; soil runoff: $F = 11.88$, $P = 6.35 \times 10^{-5}$).
‡ Significantly larger than < 0.45 µm ($P < 0.05$).
§ Significantly smaller than < 0.45 µm ($P < 0.05$).

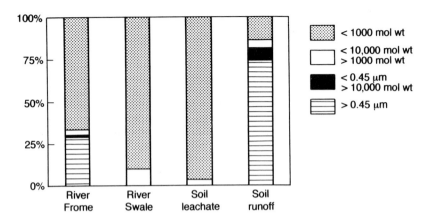

Fig. 20.1. Proportion of MRP in particle-size ranges of four natural waters.

runoff water implies that physical movement of solid material must be included in mechanistic models of soil-P mobility. Hannapel *et al.* (1964) suggested this when considering mobility of organic P attached to microorganisms moving downward through the soil profile, and these results suggest that the concept could be applied to the transport of inorganic material. The wider implication is that P could be physically mobile, by being attached to colloids and particles, while remaining chemically immobile – for example, when incorporated in/on a precipitate or large ligand – but perhaps biologically available.

Conclusions

Concentrations of MRP differed in different-sized fractions from < 1000 MW to unfiltered in four samples of natural waters. Molybdate-reactive P was not wholly associated with low MW forms of P and, in some circumstances, a considerable proportion of MRP was attached to particles > 1000 MW. The results also highlight the significance of filter size when determining MRP and suggest that the conventional filtration < 0.45 µm and determination with molybdenum blue may underestimate the total reactive P, particularly if dominated by large particulates, such as in soil runoff. Colloidal and particulate material plays a significant role in the mobility and potential bioavailability of P species.

Acknowledgements

We are grateful to the Ministry of Agriculture, Fisheries and Food, London, and the Natural Environment Research Council (NERC) Land Ocean Interaction Study (LOIS) programme for providing financial support.

References

Bosdrom, B., Persson, G. and Broberg, B. (1988) Bioavailability of different phosphorus forms in fresh water systems. *Hydrobiologia* 170, 133–155.

Hannapel, R.J., Fuller, W.H. and Fox, R.H. (1964) Phosphorus movement in a calcareous soil: II. Soil microbial activity and organic phosphorus movement. *Soil Science* 97, 421–427.

Haygarth, P.M., Ashby C.D. and Jarvis, S.C. (1995) Short-term changes in the

molybdate reactive phosphorus of stored soil waters. *Journal of Environmental Quality* 14(6) 1133–1140.
PARCOM (1993) *Nutrients in the Convention Area 1993.* Oslo and Paris Commissions, Paris, 86 pp.

20.18 Contribution of Agriculture to the Phosphorus in Surface Waters in Finland and Measures to Reduce it
S. Rekolainen
Finnish Environment Agency, Impact Research Division,
PO Box 140, FIN-00251 Helsinki, Finland

Introduction

Nutrients leached from agricultural land and lost during manure storage induce eutrophication of surface waters in Finland. Based on 16 water-quality variables, about 20% of lakes and 55% of rivers show various degrees of degradation (Finnish National Committee for UNCED, 1991). Of these, a significant fraction are affected solely or mainly by agriculture. Agricultural land covers 9% of the land area of Finland and is restricted mainly to southern and central Finland. Agriculture contributes to eutrophication mostly in south-western Finland, where the proportion of agricultural land exceeds 30% in many river basins.

For a long time, phosphorus (P) and nitrogen (N) loads from agricultural land have been larger than industrial and municipal loads combined. The proportional contribution of agriculture to P loading has increased, especially during the last two decades, due to declining loads from municipal waste waters. This has focused attention on policies for reducing non-point sources of pollution.

The Finnish government decided in 1988 (Ministry of Environment, 1988) that P loading from agriculture should be reduced by 30%, combined with a significant reduction in N loading, by 1995. Decreasing P loading was given priority, because P is usually the limiting factor for algal growth in Finnish fresh waters.

Effective eutrophication control through best management practices requires knowledge of the nutrient losses from agriculture and how they are affected by management practices. Research on P losses and their reduction by agricultural management practices is summarized here.

Results and discussion

The use of commercial P fertilizers increased from the mid-1950s to the early 1990s. Since then, a sharp decline has taken place, from 35 kg P ha^{-1} to 19 kg P ha^{-1} annually. The large input of P resulted in its accumulation in agricultural soils. Estimates suggest that total P in surface soils has increased by one-third in the last 40 years.

Estimates of P losses from agricultural areas in Finland were based on discharge and water-quality records of small representative drainage basins. At present, the monitoring network includes 13 basins, varying in size between 0.07 and 21.7 km^2, and land use ranging from zero to 100% agriculture.

In 1986–1990, P loss from agricultural land was 0.95–1.70 kg P ha^{-1} (Rekolainen *et al.*, 1995), which showed no change from 1981–1985 (Rekolainen, 1989). Based on these results, the total P load from all agricultural land in Finland was estimated to be 2000–4000 t year^{-1}. This load far exceeded that from the other main nutrient sources: industry, municipalities and fish farming (Table 20.2).

Table 20.2. Loads of total phosphorus from the different sources in Finland.

	Total phosphorus (t year^{-1})
Industry	410
Municipalities	240
Fish farming	190
Field cultivation	2000–4000
Manure storage	300

In south-western Finland, where the impact of agriculture on eutrophication is greatest, the dominant soil type is clay, with clay content exceeding 30%. Phosphorus losses can be significantly decreased by minimizing erosion, because about 75% of the total P moving to watercourses is attached to eroded soil particles.

If not under grass, most of the cropped fields in Finland are mould-board-ploughed in September or October, and very little plant residue is left on the surface to reduce the erosive impact of raindrops and running water on soil particles.

Model estimates, using a modified version of the chemicals, runoff and erosion from agricultural management systems (CREAMS) model (Knisel 1980; Rekolainen and Posch, 1993), showed that replacing mould-board ploughing with reduced tillage could decrease erosion substantially. According to these estimates, best results would be achieved in clay and sandy soils, but a great reduction in the tillage intensity would be needed to get a significant reduction in soil loss (Rekolainen and Posch, 1991). Without tillage in autumn or if tillage is postponed to the following spring, soil loss is less than half that caused by conventional tillage.

Also, according to the CREAMS model estimates, filter strips, with vegetation at the lower end of slopes, can reduce soil loss significantly in Finnish conditions, especially in sandy soils, because less soil is lost from the strip itself and eroded material is deposited within the strip.

Conclusions

Experimental studies and model evaluations suggest that an extensive establishment of vegetation filter strips, a general adoption of reduced-tillage techniques and perennial grassland on all set-aside fields may reduce the total P load to surface waters in Finland by about 40%, by minimizing soil erosion. In addition, proper storage facilities for manure could reduce total P loss by another 10–15%. Large applications of P fertilizers to some agricultural soils in Finland have increased the soluble P concentration, indicating a risk for leaching. In extreme cases, only 20% of the P adsorption capacity is unused. Losses of soluble P can probably be reduced most effectively by controlling the input of P into soil, but the quantitative effect of controlled fertilization levels requires further investigation.

References
Finnish National Committee for UNCED (1991) *National Report to UNCED 1992.* Publication of Ministry for Foreign Affairs 13/91, Helsinki.
Knisel, W.G. (ed.) (1980) *CREAMS: a Field-Scale Model for Chemicals, Runoff, and Erosion from Agricultural Management Systems.* Conservation Research Report 26, US Department of Agriculture, Washington DC.
Ministry of Environment (1988) *Water Protection Programme until 1995 in Finland.* Series B, 12/1988, Ministry of Environment, Helsinki.
Rekolainen, S. (1989) Phosphorus and nitrogen load from forest and agricultural areas in Finland. *Aqua Fennica* 19, 95–107.
Rekolainen, S. and Posch, M. (1991) Effects of conservation tillage techniques on erosion control in Finland – a model evaluation. In: *Preprints*, Vol. 1, *International Hydrology and Water Resources Symposium.* National Conference Publication No. 91/22, The Institution of Engineers, Barton, ACT, Australia, pp. 236–240.
Rekolainen, S. and Posch, M. (1993) Adapting the CREAMS model for Finnish conditions. *Nordic Hydrology* 24, 309–322.
Rekolainen, S., Pitkänen, H., Bleeker, A. and Felix, S. (1995) Nitrogen and phosphorus fluxes from Finnish agricultural areas to the Baltic Sea. *Nordic Hydrology* 26, 55–72.

20.19 Identifying Critical Sources of Phosphorus Export from Agricultural Catchments

A. Sharpley and J. Lemunyon
USDA-ARS, Pasture Systems and Watershed Management Research Laboratory, Curtin Road, University Park, Pennsylvania and USDA-NRCS, Fort Worth, Texas, USA

Introduction
Increased transport of phosphorus (P) in agricultural runoff often promotes the growth of algae in lakes, which can consume so much oxygen that fish suffocate. Phosphorus losses are of particular concern in areas of intensive cropping and livestock production, in which P inputs have exceeded outputs for several years, often leading to accumulation of P in surface soil in excess of crop requirements and P enrichment of runoff (Sharpley *et al.*, 1994). Thus, sites vulnerable to P loss must be identified to implement cost-effective remedial strategies. Site vulnerability may be assessed by field studies and/or computer simulation. Field assessments are time-consuming, costly and labour-intensive while simulation models require detailed soil and management information. A simple P index was developed as a field tool to quantify the relative importance of transport (runoff and erosion) and source factors (soil type and rate, timing, method and type of P applied) controlling P loss in runoff (Lemunyon and Gilbert, 1993). The index has had limited field testing and validation, but results from application to 26 grassed and cropped catchments in Oklahoma and Texas, USA, are compared with losses of P in runoff from these watersheds over the last 16 years.

Methods
Catchment management included native and introduced grass pastures, conventional and no-till winter wheat, reduced and no-till fallow–sorghum–wheat rotations and peanut cropping. Soils encompassed fine sandy loams to clay loams and fertilizer P applications were from 0 to 30 kg P ha^{-1} year^{-1}, based on Mehlich-3 soil-P recommendations. Catchment instrumentation, runoff collection and P analysis started in about 1976 and additional information on the catchments and

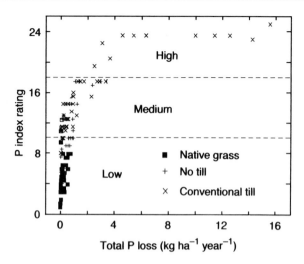

Fig. 20.2. Relationship between P index rating and mean annual total P loss in runoff as a function of catchment management.

runoff analysis are given by Sharpley *et al.* (1992).

The P index is outlined in Tables 20.3 and 20.4. Each site characteristic affecting P loss is assigned a weight, assuming that certain characteristics have a relatively greater effect on potential P loss than others. An assessment of catchment vulnerability to P loss in runoff is made by selecting the rating value for each characteristic from the index (Table 20.4). Each rating is multiplied by the appropriate weighting factor. Weighted values of all site characteristics are summed and catchment vulnerability is obtained from Table 20.4.

Results

Mean annual sediment (25–5300 kg ha^{-1} year^{-1}) and total P (TP) discharged (0.1–12 kg ha^{-1} year^{-1}) from the catchments have ranged widely over the last 16 years. The TP loss was compared with index ratings for vulnerability to P loss in runoff (Fig. 20.2) and the order of vulnerability is consistent with measured TP loss. Overall, the grassed catchments had a lower vulnerability to P loss than the cropped catchments. From Fig. 20.2, it can be seen that a TP loss in runoff exceeding 2 kg ha^{-1} year^{-1} generally classifies a catchment as having a high vulnerability to P loss. The close relationship between TP loss and P index rating shown in Fig. 20.2 indicates that the indexing procedure can give a reliable estimate of catchment vulnerability to P loss. This was true, even though the catchments had widely differing management. Widespread use of the index by extension agencies is planned as a tool for field personnel to easily identify agricultural areas or practices that have the greatest potential for P loss. It is intended that the index will identify management options available to land users that will allow them flexibility in targeting control strategies that minimize P loss, while maintaining crop productivity.

Table 20.3. The P indexing system to rate potential P loss in runoff.

Site characteristic*	Phosphorus loss potential (value)				
	None (0)	Low (1)	Medium (2)	High (4)	Very high (8)
Soil erosion (t ha^{-1})	<10	10–100	100–500	500–5000	>5000
Runoff (cm)	<0.1	0.1–1.0	1–5	5–10	>10
Soil test P (mg kg^{-1})	<10	10–20	20–40	40–65	>65
P application rate (kg P ha^{-1})	None applied	<10	10–20	20–40	>40
P application method	None applied	Placed with planter deeper than 5 cm	Incorporated immediately before crop	Incorporated >3 months before crop or surface-applied <3 months before crop	Surface-applied >3 months before crop

* Each site characteristic is weighted by the following factors: soil erosion, 1.5; runoff, 0.5; soil test, 1.0; P application rate, 0.75; P application method, 0.5.

Table 20.4. Site vulnerability to P loss as a function of total index values.

Site vulnerability	Total index rating value
Low	< 10
Medium	10–18
High	19–36
Very high	> 36

References

Lemunyon, J.L. and Gilbert, R.G. (1993) The concept and need for a phosphorus assessment tool. *Journal of Production Agriculture* 6, 483–496.

Sharpley, A.N., Smith, S.J., Jones, O.R., Berg, W.A. and Coleman, G.A. (1992) The transport of bioavailable phosphorus in agricultural runoff. *Journal of Environmental Quality* 21, 30–35.

Sharpley, A.N., Chapra, S.C., Wedepohl, R., Sims, J.T., Daniel, T.C. and Reddy, K.R. (1994) Managing agricultural phosphorus for the protection of surface waters: Issues and options. *Journal of Environmental Quality* 23, 437–451.

20.20 Improved Measurements of Phosphorus Loss to Watercourses from Agricultural Areas

L. Wiggers
Aarhus County, Environmental Division, Lyseng Alle 1, 8270 Hoejbjerg, Denmark

Introduction

Phosphorus (P) from unpurified or partly purified sewage water has been the main reason for eutrophication in Danish lowland freshwater ecosystems. Improved treatment of sewage water in Denmark has reduced P loadings from this source – for example, in Aarhus County by 75% during the period 1989–1994. In consequence, P runoff from agricultural areas, including loadings from houses without sewage systems, has become proportionally more important. Also, the contribution of P from these diffuse sources has been underestimated, due to imperfections in the sampling technique. Improved methods for estimating P transport to streams shows that the contribution of P from agricultural areas is larger than previously calculated.

Methods

The sampling frequency in most Danish streams has typically been 18–26 times a year from 1989 until now. In 1993, a new sampling technique was implemented at a number of stations representative of streams in small agricultural catchments without loadings from industrial or municipal waste water.

In Aarhus County there are two stations: Jaungyde Brook, a catchment of 10.6 km^2 with 94% agricultural land, and Horndrup Brook, a catchment area of 5.5 km^2 with 74% agricultural land. The soils of the Jaungyde Brook and Horndrup Brook catchments are dominated by clayey sand and sandy clay, respectively.

Phosphorus transport in streams is estimated from samples taken hourly throughout the year by means of an automatic sampler (ISCO 3700). This frequency ensures that storm-flow events for shorter periods are represented. The water level is automatically logged and the data can be monitored at the office via a modem.

Samples are pooled according to flow events. In weeks without sudden rises in water level, all samples are pooled. However, when the water level rises suddenly due to a storm-flow event, samples from this event are pooled separately; the smallest time interval has been 24 h. The pooled samples are analysed for total P (TP) and inorganic dissolved P (DP). Transport of TP and DP in the brooks is estimated by multiplying water flow, estimated for 1-day time intervals, by concentrations, estimated for 1-day or longer time intervals, depending on the flow events. Particulate P (PP) is calculated as the difference between the transport of TP and DP.

The sampling technique ensures that high concentrations of TP, which typically occur in conjunction with storm-flow events, are multiplied by the simultaneously large volumes of water, which are typically of short duration. In the brooks, the normal sampling strategy, with 18 or 26 single samples, has been continued also and P transport based on these samples has been estimated by the C-linear interpolation method.

Results

In Jaungyde Brook and Horndrup Brook, the discharge of TP for the last 2 years has been estimated at between 0.5 and 1.3 kg P ha^{-1} each year, using frequent sampling (Table 20.5). Losses of TP were highly dependent on water discharge, and P losses calculated by this method were between 1.5 and 2.5 times larger than those calculated from 18–26 single samples. For DP, the losses calculated by the two methods were about the same. Thus, the difference in losses of TP estimated by the different sampling methods is due to a larger contribution of PP, determined by frequent sampling. For example, in 1994, more than 75% of the TP was transported in both brooks during 10% of the time, which explains the significant difference in discharge calculated from the two sampling strategies.

The diffuse-P load in a catchment originates from a natural background load, waste water from households without sewage systems and runoff from agricultural land. The background load is estimated using TP concentrations from streams running in non-cultivated areas. The P load from households without sewage systems is estimated using a standard emission value of 2 g P per capita per day – this is decreased by 55% due to retention in the soil. The contribution from agricultural land is estimated as the difference between total transport and that from other sources. The agricultural load is by far the

Table 20.5. Loss of total (TP) and dissolved phosphorus (DP) calculated using a frequent sampling method in two catchments in Denmark.

	Jaungyde Brook			Horndrup Brook		
	TP (kg ha^{-1})	DP (kg ha^{-1})	Water (mm)	TP (kg ha^{-1})	DP (kg ha^{-1})	Water (mm)
1993	0.51	0.15	238	0.64	0.16	342
1994	1.26	0.36	435	1.34	0.30	524

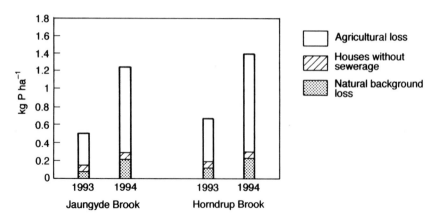

Fig. 20.3. Total phosphorus loss.

dominant source (Fig. 20.3). Estimating P transport on the basis of frequent sampling substantiates this, as do the new standard P emission values per capita, which are half the earlier values.

Particulate P, which accounts for by far the biggest part of the load, can be segregated in different fractions: easily adsorbed P, organic-bound P, iron- or aluminium-bound P and calcium- and magnesium-bound P. A significant part (more than 80%) of the PP transported during a storm-flow event in five streams was either organic-bound P or associated with iron or aluminium compounds. The first fraction readily becomes biologically available in lakes, and also, in anaerobic conditions, the second fraction is typically found in the sediment in many lakes. Consequently, the pulse of P transported during storm flow in the streams influences the downstream lakes and coastal waters by increasing eutrophication.

20.21 Phosphorus Loss to Water from Agriculture in the UK
P.J.A. Withers
ADAS Bridgets Research Centre, Martyr Worthy, Winchester SO21 1AP, UK

Introduction

The Ministry of Agriculture, Fisheries and Food (MAFF) currently funds a research programme by the Agricultural Development and Advisory Service (ADAS) aimed towards a greater understanding of the agricultural contribution to eutrophication and identification of appropriate measures to reduce agricultural phosphorus (P) loss to sensitive waters. While point sources are known to have a major influence on the P content of many major rivers (Muscutt and Withers, 1996) there is little quantitative information in the UK on the concentrations and loss of P to water from agricultural land. Preliminary results from the research programme are presented here.

Methods

Experimental catchments representative of arable farming, mixed agriculture and upland farming have been established and instrumented to continuously monitor flow and automatically collect water samples in both low- and high-flow conditions. Selective monitoring of subcatchments and field drains is used to define the pathways of P loss, while replicated field experiments are used to measure the amount, form and timing of P loss in surface and subsurface runoff from arable land receiving livestock manures. Leaching losses of P from manured soils with large amounts of bicarbonate-soluble P (Olsen P) are being assessed in lysimeter studies and also at field sites by means of Teflon-coated suction cups installed at various depths. Molybdate-reactive P (MRP) and total dissolved P (TDP) on filtered (0.45 µm) and total P (TP) on unfiltered samples are measured routinely.

Results

Catchment monitoring. Environmentally significant loads and concentrations of both TDP and particulate P (PP) were measured in surface and subsurface flow within agricultural catchments. Annual TP losses ranged from 0.7 to 3.4 kg P ha^{-1} (Table 20.6). Total P concentrations in surface runoff and drain flow were as large as 6 and 2 mg P l^{-1}, respectively, during periods of large flow. Most P loss was in particulate form, but concentrations of MRP were also above limits for eutrophication for both surface runoff (up to 0.8 mg l^{-1}) and drain flow (up to 0.2 mg l^{-1}). Most TDP occurred as MRP, and TP and MRP loss increased with flow rate.

Organic manures. Application of liquid manures to soils of moderate P status under arable cropping caused substantial increases in both the concentrations and loads of MRP and TP in both surface runoff and drain flow (Table 20.7). Maximum P concentrations occurred in the first drainage event after slurry application.

Leaching. In lysimeter studies, cumulative applications of P equivalent to a maximum of 1200 kg ha^{-1} have not yet increased concentrations of P in the leachate, but, at field sites which had accumulated > c. 60 mg Olsen P l^{-1} in the topsoil, concentrations of MRP in the soil solution extracted by suction cups to

Table 20.6. Annual P loss in small rural catchments in England.

Catchment	Size (ha)	Land use	Loss (kg P ha^{-1}) TP	MRP	Mean concentration (mg P l^{-1})
Redesdale	449	Upland grass	0.70	0.23	0.14
Jubilee	30.6	Lowland arable	2.60	0.72	0.80
Foxbridge*	5.9	Lowland arable	1.63	0.24	0.85
Cliftonthorpe	95.5	Lowland mixed	3.40	0.22	0.95
Lower Smisby	260	Mixed/urban	1.40	0.48	0.47

* Foxbridge is a field drain that feeds into the Jubilee catchment.

Table 20.7. Concentrations and loads of phosphorus following application of livestock manure at two sites in England.

	ADAS Rosemaund: surface flow		ADAS Boxworth: drain flow	
	Control (no slurry)	Cow slurry (76 m³ ha⁻¹)	Control (no slurry)	Pig slurry (59 m³ ha⁻¹)
Max. conc. (mg P l⁻¹)				
TP	0.70	27	0.16	10
MRP	0.26	12	0.03	3
Total load (kg P ha⁻¹)				
TP	0.04	1.54	0.20	1.20
MRP	< 0.01	0.74	0.04	0.45

90 cm exceeded 1 mg l⁻¹. At two sites, MRP concentrations were 0.5 mg l⁻¹ even at 2 m depth. These results suggest that significant leaching losses of P can occur at sites that have accumulated large amounts of extractable P in the soil from past applications of organic manures.

Acknowledgements

Financial support of MAFF is gratefully acknowledged. Thanks are also extended to all ADAS staff involved in this work.

Reference

Muscutt, A.D. and Withers, P.J.A (1996) The phosphorus content of rivers in England and Wales. *Water Research* 30, 1258–1268.

MODELLING AND MANAGEMENT

20.22 Phosphorus Concentrations in Surface Water and Drainage Water in the Watershed of the Poekebeek, Flanders, Belgium

R. Hartmann, H. Verplancke, P. Verschoore, M.M. Villagra
University of Ghent, Department of Soil Management and Soil Care, Division Soil Physics, Coupure Links 653, B-9000 Ghent, Belgium

Introduction

Within the project of the regional inventory of the phosphate saturation degree (PSD) of the sandy and light sandy loam soils in Flanders, a more thorough study is being made in the watershed of the Poekebeek. This watershed is partially situated in West Flanders and East Flanders and covers some 110 km². Intensive animal husbandry in this area results in an overproduction of manure and very intensively manured farmland. The phosphorus (P) input exceeds the P requirements of the crops and P accumulates in the soil. The sorption capacity of these soils for P is limited, and, when P breakthrough occurs as the sorption capacity is exceeded, P will leach during the winter to the shallow groundwatertable or drain to the surface water (De Smet *et al.*, 1990). Because the watershed is an area for the production of drinking water, monitoring the quality of the surface water is of utmost importance.

Methods

A regional inventory of the phosphate sorption capacity (PSC) and the PSD of the soils in the watershed was made. Soils were sampled on a square grid with 1000 m intervals, which, together with some short-distance (250 m) sampling locations, provided 131 sampling points. At a later stage, 59 extra samples were taken to improve the data analysis. At each sampling point, five samples were taken in an area of 30 m^2 and bulked. The soil profile was sampled in layers of 30 cm to a depth of 90 cm. The determination of the PSC and the PSD was done according to Van der Zee *et al.* (1990).

A geostatistical analysis was made of the point data, to indicate the regions where a breakthrough of P during the winter is possible. Block kriging was used as an interpolation technique. The PSD was estimated for units of 200 m × 200 m, which resulted in a spatially continuous image of the PSD. By taking into account the block kriging standard deviation for each unit, a probability map of the PSD was created.

Within the watershed, the main river, the Poekebeek, has a length of about 24 km and is fed by several tributaries. To evaluate the orthophosphate condition of the surface water in the watershed, water samples were taken regularly and orthophosphate was measured (Murphy and Riley, 1962).

To estimate orthophosphate coming from arable land, water samples were taken from drainpipes throughout the watershed.

Results

When the critical PSD was set at 30% of the PSC, then 71% of the sampled points were considered to be phosphate-saturated throughout the profile, with a median of 38.7%.

A probability map of the PSD (not shown here) in the watershed shows that there is a 95% probability that 17% of the area is phosphate-saturated throughout the profile, while 53% of the area is phosphate-saturated with a probability < 68%.

The surface water in the several sub-basins of the watershed and in the Poekebeek itself does not meet the 0.2 mg orthophosphate l^{-1} norm. The largest concentrations are in the Poekebeek, because of the high orthophosphate values (between 0.4 and 20.7 mg orthophosphate l^{-1}) resulting from the discharge of domestic and industrial waste water by the city of Tielt, located close to its source.

Orthophosphate in the Poekebeek varied between summer, 2.1 mg orthophosphate l^{-1}, and winter, 0.7 mg orthophosphate l^{-1}, probably because the discharge of the Poekebeek was ten times higher in winter than in summer. However, the dilution in winter is not enough to come below the 0.2 mg orthophosphate l^{-1} norm.

The contribution of the 17% of the area with a 95% of probability of phosphate-saturated soils to the orthophosphate load of the surface water cannot be evaluated from the analysis of the water samples, because background orthophosphate concentrations are too high.

The concentrations of orthophosphate in the drainage water often exceed 0.1 mg orthophosphate l^{-1}, while values higher than 1 mg l^{-1} are not exceptional. Orthophosphate in drainage water varied from 0.01 to 0.27 mg l^{-1}, which shows the need for frequent sampling.

At present, the orthophosphate load of the Poekebeek is largely determined by the discharge of domestic and industrial waste water and work is being done to abate these discharges. That part of the orthophosphate load which comes from agriculture cannot yet be quantified properly.

Acknowledgements
The authors wish to express their gratitude to the Flemish Land Society (VLM, Brussels) for subsidizing this research.

References
De Smet, J., Hartmann, R. and De Boodt, M. (1990) De fosfortoestand van de bodem en het grondwater in het arrondissement Tielt. *Med. Fac. Landbouwwetenschappen, Rijksuniversiteit Gent* 55(1), 17–23.
Murphy, J. and Riley, J.P. (1962) A modified single solution method for determination of phosphate in natural waters. *Analytica Chimica Acta* 27, 31–36.
Van der Zee, S.E.A.T.M., Van Riemsdijk, W.H. and de Haan, F.A.M. (1990) *Het Protokol Fosfaatverzadigde Gronden. Deel I: Toelichting.* Vakgroep Bodemkunde en Plantevoeding, Wageningen Agricultural University, Wageningen.

20.23 Evaluating a Phosphate Saturation Inventory of Soils in Northern Belgium
I. Schoeters, R. Lookman, R. Merckx and K. Vlassak
Laboratory of Soil Fertility and Soil Biology, KU Leuven, Kardinaal Mercierlaan 92, 3001 Heverlee, Belgium

Introduction
The livestock population in Belgium has increased dramatically during the past few decades. This, in combination with a gradual decrease of total available agricultural area, has led to regional excesses of animal manure. Excessive manuring of sandy soils, with rather low phosphorus (P) fixing capacities, has resulted in excessive amounts of nitrogen (N) and P being leached to shallow groundwater and surface water, leading to surface-water eutrophication. As a basis for future governmental measures to protect ground- and surface-waters from excessive P enrichment, an inventory of the PBC (phosphate binding capacity) and the PSD (phosphate saturation degree) of the sandy soils in Belgium was started, while research sought a relationship between the PSD and the P content of the soil solution (P-sol). This study presents the statistical and geostatistical results of the PBC of a region in northern Belgium with a total area of 1600 km^2 and the relationship that has been found between PSD and P-sol.

Methods
An inventory was prepared using the methodology of Van der Zee *et al.* (1990a,b). At 1050 points determined using a square grid at 1800 m interval, with additional nesting with distances of 1500, 450, 100, 50 and 20 m, soils were sampled to 90 cm. Phosphate binding capacity and PSD were calculated by analysing soil samples for oxalate-extractable aluminium (Al) (Al-ox), iron (Fe) (Fe-ox) and P (P-ox). Frequency distributions of Al-ox, Fe-ox and PBC were analysed and their moments computed. The sample semivariogram of PBC was modelled and used to produce a map of the latter by 0.5 km^2 block kriging. The result was presented as a raster-format map. An overlay of the PBC map with texture and geological maps provided information about the relationship of PBC

and the origin of the soils. The relationship between PSD and P-sol for the top 30 cm of soil was studied on a field transect with 60 points, the soil was analysed for P-sol (Lookman *et al.*, 1995) and its PSD was calculated.

Results

The results of Al-ox, Fe-ox and PBC are summarized in Fig. 20.4 and Table 20.8. Both Al-ox and, especially, Fe-ox had a skewed distribution. When samples with an extreme Al-ox and/or Fe-ox content were identified on the texture and geological maps, it was found that soils of quaternary origin were rich in Al-ox while soils of a tertiary alluvial origin had a high glauconite (Fe-holding)

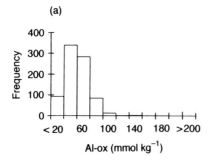
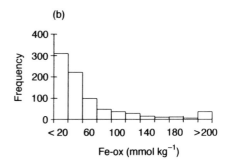
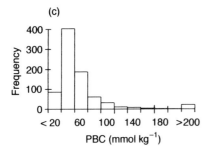

Fig. 20.4. Frequency distribution of the ammonium oxalate-extractable aluminium (Al-ox) (a) and iron (Fe-ox) (b) and phosphate binding capacity PBC (c).

Table 20.8. Basic statistics of the soils sampled.

	Al-ox (mmol kg^{-1})	Fe-ox (mmol kg^{-1})	PBC (mmol kg^{-1})
Sample size	1050	1050	1050
Mean	40.3	62.1	51.2
Standard deviation	17.3	140.9	70.2
Skewness	0.8	7.1	7.1
Kurtosis	4.7	64.4	65.3
Median	38.9	27	36

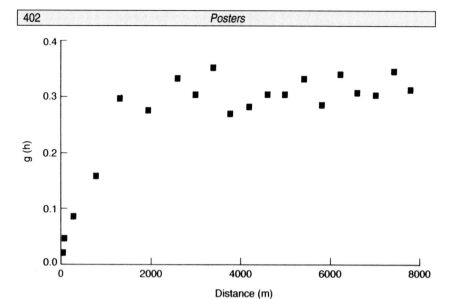

Fig. 20.5. Semivariogram for log PBC × g(h) = semivariance at lag distance h.

content. Soils with a texture of loamy sand, light sandy loam and some alluvial soils have a higher PBC than sandy soils, which are poorer in Al-ox and Fe-ox. Data of PBC were log-transformed to stabilize the variance when calculating the sample variogram. The sample variogram (Fig. 20.5) of the latter combine two components of variation: a short-range or field-to-field component, related to an agricultural effect and extending up to 250 m, and a long-range component, extending to 2300 m, which is probably the result of the textural and geological effect. Maps of PBC, made by block kriging, showed that regions with the lowest PBC also have the highest PSD and so present the highest risk for P leaching. The results of the relationship found between the PSD and the P-sol are presented in Fig. 20.6. Both an exponential and a power fit had a similar correlation coefficient. Thus PSD is a valuable tool to assess P leaching risk. From the regional inventory of PSD and its relationship with P-sol, P losses to groundwater can be estimated.

References

Lookman, R., Jansen, K., Merckx, R. and Vlassak, K. (1996) Relationship between soil properties and phosphate saturation parameters – a transect study in northern Belgium. *Geoderma* 69, 265–274.

Van der Zee, S.E.A.T.M., van Riemsdijk, W.H. and de Haan, F.A.M. (1990a) *Het protocol fosfaatverzadigde gronden. Deel I.: Toelichting.* Vakgroep bodemkunde en plantenvoeding, Wageningen Agricultural University, Wageningen, 69 pp.

Van der Zee, S.E.A.T.M., van Riemsdijk, W.H. and de Haan, F.A.M. (1990b) *Het protocol fosfaatverzadigde gronden. Deel II.: Technische uitwerking.* Vakgroep bodemkunde en plantenvoeding, Wageningen Agricultural University, Wageningen, 25 pp.

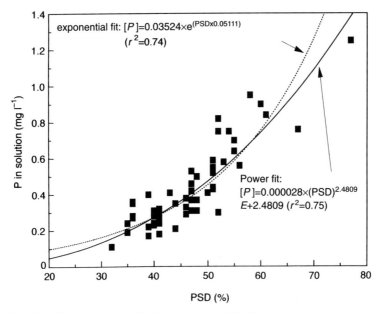

Fig. 20.6. Relationship between P in the soil solution (P-sol) and phosphate saturation degree (PSD).

20.24 Developing an Expert System for the Evaluation of Nutrient Losses from Agriculture to Water in Belgium

P. Scokart, P. Nyssen and P. De Cooman
Institute of Chemical Research, Ministry of Agriculture, Administration of Agricultural Research, Leuvensesteenweg, 17 B-3080 Tervuren, Belgium

Introduction

Before nutrients in agriculture can be managed by legislation, it is necessary to evaluate the share of all pathways to the total loading of surface water. For phosphorus (P), losses are mainly due to erosion and runoff (Nolte and Werner, 1991) and many models have been developed to evaluate such losses at a small scale. Using these models is often difficult and not user-friendly for managerial purposes when making decisions at a larger scale, when all pathways of loss must be taken into account. This study describes a more functional, pragmatic, empirical method for the calculation and estimation of P losses on a macro-economic level and its evaluation by field studies in three representative catchment areas in Belgium.

Methods

The total load to water from the various sources of P and nitrogen (N) have been divided into six categories, each of which can be estimated in a very simple way from official statistics and current agricultural practices. They are as follows.

1. Direct losses: input of nutrients without intermediate transport divided into the following.

- Direct losses of mineral fertilizers during storage, transport and application, estimated as a percentage of the total consumption of fertilizers.
- Direct losses of animal manure during grazing, calculated as a percentage of the nutrient excretion by the total number of animals during the grazing period (6 months).
- Direct losses of animal manure in stables, expressed as a percentage of the nutrient excretion by animals during housing and modulated by a function of the degree of modernization of livestock buildings.
- Direct losses from silage and manure storage capacity, calculated by a function of the total production of manure and yields of ensiled fodder crops.

2. Runoff losses: direct input of nutrients after spreading of slurry and runoff on country roads, expressed as a percentage of the total production and modulated in accordance with codes of good agricultural practice.

3. Drainage losses: input of nutrients via drainage water, calculated by the product of the mean nutrient concentration measured in drain water in various fields cultivated with normal agricultural practices, and the mean volume of drain water, estimated by establishing an approximate, annual hydrological balance.

4. Groundwater losses: input of nutrients via transfer of groundwater to surface waters with normal agricultural practices, calculated in the same way as drainage losses.

5. Erosion losses: losses of nutrients via soil erosion and runoff during large rainfall events. Soil losses are estimated by the universal soil loss equation (Gabriels et al., 1986) and nutrient enrichment by the Auerswald equation (Auerswald, 1989) and an enrichment coefficient. Losses in soluble form were calculated by the volume of water coming from rain causing erosion (Rogler, 1981), assuming that only 30% of water coming from arable land is discharged to surface waters.

6. Excess losses: input of nutrients from all 'nutrient streams' due to excessive application of livestock manures, estimated by an exponential function of the manure production in excess of a limit corresponding to good agricultural practices.

Each type of loss is calculated at the smallest available scale (municipality) and totalled for decision-making systems at the appropriate scale (catchment or regional scale).

Results and discussion

Figure 20.7 gives results of the calculation described above for the Flanders and Wallonia regions of Belgium in 1992.

Total P losses were very much larger in Flanders (1084 t year^{-1}) than in Wallonia (606 t year^{-1}), corresponding, respectively, to 1.87 kg and 0.81 kg P ha^{-1} of agricultural area in each region. Direct losses represent 51% of the total losses in Flanders and 39% in Wallonia. Runoff and excess losses due to bad agricultural practices can be estimated at 355 and 147 t P year^{-1} or 0.6 and 0.19 kg P ha^{-1}, respectively, for Flanders and Wallonia.

Direct losses are localized in only three catchments in Flanders with a very high density of livestock. Policies to reduce direct P losses to water would thus be focused on these high-livestock-density areas. In Wallonia losses originate from more diffuse sources and their reduction by direct control is more difficult at this small scale.

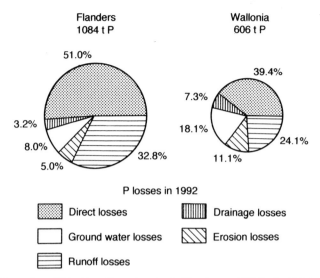

Fig. 20.7. Proportion of P lost by different routes in Flanders and Wallonia, Belgium in 1992.

The model has been tested and validated for P and N in three catchments. In general, good agreement was found between calculated and measured losses. The only discrepancies were observed where direct losses were very high and thus very variable with time.

The model can also be used to predict the reduction of nutrient losses which may be expected when legal measures or action programmes are implemented by the authorities. For example, a better use efficiency of P in pig and poultry feeds by using the enzyme phytase could give a reduction of P excretion of 20% and a reduction of total losses of approximately 50 t P year^{-1} in Flanders but only 5 t P year^{-1} in Wallonia.

Improvement in the storage capacity and silage installations could decrease losses by about 500 t P year^{-1}. Reductions of this type, representing 30% of total losses, have already been achieved between 1985 and 1992 by the implementation of agroenvironmental policies in the two regions of Belgium. The large reduction obtained in this case justified the choice made by the authorities to favour this type of measure by regulation and/or subsidy. The cost of this policy can then be compared with environmental benefits from reduced nutrient losses.

Conclusion

The model seems to be an adequate tool for predicting the effect of regulations on nutrient losses in agriculture. It can be integrated in a geographical information system to follow annual changes in nutrient losses from agriculture to water and to estimate the results of the implemented regulations. It can become a management tool for promoting good agricultural practices.

References

Auerswald, K. (1989) Predicting nutrient enrichment from long term average soil loss. *Soil Technology* 2, 271–277.
Gabriels, D., Van Molle, M., Callebaut, F., Cnockaert, L., Corthals, K., Devos, K.,

Poesen, J. and Van Ghelue, P. (1986) Erosie en verontreiniging. In: *IVe Congres voor Groenvoorziening, Water voor Groen*, pp. 253–258.

Nolte, C. and Werner, W. (1991) Stickstoff und Phosphateintrag über diffuse Quellen in Fließgewässer des Elbeeinzugsgebietes im Bereich der ehemaligen DDR. *Agrarspectrum* 19, 118S.

Rogler, H. (1981) Die Erosivität der Niederslage in Bayern. Dipl. Arbeit Lehrst. f. Bodenk., TU München, Freising-Weihenstephan.

20.25 Phosphorus Loads from Agricultural Areas in an Austrian Watershed: Measurements and Estimation Using Geographical Information System Technology

P. Struß and W.E.H. Blum
Institute for Soil Science and Engineering Geology, University of Agriculture, Forestry and Renewable Resources, Gregor-Mendelstrasse 33, A-1180, Vienna, Austria

Introduction

The watershed of the River Kamp ($1800\ km^2$), situated in the eastern part of Austria, is faced with increasing problems of macrophyte growth. To find measures to reduce nutrient loadings, phosphorus (P) inputs from agricultural areas were evaluated.

Methods and results

Three representative agricultural subwatersheds (two of them under intensive agricultural use) were chosen and P loadings for dissolved (molybdate-reactive P(MRP)) and particulate P (PP) were measured for 1 year (Struß *et al.*, 1994). For the intensively used watersheds, nearly all the annual P load was caused by a very few heavy rainfall events, whereas, for the extensively used watershed, one event of snow melt, in combination with light rainfall, was the main source of the annual P load. Annual P loadings ranged between 0.27 and 0.32 kg P ha^{-1} year^{-1}. Due to climatic conditions of that year, this is probably a 30% underestimate compared with other results under similar climatic conditions (Bayrisches Landesamt für Wasserwirtschaft, 1984). Although the third watershed was extensively used, total P loading did not differ, but MRP concentrations were about 30% higher than in the intensively used watersheds. This can be explained by the special geomorphological conditions of the third watershed where the soils are partly influenced by hydromorphic conditions.

Based on the results of the measurements for the subwatersheds, the P loading for the watershed of the River Kamp was estimated using a combination of the universal soil loss equation of Wischmeier and Smith (1978) and geographical information system (GIS) techniques. Parameter values for the estimation of soil-loss rates for agricultural land were derived from high-resolution digital maps. Soil losses were added up for 372 subwatersheds and subsequently a total P delivery rate was estimated for each of them. For forests and pasture land, P loading coeffients were taken from the literature. Using this estimation procedure, it was possible to identify subwatersheds with high P loading (Fig. 20.8). In general, loadings increased from the western to the eastern part of the watershed. A comparison between estimated non-point-induced P loading and point-induced P loading (Wasserwirtschaftskataster, 1992) revealed that, for the whole watershed, non-point P loading accounted for 30% of the total P loading, whereas, for single subwatersheds, non-point P loading sources could account for up to 100% of the total. The mean contribution of P from intensively used agricultural areas (without forests and

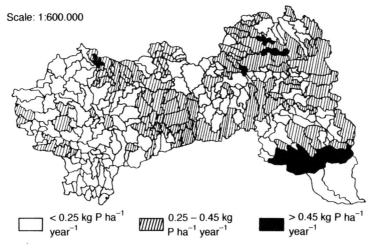

Fig. 20.8. Distribution of non-point-source-induced phosphorus loadings for 372 subwatersheds of the River Kamp (in kg P ha^{-1} year^{-1}).

pastures) for the Kamp watershed was estimated to be about 0.53 kg P ha^{-1} year^{-1}.

References
Bayerisches Landesamt für Wasserwirtschaft (1984) *Nährstoffaustrag aus landwirtschaftlich genutzten Flächen*. Informationsberichte Bayerisches Landesamt für Wasserwirtschaft 2, Munich.
Struß, P., Apschner, C. and Blum, W.E.H. (1994) Stoffaustrag aus drei landwirtschaftlich genutzten Einzugsgebieten im niederösterreichischen Waldviertel, Teil 1 – Schwebstoff und Phosphor. *Zeitschrift für Kulturtechnik und Landentwicklung* 35(5), 292–299.
Wasserwirtschaftskataster (1992) *Gewässergütestudie Kamp*. Bundesministerium für Land- und Forstwirtschaft, Vienna.
Wischmeier, W.H. and Smith, D.D. (1978) *Predicting Rainfall Erosion Losses – a Guide to Conservation Planning*. Agricultural Handbook No. 537, US Department of Agriculture, Washington DC.

20.26 Dutch Policy Towards Phosphorus Losses in Agriculture
D.T. van der Molen,[1] A. Breeuwsma,[2] P.C.M. Boers[1] and C.W.J. Roest[2]
[1]*Institute for Inland Water Management and Waste Water Treatment, PO Box 17, 8200 AA Lelystad, the Netherlands;* [2]*The Winand Staring Centre for Integrated Land, Soil and Water Research, PO Box 125, 6700 AC Wageningen, the Netherlands*

Phosphorus (P) is the main growth-limiting factor for algae in Dutch surface waters (Boers and van der Molen, 1994). Since the mid-1980s, the P concentration has decreased (Fig. 20.9) because of improved waste-water treatment, the use of P-free detergents and lower losses from industrial sources. Also, the P load arriving in the main rivers entering the Netherlands, especially the River Rhine, and consequently to the North Sea, decreased significantly. However, most lakes

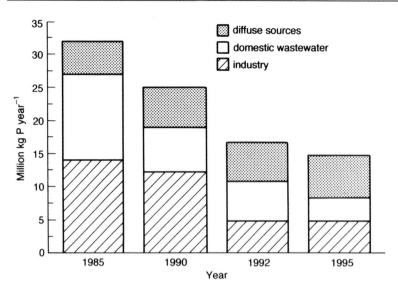

Fig. 20.9. Phosphorus emissions to surface waters in the Netherlands, 1985–1995.

in the Netherlands are turbid (Secchi disc transparency less than 0.5 m) and the algal species composition is dominated by noxious cyanobacteria.

Agriculture, the main source of the 'diffuse' load in Fig. 20.9, causes about 40–50% of the total P emission to the surface waters in the Netherlands, and this contribution has changed little over the 10-year period. These P losses are mainly caused by the overuse of manure and fertilizers; on average, 39 kg P ha^{-1} is given in excess of crop uptake. Regionally, the impact of agricultural P emission can be even larger (Boers, 1996). To restore quality to the Dutch surface waters, a significant decrease in the nutrient emission from the agricultural sector is necessary.

The agricultural nitrogen model (ANIMO) is used to calculate the nutrient loss to surface waters to aid making decisions in the Netherlands. The model has been calibrated and validated by laboratory experiments and field studies. Table 20.9 presents some typical results for the P load to surface water, based on calculations for a regional study in a sandy area with both intensive agriculture and land that receives no fertilizers. The same model has been used to calculate future loads to surface waters, for areas with a manure surplus and P-saturated soils, for P surpluses of 5–20 kg P ha^{-1} year^{-1}. Mean values for catchments vary between 1.5 and 6.0 kg P ha^{-1} year^{-1} (Groenenberg et al., Personnal Communication).

The Dutch policy to combat P emissions from agriculture is a stepwise reduction of the nutrient supply towards an equilibrium fertilization. Equilibrium fertilization is defined as the sum of the manure and fertilizer supply that meets the need of the crops and compensates for inevitable losses. In 1994/95, desk studies defined inevitable losses under 'good agricultural practices' and the acceptable losses needed to meet environmental standards (Oenema and van Dijk, 1994). The level of the losses is determined by the desired soil-P level, soil type and hydrology. Based on field data, a P supply in excess of the crop demand of 10–30 kg P ha^{-1} year^{-1} is needed to maintain an optimal soil P. Based on average net precipitation and Dutch surface-water quality standards, a loss of less

Table 20.9. Phosphorus emission (kg P ha^{-1} year^{-1}) to surface water in a sandy area, calculated with the model ANIMO for 1988–1993.

Land use	Average groundwater level	Surface runoff (kg P ha^{-1} year^{-1})	Subsurface runoff to drains (kg P ha^{-1} year^{-1})	Subsurface runoff to ditches (kg P ha^{-1} year^{-1})	Total emission (kg P ha^{-1} year^{-1})
Grass	< 1 m	0.37	1.38	0.27	2.02
Grass	> 1 m	0.01	0.02	0.09	0.12
Maize	< 1 m	0.24	3.25	0.34	3.83
Maize	> 1 m	0.01	0.01	0.07	0.09
Arable farming		0.05	0.16	0.05	0.26
Heather, forest		0.00	0.01	0.02	0.03

than 1 kg P ha^{-1} year^{-1} is acceptable. There is a significant gap between inevitable nutrient losses from agriculture and acceptable nutrient losses based on environmental standards.

The expected P losses from agriculture exceed the losses based on environmental standards. Further, no exception is made for soils that are saturated with P, which are especially responsible for the P emission to surface water. With a P excess of about 5 kg P ha^{-1} year^{-1}, it is expected that the area which is defined as 'strongly P-saturated' (P concentration in drainage water > 0.5 mg l^{-1}) will double by 2050.

Reconsideration of the optimum soil-P level and nutrient application and the use of modern application techniques may reduce the gap between current and acceptable nutrient losses from agriculture. Additional measures are needed to reduce the environmental pollution from P-saturated soils. Examples are restoration of buffer strips and wetlands, lowering of groundwater levels and applying ferric and aluminium compounds to the soil. The restoration of freshwater ecosystems cannot be a water-quality target without a significant reduction in manure application and without additional measures.

References
Boers, P.C.M. (1996) *Water Science and Technology* 33, 183–189. Nutrient emissions from agriculture in the Netherlands, causes and remedies. In: 14–18 August 1995, Prague. *Proceedings IAWQ Conference on Diffuse Pollution.*
Boers, P.C.M. and van der Molen, D.T. (1994) Control of eutrophication in lakes: the state of the art in Europe. *European Water Pollution Control* 3, 19–25.
Oenema, O. and van Dijk, T.A. (1994) *Fosfaatverliezen en fosfaatoverschotten in de Nederlandse landbouw.* (In Dutch.) Rapport van de technische projectgroep 'P-deskstudie'.

20.27 Present and Future Dutch Regulations to Reduce Phosphorus Loss to Water from Agriculture

P. Hotsma

National Reference Centre for Agriculture, PO Box 474, 6710 BL Ede, Netherlands

Introduction

Agriculture is of major importance to the Netherlands economy and, since the 1950s, its share has increased. Between 1950 and 1984, production of milk doubled, eggs trebled, pigs quadrupled and poultry production increased tenfold. Yet, in the same period, the net area of land devoted to agriculture declined by about 15%. Because of the measures that were taken to protect the environment and the European Community (EC) Common Agricultural Policy, production has remained at about the same level since the mid-1980s.

Due to the combination of intensive agriculture and a limited land base, which is situated in the delta of three major rivers discharging into the North Sea, the Netherlands was confronted fairly early with the drawback of intensive production. Feed imports, which allowed livestock numbers and thus manure production to increase, and the use of inorganic fertilizers resulted in intensive nutrient applications on small fields. One of the results was eutrophication of surface waters.

In the early 1980s, awareness arose that, in the short term, nature and the environment were being affected by a too intensive agriculture, as well as by industries, traffic and transportation, and that the quality of life itself would be affected in the long term. Environmental quality became a public and political issue and it was decided to take steps to prevent further deterioration. In 1989 the government presented a large number of preconditions and objectives. In relation to phosphorus (P) and nitrogen (N) loss to water from agriculture, the following goals were set.

- Total nutrient emission to surface waters from all sources was to be reduced by 50% in 1995 relative to 1985.
- The standards for groundwater, surface waters and drinking-water set by the Netherlands and the EC were to be realized.
- A balanced application of manure and fertilizers, which should not exceed crop demand ('equilibrium fertilization') for P and N, was to be realized by the year 2000. In addition, there must be no structural accumulation of P in soil, and N losses are not to exceed those required to meet the water-quality standards mentioned above.

Present regulations

The nutrient problem is mainly caused by excessive animal-manure production and application. The first action to tackle this problem was to freeze the size of the herds on each farm and, since November 1984, numbers have not been allowed to increase relative to those in 1982. From 1987, the Act on Manures and Fertilizers set an upper limit on manure-P production to prevent any further increase. Consequently, a manure-production quota was defined for each farm. A farmer can nowadays only extend his/her herd by buying 'manure production' quota from farmers who stop farming, but a farmer can sell only 75% of his/her rights; the rest must be given to the government.

In 1986, the government began with a three-phase approach to the nutrient problem, to give farmers and the agricultural industry sufficient time to develop and introduce solutions designed to reduce the nutrient burden on the environment to acceptable levels.

Phase 1 (1987–1990) was aimed at stabilizing the problem and preventing animal-manure production from increasing further. Under the Soil Protection Act, upper limits were set for the maximum amount of P to be applied per hectare in the form of animal manure, compost and/or sewage sludge. Initially, these limits were set at such a level as to ensure that all the manure produced in the Netherlands could be disposed of within the country. The standards were based primarily on P because of the relatively constant amount produced per animal and its non-volatile and stable character, which facilitates the control and enforcement of the regulations.

In addition, the time when manure could be spread was restricted and the use of improved manure-application techniques was made obligatory in order to improve the utilization of nutrients in manure, to reduce ammonia volatilization and to prevent surface-water pollution by agricultural runoff and leaching.

Phase 2 (1991–1994) was aimed at a gradual reduction of the burden placed on the environment by lowering the maximum surpluses. In this period, the extent and rate at which these limits were lowered were tailored to enable the development of solutions for either a decrease in the surplus manure or environmentally acceptable methods for its disposal. For example, reducing mineral excretion through reduced mineral intake via feedstuffs, e.g. by adding enzymes which would assist the animal to digest phosphate better and thus eliminate the need for inorganic P supplements, and through manure distribution and processing. Farmers producing manure in excess of the equivalent of 56 kg P ha^{-1} year^{-1} have to pay a levy. Intensive advisory and extension campaigns were held to help farmers improve their use of manure as a source of nutrients, improve their fertilizer techniques and reduce fertilizer application levels. The period during which manure spreading on land is not permitted was extended from September to February inclusive.

Future regulations

Phase 3 (1995–2000) is aimed at a further reduction of manure and fertilizer applications to realize a balanced application of fertilizers and manure as regards both P and N by the year 2000. This means that the amount of nutrients applied will have to match crop demand, thus preventing accumulation of P in the soil. Moreover, the unavoidable P and N losses are not to exceed the levels indicated in the water-quality standards for sustainable environmental quality. Initially, the upper limits for application of animal manure will be further reduced. In 1995, these upper limits are set at 49 kg P ha^{-1} year^{-1} for arable crops and 67 kg P ha^{-1} year^{-1} for grassland. The regulation of inorganic fertilizer use, however, requires methods that are more sophisticated than generic application standards.

To make allowances for the variation in, for example, soil conditions and water availability and thus differences in crop production and uptake, another type of regulation is required. Therefore a nutrient accounting system is being developed. In 1993, the government proposed that, before the end of phase 3, every Dutch farmer will be obliged to keep a nutrient ledger, with a yearly nutrient assessment. If input of nutrients is higher than the output, the surplus that is not acceptable will be subject to a levy. The levy will force farmers to lower their nutrient surplus, eventually to the level of the maximum surplus that is acceptable from the environmental point of view. Separate levels will be set for P and N.

During the development of this proposal, it became clear that solutions to the problem are even more complicated than already recognized. Since

November 1994, the new government started fresh discussions about the right course forward and at the end of 1995 will discuss new proposals on the details for phase 3. The Minister of Agriculture and the Minister of Environment will together present these proposals in October. After the political discussion and decisions have taken place, a brochure with more detailed information on the regulations in phase 3 will be available – in English – at the 'infotheek' of the Ministry of Agriculture, Nature Management and Fisheries (tel. + 31 70 379 2062).

20.28 Hydrological and Chemical Controls on Phosphorus Losses from Catchments – Coordination of Field Research, Geographical Information Systems and Modelling

J.A. Zollweg, W.J. Gburek, A.N. Sharpley and H.B. Pionke
Pasture Systems and Watershed Management Research Laboratory, USDA-ARS, University Park, Pennsylvania 16802-3702 USA

Introduction

The processes governing phosphorus (P) losses from agricultural catchments are diverse and interrelated, and research in this area requires a coordinated application of a set of research tools. The integration of geographical information systems (GIS) technology, physical-process-based modelling and field studies to determine critical source areas of P export from agricultural catchments is discussed here.

In humid temperate areas, the variable-source-area (VSA) concept of watershed hydrology usually describes runoff production. This concept proposes that runoff production is dominated by saturation-induced overland flow from source areas which expand and contract seasonally, as well as during storm events. Phosphorus is transported within and from the source areas in both soluble and adsorbed forms. The primary vectors for P movement are water and water-borne sediment in the surface (storm) runoff components of the hydrological cycle. Therefore, the key to understanding P dynamics at the catchment scale is to quantify the processes controlling the transfer of soil P to runoff and the potential for transporting that P to receiving waters. Taking a catchment-scale view is critical, because spatially variable P sources, sinks, temporary storages and transport processes are linked by the catchment's flow system.

Methodology and discussion

A GIS, by its design, is well suited to handle the spatial variability of VSA-dominated hydrological systems. It also offers distinct advantages in environmental modelling by providing the following: preservation of physical meaning, internal validation, data visualization, programming productivity, metalanguage scripting, flexibility, amenability to non-point-source pollution modelling and ability to handle variability/uncertainty.

The fundamental input data for hydrological modelling (topography, soils distribution and land use) are becoming much more readily available as GIS data layers. Because of these features, a GIS specifically Geographical Resources Analysis Support System (GRASS) was selected as the foundation of the proposed computer model. In addition, the Soil Moisture-based Runoff Model (SMoRMod), a physical-process-based, VSA-runoff model, which is supplemented with algorithms for P availability and transport by surface runoff, was incorporated. The GIS-embedded, distributed modelling approach employed by SMoRMod enables identification of source areas of runoff and P, as well as generation of the runoff hydrograph and calculation of the total delivery of P to the outlet.

The modelling is supported by field studies at several sites within WE-38, an upland catchment used for research in east-central Pennsylvania, USA. The climate is humid temperate, with approximately 1050 mm precipitation per year. The area is predominantly cropland, with some pasture and a few woodlots. Previous studies in this area have shown that runoff and storm flow contribute about 10% of annual stream flow but as much as 90% of annual P export. During storm events, average dissolved P concentrations considerably exceed threshold levels associated with accelerated eutrophication of fresh waters. Thus, identification of the critical source areas of runoff and P export is essential for effectively targeting future remedial efforts.

Sufficient instrumentation has been developed to generate a continuous record of the extent and behaviour of runoff-producing ('critical') areas. Three types of instruments are being used. Very dense arrays (1 to 6 m spacing) of simple yes/no sensors are used to determine the extent of saturated-runoff source areas before, during and after storm events, as well as the timing and extent of runoff generation during events. Also, numerous interval-runoff samplers are employed to measure, on a spatially distributed basis, the amounts of runoff generated. The samples collected can be chemically analysed to determine the spatial distribution and variability of P in the runoff, in order to characterize the behaviour of P with respect to the hydrology, soils and land use of the site. Finally, several continuously recording flow-measurement and runoff-collection devices are used to record the aggregated output from each of several sectors of the study site. The combination of the three types of instruments creates a 'layered' data-collection system that provides a complete view of the hydrology and P dynamics of the catchment system.

Modelling results suggest that zones of runoff production and P export are a relatively small and delineable portion of the catchment. Highest P losses from the catchment are a result of moderate to high runoff zones intersecting areas of high soil-P status. Areas of the catchment having characteristically high runoff rates but only limited P contribute major amounts of flow to the storm hydrograph but only small to moderate amounts of P loss. So far, field studies have shown that the runoff and P source areas expand rapidly during storm events and contract rapidly after the end of the event. These areas also vary significantly on a seasonal basis and the physical extent of the source areas shows a great deal of spatial variability.

Acknowledgements

This study is a contribution from the US Department of Agriculture, Agricultural Research Service, in cooperation with the Pennsylvania Agricultural Experiment Station, Pennsylvania State University, University Park, Pennsylvania, USA.

21 Phosphorus Status of Soils and Fertilizer Recommendations

PHOSPHORUS STATUS OF SOILS

21.29 Phosphorus Composition of Soil Solution: Effects of Sample Preparation and Soil Storage

P.J. Chapman, C.A. Shand, A.C. Edwards and S. Smith
Macaulay Land Use Research Institute, Craigiebuckler, Aberdeen AB9 2QJ, UK

Introduction

Obtaining representative samples of soil solutions is difficult due to the spatial variability of soil and the number of different methods and conditions used to obtain solutions. Changes in chemical composition of soil solution can also result from: (i) time of soil storage; (ii) pretreatment of soil; and (iii) storage of isolated soil solutions. The objectives of this study were to: (i) compare the phosphorus (P) composition of solutions obtained by centrifugation of sieved and intact soil cores; and (ii) investigate changes in the P composition of soil solutions following storage of intact and sieved soil cores.

Method

Four pairs of intact soil cores were taken side by side from 16 locations on improved grassland. The soil was freely draining iron podzol of the Countesswells Association developed from granitic till. After removing the surface vegetation, one of the cores from each pair was sieved (< 6 mm) to remove stones and large roots. Sieved and intact cores were stored at soil temperature (4°C) for 0, 1, 3 or 8 days before being centrifuged at 1000 g for 1 h to obtain soil solutions. The soil solution was further centrifuged at 25,000 g for 30 min and filtered through a 0.45 µm membrane filter. The solutions were analysed for total dissolved P (TDP), molybdate-reactive P (MRP), dissolved organic P (DOP) and dissolved condensed P (DCP) (Ron Vaz *et al.*, 1992) immediately after obtaining the solution.

Results

Overall, concentrations of TDP and MRP were significantly ($P < 0.01$) different between solutions from intact and sieved soils, although the relative difference was dependent on period of storage (Fig. 21.1). No overall significant difference in DOP and DCP concentrations was observed between solutions from sieved and intact cores, although significant ($P < 0.01$) differences were observed between solutions on individual days. Although the actual concentrations of P fractions were different between solutions from sieved and intact cores, particularly on day 0, the pattern of distribution of MRP, DOP and DCP making up TDP was similar (Fig. 21.2). However, the MRP proportion was consistently larger in solutions from intact cores on all days (Fig. 21.2).

Storage of both sieved and intact soil prior to centrifugation significantly ($P < 0.01$) affected the P composition of soil solutions (Fig. 21.1). Concentrations of TDP, MRP and DOP all declined rapidly with increase in storage time, whereas DCP concentrations, which were initially very low, increased in solutions from both sieved and intact cores. The contribution of the MRP component remained relatively constant with length of soil storage at an average of 44% for solutions from intact soil and an average of 38% for solutions from sieved soil (Fig. 21.2). In contrast, the proportion of DOP decreased substantially, while the DCP proportion increased significantly ($P < 0.01$) in solutions from both sieved and intact cores as length of soil storage increased (Fig. 21.2).

Discussion

The difference in concentrations of P fractions in solutions from sieved and intact cores is attributed to the destruction of the natural soil structure during sieving. Sieving increases the proportion of exposed soil surfaces, which influences adsorption/desorption processes, causes physical disruption to the pore-size distribution and continuity of the pores, which affects the water retention properties of the soil, and aerates the soil, thereby modifying biological activity. However, the extent and relative importance of these processes may have

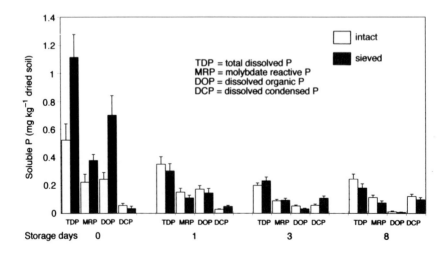

Fig. 21.1. Effect of storage on soluble P concentrations in soil solutions from sieved and intact soil cores (error bars represent standard error).

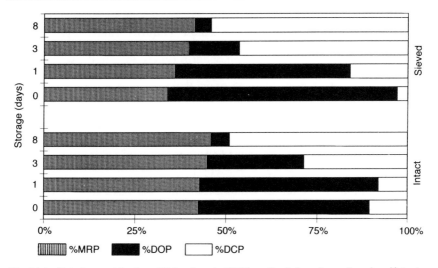

Fig. 21.2. Relative contribution of P fractions to TDP in soil solutions from sieved and intact soil cores.

opposing influences on the P composition of soil solutions. In addition, the individual importance of a particular process will vary temporally. The results must, therefore, be interpreted in terms of a net effect of these processes.

Storage of both sieved and intact soil samples resulted in a reduction in soluble TDP and a change in the relative contribution of DOP and DCP to TDP. However, the relative changes were more pronounced in solutions from sieved soils.

The results highlight the need for caution when interpreting and comparing soil-solution data and the need for consistency in soil-solution sampling.

Acknowledgements
This work was jointly funded by the European Community (EC) and the Scottish Office Agricultural and Fisheries Department.

Reference
Ron Vaz, M.D., Edwards, A.C., Shand, C.A. and Cresser, M. (1992) Determination of dissolved organic phosphorus in soil solutions by an improved automated photo-oxidation procedure. *Talanta* 39, 1479–1487.

21.30 Inventory of the Phosphate Saturation Degree of the Light-Textured Soils in West Flanders, Belgium

J. De Smet,[1] K. Scheldeman,[1] G. Hofman,[1] M. Van Meirvenne,[1] J. Vanderdeelen[2] and L. Baert[2]

[1]*University of Ghent, Dept. Soil Management and Soil Care, Coupure Links 653, B-9000 Ghent, Belgium;* [2]*University of Ghent, Dept. Analytical and Physical Chemistry, Coupure Links 653, B-9000 Ghent, Belgium*

Introduction

During the past three decades, an almost industrial, land-independent pig husbandry, on the one hand, and intensive vegetable growing, on the other, have resulted in the nitrogen (N) and phosphorus (P) economy in the soil being out of balance in the province of West Flanders. Maize fields, especially, were used to get rid of the excessive amounts of pig slurry. In some regions, the P added to the fields, as animal manure and/or mineral fertilizer, regularly exceeded 110 kg P ha^{-1} year^{-1} (VLM, 1991). As the annual P needs of the crops ranged from 27 kg P ha^{-1} (potatoes) to 53 kg P ha^{-1} (mowed grassland) (Hofman et al., 1984), P accumulated in the soil profile.

Recently, in Flanders, the policy-makers have promulgated constraints about fertilizer use on cultivated land. The scope of this study was to delimit, by means of a regional inventory of the P status of the soil, those areas where the risk for eutrophication of the surface waters is acute. In these 'black areas', the use of P will be restricted in a more rigid way in the near future.

Materials and methods

In the province, the sandy-loam soil region (± 1000 km^2) and the sandy region (± 600 km^2) were sampled separately on a square grid with unit dimensions of 2000 m and 1500 m, respectively. At random, in between two sampling points, samples were taken at a shorter distance to be able to quantify the short-distance variability. In total, 296 locations were sampled in the sandy loam soils and 316 locations in the sandy soils. At each location, the soil was sampled five times in a circular area of 30 m^2. The soil profile was sampled in 30 cm layers to a depth of 90 cm. For each layer, a mixed sample of the five subsamples was air-dried and ground (< 2 mm), before being extracted with ammonium oxalate (Schwertmann, 1964). In the extract, the quantities of iron (Fe) and aluminium (Al) were measured by atomic absorption spectrophotometry and P in solution was determined according to Scheel (1936). Finally, for the different locations, the mean phosphate saturation degree (PSD) of the total soil profile was calculated (Van der Zee et al., 1990). The data were statistically and geostatistically analysed.

Results

In the upper 0–30 cm soil layer, the median ammonium-oxalate-extractable P (P-ox) was found to be 620 mg P kg^{-1} in the sandy loam region and 810 mg P kg^{-1} in the sandy region. In both regions, the distribution of P-ox down the profile showed an identical pattern, with 65, 25 and 10% of the total P-ox content in the 0–30, 30–60 and 60–90 cm layers, respectively.

The PSD of the total soil profile ranged from 8 to 75% in the sandy loam region and from 9 to 110% in the sandy region. Postulating that a critical PSD of 30% could result in the P concentration in the upper groundwater exceeding 0.1 mg orthophosphate l^{-1}, 43% of the soils in the sandy loam region and 71% of the soils sampled in the sandy region were P-saturated.

In order to make an unbiased spatial interpolation of the point data with a minimum estimation variance, the global structure of the spatial variability of the PSD was investigated (Van Meirvenne, 1991). The autocorrelation of the PSD reached to a distance of 2400 m in the sandy loam region and 2200 m in the sandy region. The maximum variance of the PSD in the sandy region was about 2.15 times the maximum PSD variance in the sandy loam region.

Block kriging (Burgess and Webster, 1980) was used as a spatial interpolation technique to estimate the mean PSD for units 500 m × 500 m. Taking into account the kriging standard deviation, a probability map of P saturation was composed (68% and 95% probability level) (Fig. 21.3). In this way, 6% of the total surface in the sandy loam region and 13% of the area in the sandy region was found to be P-saturated with a probability of 95%.

Acknowledgement

The authors wish to express their gratitude to the Institute for the Encouragement of Scientific Research in Industry and Agriculture (IWONL, Brussels) and the Flemish Land Society (VLM, Brussels) for subsidizing this research.

References

Burgess, T.M. and Webster, R. (1980) Optimal interpolation and isarithmic mapping of soil properties. II. Block kriging. *Journal of Soil Science* 31, 333–341.

Hofman, G., Ossemerct, C., Ide, G. and Van Ruymbeke, M. (1984) Stikstof-, fosfor- en kalibehoeften van de belangrijkste akkerbouwteelten. In: KVIV (ed.) *Bemesting van intensieve akkerbouwteelten.* Antwerp, pp. 4.1–4.22.

Scheel, K.C. (1936) Colorimetric determination of phosphoric acid in fertilizers with the Aulfrich photometer. *Zeitschrift für Analytische Chemie* 105, 256–269.

Schwertmann, U. (1964) Differenzierung der Eisenoxide des Bodens durch Extraktion mit Ammoniumoxalat-Lösung. *Zeitschrift für Pflanzenernährung, Düngung und Bodenkunde* 105, 194–202.

Van der Zee, S.E.A.T.M., Van Riemsdijk, W.H. and De Haan, F.A.M. (1990) Het

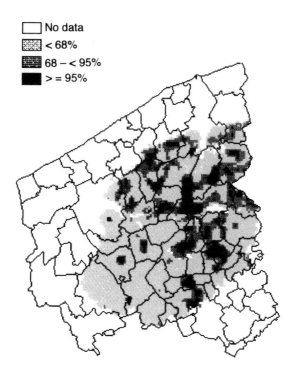

Fig. 21.3. Probability map (68% and 95% probability level) of phosphate saturation (PSD > 30%) in the sandy loam and sandy region of West Flanders.

protokol fosfaatverzadigde gronden. Deel II: Technische uitwerking. Vakgroep Bodemkunde en Plantevoeding, Wageningen Agricultural University, Wageningen, the Netherlands, 25 pp.

Van Meirvenne, M. (1991) Characterization of soil spatial variation using geostatistics. PhD thesis, University of Ghent, Faculty of Agricultural Sciences, Ghent, Belgium, 168 pp.

VLM (1991) *Mest: problemen en oplossingen.* Flemish Land Society, Brussels, Belgium, 36 pp.

21.31 Changes in Two Transport- and Retention-Related Soil-Phosphate Parameters Following Phosphate Addition

R. Indiati[1] and A.N. Sharpley[2]

[1]*Istituto Sperimentale per la Nutrizione delle Piante, Via della Navicella 2, 00184 Rome, Italy;*
[2]*USDA-ARS, Pasture Systems and Watershed Management Research Laboratory, Curtin Road, University Park, PA 16802-3702, USA*

Introduction

Following continuous large additions of phosphorus (P) in fertilizer or manure, the available P content of most soils can increase to levels such that the loss of P to surface and groundwater accelerates freshwater eutrophication. Current soil P-tests were designed to estimate plant availability, not potential release of soil P to runoff or drainage water. Thus, we investigated the use of the P sorption index (SI) to predict changes in two environmental soil-P parameters – equilibrium P concentration (EPC) and bioavailable P (BAP) – following P application (100 mg P kg^{-1}) to 30 Italian soils. The P index is a quick-test, single-point isotherm, used to estimate the P sorption maxima of soils (Bache and Williams, 1971); EPC evaluates whether soils and sediments will gain or lose P when in contact with runoff waters (Taylor and Kunishi, 1971); and BAP is defined as the amount of inorganic P a P-deficient algal population can utilize over a period of 24 h or longer (Sonzogni *et al.*, 1982). The estimation of both EPC and BAP in runoff is needed to evaluate more accurately the impact of agricultural management practices on the biological productivity of surface waters.

Methods

Thirty soils (0–10 cm depth) were collected from several regions in the southern Latium area of Italy, air-dried and sieved (2 mm) (Table 21.1). The soils were amended with soluble P (as potassium dihydrogen phosphate (KH_2PO_4)) at amounts equivalent to 0 and 100 mg P kg^{-1} soil and incubated aerobically for 90 days at 20 ± 3°C. Soils were watered to field capacity twice a week. Soils incubated with and without added P were air-dried, sieved (2 mm) and analysed for EPC and BAP. Equilibrium P concentration was calculated according to Taylor and Kunishi (1971). Bioavailable P, or iron (Fe)-oxide strip P, from simulated runoff samples was determined according to Sharpley *et al.* (1994). The amount of P sorbed, X (mg 100 g^{-1}), from one addition of 1.5 g P kg^{-1} soil was determined after shaking for 24 h at a water-to-soil ratio of 10:1. The P SI was calculated using the quotient $X/\log C$, where C is the solution P concentration expressed as µmol l^{-1} (Bache and Williams, 1971).

Results

Phosphorus application increased soil EPC and BAP and the degree of change was dependent on soil-P properties. Increases for EPC ranged from 0.01 to

Table 21.1. The range of values for selected properties of the 32 Italian soils.

Property	Minimum	Maximum	Mean	Standard deviation
pH	5.6	8.0	7.0	0.7
Clay (mg kg^{-1})	40	780	265	194
Organic C (mg kg^{-1})	2	58	15	13
CDB-Al$_2$O$_3$ (mg kg^{-1})	1	14	6	4
CDB-Fe$_2$O$_3$ (mg kg^{-1})	5	52	21	15
CaCO$_3$ (mg kg^{-1})	0	344	44	91
NaHCO$_3$-extracted P (mg kg^{-1})	0.1	31.1	7.3	6.7
P sorption index	4	54	27	15

C, carbon; Al$_2$O$_3$, aluminium oxide; Fe$_2$O$_3$, ferric oxide; CaCO$_3$, calcium carbonate; CDB, citrate dithionite bicarbonate; NaHCO$_3$, sodium bicarbonate.

1.10 mg P l^{-1}. The mean increase for low-fixing P soils (SI < 10, $n = 6$) was 0.87 mg P l^{-1}, while for high-fixing P soils (SI > 30, $n = 13$) EPC was 0.02 mg P l^{-1}. Bioavailable P increases ranged from 0.01 to 0.50 mg P l^{-1}. The mean increase for soils with SI < 10 was 0.39 mg P l^{-1} and for soils with SI > 30 0.14 mg P l^{-1}. The relationships between change in EPC and change in BAP and soil-P SI values are shown in Fig. 21.4. These relationships could be utilized in estimating the potential increases in EPC and BAP parameters caused by P fertilizer application for soils ranging widely in P sorption capacity.

Conclusions

Expansion of the traditional role of soil-P testing from agronomic to 'environmental' tests is needed to identify soils that have the potential to enrich surface and groundwater with P. It is suggested that this can be obtained by including measures of other pools of P that offer an environmental interpretation of the analytical results. In this study, SI appears to be a simple soil-P parameter to

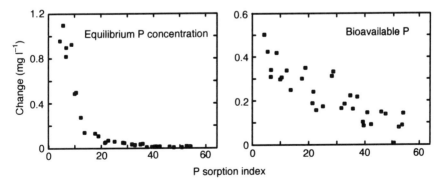

Fig. 21.4. Changes in equilibrium P concentration and bioavailable P following P application (100 mg P kg^{-1}) to 32 Italian soils as a function of soil-P sorption index.

characterize the potential of a given soil for change in transport in runoff or retention by soil of P following fertilizer or manure additions.

References

Bache, B.W. and Williams, E.G. (1971) A phosphate sorption index for soils. *Journal of Soil Science* 22, 289–301.

Sharpley, A.N., Indiati, R., Ciavatta, C., Rossi, N. and Sequi, P. (1994) Interlaboratory comparison of iron oxide-impregnated paper to estimate bioavailable phosphorus. *Journal of Environmental Quality* 23, 14–18.

Sonzogni, W.C., Chapra, S.C., Armstrong, D.E. and Logan, T.J. (1982) Bioavailability of phosphorus inputs to lakes. *Journal of Environmental Quality* 11, 555–563.

Taylor, A.W. and Kunishi, H.H. (1971) Phosphate equilibria on stream sediment and soil in a watershed draining an agricultural region. *Journal of Agricultural Food Chemistry* 19, 827–831.

21.32 The Downward Movement and Retention of Phosphorus in Agricultural Soils

A.E. Johnston and P.R. Poulton

Soil Science Department, IACR-Rothamsted, Harpenden, Hertfordshire AL5 2JQ, UK

Introduction

It has been generally accepted that there are only very small losses of phosphorus (P) from agricultural soils treated with phosphate fertilizers. This is based both on soil-P chemistry and results from field experiments. Estimated losses have been about 1–2 kg P ha^{-1} annually, agriculturally an insignificant amount. Recent concern about the increasing occurrence of algal blooms in fresh waters has considered P to be the limiting nutrient. This P can come from point sources, such as sewage-treatment works, or diffuse sources, of which agricultural land is one. Concentrations as low as 20 µg P l^{-1} can be sufficient for the development of algal blooms, but these equate to only very small amounts of P being lost in drainage from soils. For example, 20 µg P l^{-1} in 250 mm through drainage represents a loss of only 0.05 kg P ha^{-1}.

Estimates of losses of P by leaching and erosion can be got from carefully monitored experiments (Howse *et al.* and Catt *et al.*, Chapter 19, this volume). However, some estimate of the risk of P leaching from differently treated surface soils can be got from long-term experiments where P has been retained in the subsoil.

Methods

Phosphorus balances have been prepared for long-term experiments on a silty clay loam (Rothamsted) and a sandy loam (Woburn) and related to total P, Olsen P and P soluble in 0.01 M calcium chloride (CaCl$_2$) (CaCl$_2$-P) in surface and subsoils.

Results

There are problems in preparing an accurate P balance when all crops have not been analysed. However, using reasonable estimates of P off-take in long-term experiments, the change in total soil P in the plough layer (0–23 cm) is usually within about 5% of the estimated P balance at Rothamsted (Johnston and Poulton, 1977). Unfortunately, if lost in drainage, the amount of P unaccounted

Table 21.2. Vertical distribution of total phosphorus in two long-term experiments on a silty clay loam soil at Rothamsted.

	Barnfield, 1.75% SOM*				Park grass, 5% SOM	
Depth (cm)	None	P	FYM	FYM + P	None	P
0–23	780	1295	1375	1970	575	1425
23–30	465	525	650	780	555	785
30–46	415	450	525	580	500	600
Below 46	400	395	440	410	–	–

* Mean of five nitrogen treatments, including rape cake, which added some organic matter.
P, superphosphate; FYM, farmyard manure; FYM + P, farmyard manure plus superphosphate.

for could raise P concentrations in the average 250 mm drainage above the critical value for algal growth.

On the Barnfield experiment, started in 1843 at Rothamsted, there had been, by 1958, some slight P enrichment of the 23–46 cm horizon where superphosphate, 33 kg P ha^{-1} each year, had been applied to arable crops grown on a soil with 1.8% soil organic matter (SOM). This horizon had been enriched with more P where farmyard manure (FYM), containing about 40 kg P ha^{-1} each year, was given. In the Park Grass experiment, the same amount of superphosphate has been applied annually since 1856 to permanent grassland with about 5% SOM; by 1959 much P had moved down the profile (Table 21.2) (Johnston and Poulton, 1992).

The observed P enrichment of subsoils appears to be related to an enhanced solubility of P in 0.01 M CaCl$_2$ (CaCl$_2$-P) in the surface soils (0–23 cm). For example, total, Olsen P and CaCl$_2$-P in 1958 Barnfield soils treated with superphosphate (1.6% SOM, mean of three of the nitrogen treatments, excluding those with rape cake) and with FYM (4.2 % SOM) are in Table 21.3.

The increase in both total and Olsen P was much the same for both superphosphate- and FYM-treated soils and the combined effect of FYM +

Table 21.3. Total, Olsen P and CaCl$_2$-P in a silty clay loam surface soil, 0–23 cm, given four different P treatments since 1843, Barnfield, Rothamsted.

		P soluble in	
Treatment*	Total P (mg P kg^{-1})	0.5 M NaHCO$_3$ (mg P kg^{-1})	0.01 M CaCl$_2$ (µg P l^{-1})
None	670	18	15
Superphosphate	1215	69	93
FYM	1265	86	396
FYM + superphosphate	1875	145	690

* Superphosphate supplying 33 kg P ha^{-1}; FYM about 40 kg P ha^{-1} each year since 1843.
NaHCO$_3$, sodium bicarbonate.

superphosphate was additive. Much more P was soluble in 0.01 M $CaCl_2$ on FYM-treated surface soils than those given superphosphate and its solubility in the FYM + superphosphate treatment was more than additive (Johnston and Poulton, 1992). This enhanced solubility of P in 0.01 M $CaCl_2$ in FYM-treated surface soils may be related to the greater downward movement of P where FYM has been applied.

A complete P balance for 20 years for the Woburn Reference experiment was related to changes in total soil P, 0–25 cm depth. The 1:1 line went through the origin and accounted for 98% of the variance for all but the FYM-treated soils, which were below the line (Fig. 21.5). On average, about 7 kg P ha^{-1} each year, had moved below 25 cm on FYM-treated soil. This suggests that little or no P applied as inorganic fertilizer had leached below 23 cm, which is at variance with the previous data. It may be that there were sufficient P sorption sites in the topsoil capable of retaining P during this 20-year period. Losses, as in the other experiments, might well have been measured later when all appropriate P sorption sites had been occupied.

Conclusions

Phosphorus enrichment of subsoils below the depth of soil cultivation in long-term experiments can indicate treatments with a greater risk of P leaching. Decreasing enrichment with increasing depth or only very slight or no enrichment is not necessarily indicative of no leaching. This is because the residence time of drainage water in the subsoil may have been too short, perhaps because of preferential flow, to allow for adsorption of P on to soil particles. Some subsoils, e.g. coarse sands, may have no capacity to retain P. Experiments can last for too short a period for leaching losses from the topsoil to be measured.

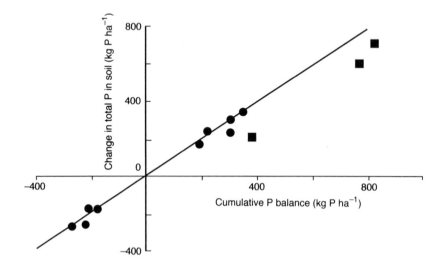

Fig. 21.5. Change in total P in surface soil, 0–25 cm, converted to kg P ha^{-1} using the known soil weight, and the cumulative P balance, kg P ha^{-1}, after 20 years, Woburn Reference experiment.

References

Johnston, A.E. and Poulton, P.R. (1977) Yields on the Exhaustion Land and changes in the NPK contents of the soils due to cropping and manuring, 1852–1975. In *Rothamsted Experimental Station Report for 1976*, Rothamsted Experimental Station, Harpenden, Part 2, pp. 53–85.

Johnston, A.E. and Poulton, P.R. (1992) The role of phosphorus in crop production and soil fertility: 150 years of field experiments at Rothamsted, United Kingdom. In: Schultz, J.J. (ed.) *Phosphate Fertilizers and the Environment.* International Fertilizer Development Center, Muscle Shoals, USA, pp. 45–64.

21.33 The Concentrations and Forms of Phosphorus in Manures and Soils from the Densely Populated Livestock Area in North-West Germany

P. Leinweber
Institute for Spatial Analysis and Planning in Areas of Intensive Agriculture (ISPA), University of Vechta, PO Box 1553, D-49364 Vechta, Germany

Introduction

The counties of Vechta and Cloppenburg, located in north-western Lower Saxony (Germany), comprise one of the regions with the highest livestock densities in Europe. The annual amounts of phosphorus (P) in excrement reach 9243 t and yet the agricultural land covers only 166,084 ha. This amount of P is, on average, 1.4 times larger than the P requirement of the plants grown (Von Hammel, 1994). In addition to excessive application of organic manures, in particular from pigs and poultry, mineral P fertilizers are also used. Little is known about the concentrations and forms of P in the organic manures and in the soils from this region, which may explain why the ecological risks from P enrichment in soil are widely neglected. This work presents data on the total P (TP) and various soluble P forms in typical organic manures and in soils from Vechta and Cloppenburg.

Materials and methods

There are four main organic manures: liquid pig manure containing 4473 t P year^{-1}; dry poultry manure from chickens, 2598 t P year^{-1}; turkey manure, 564 t P year^{-1}; and duck manure, 45 t P year^{-1}.

A total of 1094 soil samples from the Ap (Ah) horizons from gleyic and cambic podzols, arenic gleyzols, gleyic luvisols and cumulic anthrosols, were taken from a range of land uses – forest, arable, grassland and soils with special crops or management, farms with certified alternative management, e.g. 'Demeter', 'Bioland', and farms with known high livestock populations per hectare. For most of the sampling sites, there was no information about fertilizer application rates.

Water-extractable P (H_2O-P, soil:water ratio of 1:25) was determined on 256 samples and double lactate P (DL-P) on all samples; both are soil-P tests used for fertilizer recommendations. In addition, sequential extraction of P (Tiessen *et al.*, 1983) was determined on 75 samples (DOWEX-P, sodium bicarbonate ($NaHCO_3$)-P, sodium hydroxide (NaOH)-P, sulphuric acid (H_2SO_4)-P, residual P) and TP on 260 samples after microwave digestion. Phosphorus concentrations in the solutions were measured by inductively coupled plasma (ICP) spectroscopy (JY 24, Jobin Yvon, France) at 214.914 nm.

Results

The frequency distributions of the DL-P concentrations showed that samples with 0.13–0.18 and 0.22–0.26 g P kg^{-1} were most frequent in grassland and special-crop soils and in arable soils, respectively. These data and the H$_2$O-P concentrations in the range 3–150 mg kg^{-1} (not given) showed that no further P fertilizer application can be recommended for many of these soils (Fig. 21.6) (Leinweber *et al.*, 1994). The organic manures contained up to 42,103 mg P kg^{-1} and were characterized by high proportions of easily soluble P forms. Dry poultry manures contained less total P but higher proportions of H$_2$O-P than

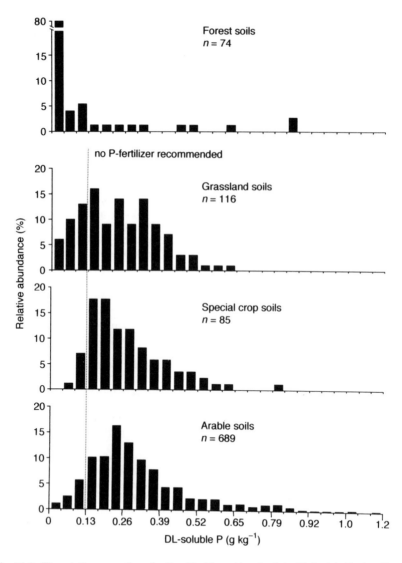

Fig. 21.6. The relative proportion of soils with different levels of double lactate P when the soils are grouped by cropping system.

liquid pig manure; the proportions of resin P were very similar in all manures (Table 21.4). It was to be expected that excessive use of these manures increased not only the total P concentrations in soils but also the proportions of easily soluble P. Arable soils, which receive the highest rates of manure, were largest in resin P and $NaHCO_3$-P (Table 21.4).

Sequential extraction of the soil P showed large proportions of resin P and $NaHCO_3$-P, up to 17% and 33% of TP, respectively, in the agricultural soils of this densely populated livestock area. The highest values were larger than those from other studies reported in the literature. Correspondingly, the small proportions of NaOH-P and H_2SO_4-P have not been reported from other locations (Table 21.4). In conclusion, significant P losses can be expected from these soils.

References
Leinweber, P., Geyer-Wedell, K. and Jordan, E. (1994) Phosphorgehalte von Böden in einem Landkreis mit hoher Konzentration des Viehbesatzes. *Zeitschrift für Pflanzenernährung und Bodenkunde* 157, 383–385.
Magid, J. (1993) Vegetation effects on phosphorus fractions in set-aside soils. *Plant and Soil* 144, 155–165.
Oberson, A., Fardeu, J.C., Besson, J.M. and Sticher, H. (1993) Soil phosphorus dynamics in cropping systems managed to conventional and biological agricultural methods. *Biology and Fertility of Soils* 16, 111–117.
Tiessen, H., Stewart, J.B. and Moir, J. (1983) Changes in organic and inorganic phosphorus composition of two grassland soils and their particle-size fractions during 60–90 years of cultivation. *Journal of Soil Science* 34, 815–823.
Von Hammel, M.-L. (1994) *Mengen- und Nährstoffbilanz organischer Rest- und Abfallstoffe für die Landkreise Vechta und Cloppenburg und das Oldenburger Münsterland.* Vechtaer Druckerei und Verlag, Vechta, 164 pp.

21.34 Comparison of Chemical Forms and Distribution of Phosphorus within Cultivated and Uncultivated Soils: Some Implications for Losses

R.O. Maguire, A.C. Edwards and M.J. Wilson
The Macaulay Land Use Research Institute, Craigiebuckler, Aberdeen AB9 2QJ, UK

Introduction
In intensively cultivated soils, phosphorus (P) availability has increased, as measured by soil-test methods. Input of P into rivers from point sources, such as sewage-treatment works, has been reduced, so that non-point sources, such as agriculture, are now likely to become the main contributors. Some P reaches rivers through leaching in soil water, but most travels into watercourses in the form of suspended particles from erosion of soil constituents and in surface runoff water. Erosion is a selective process; smaller soil particles/aggregates are eroded and transported to a greater extent than larger ones because they take longer to sediment out of suspension (Walling, 1990). If these small particles behave differently with respect to P than does the soil as a whole, this has implications for the amount of P lost. Once removed from the field, the particles enter different environments, which alter their P retention/release characteristics.

Methods
Three pairs of soils, one uncultivated (U) the other cultivated (C), from northeastern Scotland were sampled, dried and sieved (< 2 mm). Subsamples of the

Table 21.4. Mean total P concentrations and its distribution in P fractions in organic manures and differently used soils from the county of Vechta and ranges of the corresponding data in soils of other geographical origin.

Samples	Soil-test P (% TP)		Sequentially extracted P (% TP)					TP (mg P kg⁻¹)
	H_2O	DL	Resin	$NaHCO_3$	NaOH	H_2SO_4	Residual	
Manures	24	44	34	20	7	21	19	27,651
Grassland soils	1	7	5 (11)*	9 (22)	16	4	66	2,693
Special-crop soils	2	18	8 (12)	10 (16)	39	9	34	1,514
Arable soils	2	10	10 (17)	19 (33)	26	10	36	1,335
Canada†			2–8	4–8	17–24	20–40	33–46	539–893
Denmark‡			4–8	10–24	35–57	12–38	22–68	381–1113
Switzerland§			5–8		23–25		36–44	732–750

* Maximum values in parentheses.
† Tiessen et al. (1983).
‡ Magid (1993).
§ Oberson et al. (1993).

soils were separated into three size fractions by wet sieving without dispersion: silt (2–52 μm), fine sand (53–150 μm) and coarse sand (151–2000 μm). Because the soil remained aggregated, the amount of clay collected was too small for complete analysis. The whole soils and the three aggregated size fractions were analysed for their total P content by fusion with sodium hydroxide and for their mixed resin-extractable P using resins in ammonium (NH_4^+) and chloride (Cl^-) forms. Bar-lines on the graphs are standard errors.

Results

Both total and mixed resin-extractable P from the whole soils showed an increase in P due to fertilization. The increase from 1300 to 1615 mg P kg^{-1} for total P is proportionally smaller than the relatively large increase for resin P, from 30 to 85 mg P kg^{-1}. Therefore, judged by resin-extractable P, fertilization had increased the availability of soil P. On average, resin P was increased from 2.2% to 5.4% of total P for the U and C soils, respectively. For the aggregate sizes, there was an inverse relationship between size and both total and mixed resin-extractable P in both U and C soils (Fig. 21.7). The aggregate size fractions also reflect the increase in P due to fertilization as seen in the whole soils. If the increase in resin P is calculated by subtracting the U values from the equivalent C ones, the distribution of fertilizer P is not the same for each size fraction; there was more resin-extractable P in the silt fraction than in the larger-sized fractions (Fig. 21.8).

Discussion

Fertilization had increased the total soil P, but it had increased the availability of P to a greater extent, as measured by mixed resin extraction. There was more total (Fig. 21.7) and mixed resin-extractable P in the silt fraction with decreasing amounts in the fine and coarse sand, respectively, using this methodology, which did not separate out the clay fraction. The largest effect of fertilizer on the particle sizes is the increase in resin-extractable P, which is most pronounced in the silt fraction (Fig. 21.8). This has implications for calculating erosion losses, where small particles are preferentially eroded (Walling, 1990). If the amount of P lost in 1 tonne of each soil fraction is calculated, then for the C soil the loss,

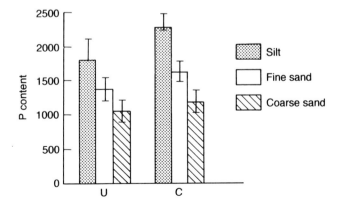

Fig. 21.7. Total P in aggregate sizes (mg P kg^{-1}) in paired uncultivated and cultivated soils.

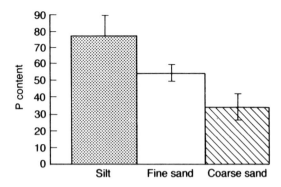

Fig. 21.8. Distribution of fertilizer-derived resin P among aggregate sizes (mg P kg^{-1}).

on average, for the whole soil was 1615 kg total P and 85 kg of extractable P, but for the silt fraction the average loss would be 2272 kg total P and 118 kg extractable P. Therefore, selective erosion causes the loss of more available and unavailable P than would be expected by only analysing the whole soil. Fertilization has increased available P to the largest extent in the fraction most likely to be eroded.

Acknowledgements
This work was funded by the European Community (EC) and the Scottish Office Agriculture and Fisheries Department.

Reference
Walling, D.E. (1990) Linking the field to the river: sediment delivery from agricultural land. In: Boardman, J., Foster, I.D.L. and Dearing, J.A. (eds) *Soil Erosion on Agricultural Land.* J. Wiley and Sons, Chichester, pp. 129–151.

21.35 Threat of Phosphorus Leaching from Intensively Farmed Agricultural Soils in the Central Reaches of the River Elbe
R. Meissner, H. Rupp and J. Seeger
Centre for Environmental Research Leipzig-Halle, Department of Soil Research, Field Research Station Falkenberg, Dorfstrasse 55, 39615 Falkenberg, Germany

Introduction
It is well known that the diffuse phosphorus (P) input in surface waters is primarily due to soil erosion. The leaching of P with seepage water and the subsequent input into surface-water systems by groundwater flows is either unknown or described as negligible in most textbooks on soil and water science, although the sorption capacity of soils for P is limited. Under the conditions of the eastern part of Germany, a surplus of 25 kg P ha^{-1} year^{-1} was determined in the fertilizer balance. Therefore, it is important to investigate the recent accumulation of P in soils and to estimate the potential threat for leaching (Behrendt and Boekhold, 1993).

Material and methods
During 1992–1995, *in situ* investigations of the top 30 cm of soil were carried out in the central stretches of the River Elbe (German Federal State of Saxony-Anhalt) on approximately 4700 ha of intensively farmed mineral and fen soils. The soil texture (particle-size analysis) and the percentage double-lactate-soluble P (PDL, which is related to plant-available P) were measured. In addition, a computer-aided information system was used to characterize the distribution of different soil textures in this region. The relation between P accumulation in soils and P leaching was investigated in lysimeter experiments.

Results and discussion
Figure 21.9 shows that the P content of 41% of soils exceed the optimal range for plant growth, 60–80 mg PDL kg^{-1} soil, and can be regarded as overfertilized with P; crop yields are not increased at a P content in excess of 70 mg PDL kg^{-1} soil.

Because of their relatively low sorption capacity, there is a risk of P leaching on sandy soils. The percentage of sandy soils in the central reaches of the River Elbe is about 27% and 19% of these soils have PDL contents of more than 80 mg PDL kg^{-1} soil. Therefore, 5% of the intensively farmed arable land in this region is characterized by a high risk of P leaching. In addition, lysimeter experiments showed a statistically significant relation between PDL soil content and P leaching for sandy soils when there is increased slurry application (Meissner *et al.*, 1992). Results of the lysimeter experiment suggest that there was an appreciable increase in P leaching above 300 mg PDL kg^{-1} (Fig. 21.10). Because intensively cultivated sandy soils are characterized by a high risk of P leaching, changing fertilization strategies is the first step to reducing the continuing P accumulation in these soils.

When fen soils are rewetted, they have relatively high PDL contents (up to 570 mg PDL kg^{-1} soil in the top layer), but PDL contents of fen soils are usually evaluated on dry bulk density. In fen soils, the soluble and plant-available portion of P in deeper soil layers depends on the groundwater level and therefore on the redox potential. The current investigations into the danger of P leaching during the rewetting of fen soils are to be continued.

References
Behrendt, H. and Boekhold, A. (1993) Phosphorus saturation in soils and groundwaters. *Land Degradation and Rehabilitation* 4, 233–243.
Meissner, R., Klapper, H. and Seeger, J. (1992) Wirkungen einer erhoehten Phosphatduengung auf Boden und Gewaesser. *Zeitschrift Wasser und Boden* 44, 217–220.

21.36 The Availability in Soil of Phosphorus Released from Poultry Litter
J.S. Robinson[1] and A.N. Sharpley[2]
[1]*Department of Soil Science, University of Reading, Whiteknights, Reading RG6 6DW, UK;*
[2]*USDA-ARS, Pasture Systems and Watershed Management Research Laboratory, Curtin Road, University Park, PA 16802-3702, USA*

Introduction
The litter (manure plus bedding material) produced during poultry production can be a valuable source of plant nutrients. However, repeated litter applications with limits based on soil and crop nitrogen (N) can lead to the accumulation of large amounts of phosphorus (P) in the surface soil, increase P loss in runoff and accelerate freshwater eutrophication (Sharpley *et al.*, 1994). Thus, for soils with

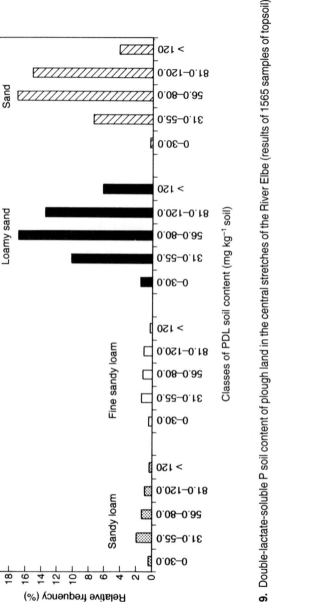

Fig. 21.9. Double-lactate-soluble P soil content of plough land in the central stretches of the River Elbe (results of 1565 samples of topsoil).

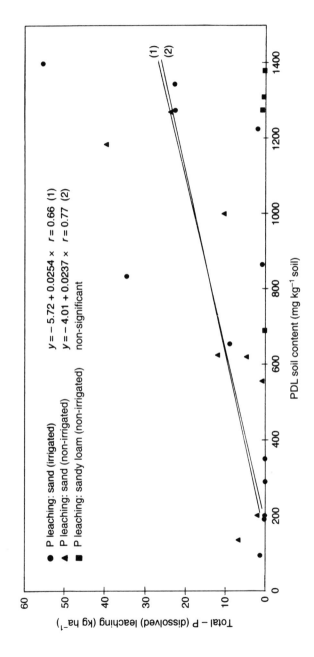

Fig. 21.10. Results of lysimeter experiments of P leaching as a function of P soil content as a result of increased slurry application.

high P contents, farmers are being encouraged to apply animal manures based on crop P requirements. Current manure-application guidelines assume that the effectiveness of manure P for plant growth is similar to that of fertilizer P. However, to develop reliable guidelines, the fate of manure P in soil should be considered. This study discusses the fate in soil of P from poultry litter compared with P from a mineral fertilizer.

Methods

Six soils (0–5 cm) that had not received fertilizer or manure were collected from eastern Oklahoma (Table 21.5) and air-dried and sieved (< 2 mm). Poultry litter (bedding material of pine-bark shavings) was collected from a broiler house in south-eastern Oklahoma. Simulated rainfall equivalent to 2.54 cm h^{-1} was applied for 20 min to 20 g of fresh, thoroughly mixed litter, spread uniformly over a 15 cm diameter by 10 cm high plexiglass tube and approximately 200 ml leachate collected. The leachate contained 43 mg l^{-1} inorganic P (IP), which was approximately 84% of the total P. A stock solution of potassium dihydrogen phosphate (KH_2PO_4) was prepared with the same P concentration. A 2.5 ml aliquot of poultry litter leachate or KH_2PO_4 (to provide approximately 108 µg IP) and water were mixed with 1 g of soil and incubated at 2°C for up to 28 days. After incubation, iron (Fe)-oxide strip P was determined (Sharpley, 1993). The Fe-oxide strip removes soil-solution P plus P that can be readily desorbed, which is related to plant-available P in soil and potentially algal-available P in runoff. Langmuir P sorption isotherms were constructed with poultry litter leachate or KH_2PO_4 as the P source, using the method of Nair *et al.* (1984). The P sorption maximum was calculated as the reciprocal of the slope of the plot of C/X and C and binding energy as the slope/intercept of the same plot, where C is P in solution, mg l^{-1}, and x is P sorbed on to the soil in mg kg^{-1}.

Results

The Fe-oxide strip P content of each soil incubated with KH_2PO_4 for 1 day was greater ($P < 0.01$) than with litter leachate (Fig. 21.11). Averaged over the six soils and after incubation for 28 days, strip P was still greater ($P < 0.01$) in KH_2PO_4 (28 mg P kg^{-1}) than in leachate-treated soils (19 mg P kg^{-1}). Thus, more P was sorbed by soil from litter leachate than from KH_2PO_4 solutions (Table

Table 21.5. Physical and chemical properties of the selected soils.

Soil	Clay (%)	pH	Organic C (g kg^{-1})	Sorption maxima (mg kg^{-1})		Binding energy (L mg^{-1})	
				Litter	KH_2PO_4	Litter	KH_2PO_4
Baxter	27.8	5.9	30.4	512	167	0.22	0.18
Dickson	31.8	5.4	16.1	582	328	0.57	0.53
Neff	8.0	5.4	21.4	536	290	0.81	0.68
Newtonia	26.2	5.9	28.9	578	306	0.52	0.42
Okemah	34.0	6.1	48.5	658	491	0.63	0.61
Tiak	14.0	5.6	14.9	419	243	0.40	0.32

C, carbon; KH_2PO_4, potassium dihydrogen phosphate.

21.5). Averaged for all soils, the P sorption maximum was 548 mg P kg^{-1} for leachate- and 304 mg P kg^{-1} for KH_2PO_4-treated soils, while respective binding energies averaged 0.527 and 0.456 L mg^{-1}. The consistently lower strip P in leachate- than in KH_2PO_4-treated soils (Fig. 21.11) could be due to a stronger P sorption mechanism and/or a larger number of sorption sites in soil treated with leachate (Table 21.5).

Conclusion

The extractability from soil of P added as poultry-litter leachate was different from that added as KH_2PO_4. The P added in poultry litter may have a lower immediate availability for plants than P added in inorganic fertilizers, but this needs to be tested under field conditions. Runoff from fields treated with mineral fertilizer may initially contain more bioavailable P than runoff from land treated with poultry litter. However, in the long term, the larger amounts of P sorbed from poultry litter may support greater bioavailable P concentrations in runoff than from mineral fertilizer treatments. Manure-application rates for agronomically and environmentally sound P management guidelines should not be based solely on data from mineral fertilizer trials.

References

Nair, P.S., Logan, T.J., Sharpley, A.N., Sommers, L.E., Tabatabai, M.A. and Yuan T.L. (1984) Interlaboratory comparison of a standardized phosphorus adsorption procedure. *Journal of Environmental Quality* 13, 591–595.

Sharpley, A.N. (1993) An innovative approach to estimate bioavailable phosphorus in agricultural runoff using iron oxide-impregnated filter paper. *Journal of Environmental Quality* 22, 597–601.

Sharpley, A.N., Chapra, S.C., Wedepohl, R., Sims, J.T., Daniel, T.C. and Reddy, K.R. (1994) Managing agricultural phosphorus for protection of surface waters: issues and options. *Journal of Environmental Quality* 23, 437–451.

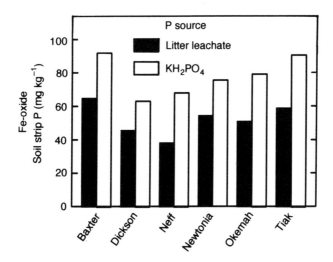

Fig. 21.11. Iron (Fe)-oxide strip soil P content after incubation for 1 day with poultry litter leachate or KH_2PO_4 when 108 mg inorganic P was added per kg of soil.

21.37 Reducing Soil-Phosphorus Availability with By-products from Power-Generation Plants

W.L. Stout, A.N. Sharpley and H.B. Pionke
USDA-ARS, Pasture Systems and Watershed Management Research Laboratory, Curtin Road, University Park, Pennsylvania 16802-3702, USA

Introduction

To meet the United States Environmental Protection Agency (USEPA) clean-air standards, electricity-generating plants must adopt methods to remove sulphur dioxide (SO_2) from exhaust gas. Two of these methods are flue-gas desulphurization (FGD) and fluidized-bed combustion (FBC). Their adoption is projected to produce over 50 million t year^{-1} of by-product gypsum by the year 2000. Research has shown that this material presents little heavy metal contamination risk when used as an amendment to cropland.

Both power-generation plants and many farms with high soil phosphorus (P) are located in north-eastern US. High soil P has resulted from long-term application of P as animal manure. In the north-east, the majority of soil (Delaware 64%, Ohio 68% and Pennsylvania 56%) analysed for P in 1990 by state soil-testing laboratories exceeded P levels needed for agricultural production (Sharpley *et al.*, 1994). In this region, the Chesapeake Bay programme has targeted a 40% reduction from the 1985 P inputs to the bay by the year 2000. To accomplish this, both the potential transport of P in runoff and the availability for soil-P release to runoff must be reduced.

By-product gypsum from power-generation plants represents a source of calcium (Ca^{2+}) that could safely and economically be used on agricultural land to precipitate water-soluble P in the soil. Our objective was to determine the effectiveness of by-product gypsum in reducing soil-P availability (Bray-I P) and solubility (equilibrium P concentration (EPC)), thereby decreasing the potential for P export through runoff to surface waters. Bray-I P estimates of plant-available soil P and EPC represent the concentration of soil P at which neither P sorption nor desorption occurs and the solution P in runoff that may be maintained by a soil.

Methods

The Ap horizon of a Berks shaly silt loam (typic dystrochrept, loamy skeletal, mixed, mesic) was collected from a central Pennsylvania farm, air-dried and sieved (2 mm). Because of a long history of receiving animal manures, this soil had an excessive Bray-I P (240 mg P kg^{-1}). Various by-product amendments, at amounts ranging from 0 to 80 g kg^{-1}, and soil were incubated for 21 days at 25°C and field moisture (about 30% w/v). The by-products used were FBC ash, containing large amounts of calcium hydroxide ($Ca(OH)_2$) and about 10% gypsum; pulverized coal (PC) ash, commonly referred to as fly ash; FGD calcium sulphate ($CaSO_4$), a very high-grade gypsum (about 95% $CaSO_4$); and FGD sludge, which was greater than 95% calcium sulphide (CaS). Following incubation, Bray-I P was determined (Bray and Kurtz, 1947). Phosphorus sorption isotherms were also constructed for each amendment and level, using the procedure of Nair *et al.* (1984).

Results

Fluidized-bed-combustion ash and CaS by-product reduced the Bray-I soil P content (Fig. 21.12). For FBC ash, significant reductions ($P > 0.05$) in Bray-I P occurred at 20 g kg^{-1} or greater. The highest FBC ash amendment (80 g kg^{-1})

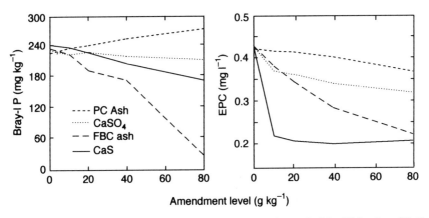

Fig. 21.12. The effect of addition of by-product materials PC ash, $CaSO_4$, FBC ash and CaS on the Bray-I P and equilibrium P concentration (EPC) in a Berks, shaly silt loam from Pennsylvania.

reduced Bray-I P from 233 to 25 mg P kg^{-1}. For CaS, significant reductions occurred at > 45 g kg^{-1}. Pulverized-coal ash or $CaSO_4$ had no effect on Bray-I P; in fact, there was a slight increase in Bray-I P with PC ash.

At the largest addition, all amendments reduced EPC (Fig. 21.12). Pulverized-coal ash reduced EPC from 0.422 to 0.368, $CaSO_4$ to 0.319, FBC ash to 0.223 and CaS to 0.207 mg l^{-1}. The greatest reduction was by CaS, where the smallest amount, 10 g kg^{-1}, decreased EPC by almost 50% compared with unamended soil.

Discussion

Although the relative effects of amendment type on soil Bray-I P and EPC were similar, small amounts of CaS had an appreciably greater effect on EPC than on Bray-I P. Even though CaS reduced plant availability of soil P, there was a greater reduction in the potential for soil-P release to runoff. Thus, CaS had a greater effect on water-soluble P forms in soil than more tightly bound aluminium (Al) and iron (Fe) forms extracted by the ammonium fluoride (NH_4F) and hydrochloric acid (HCl) of the Bray extractant.

Clearly, certain by-product materials can dramatically reduce soil-P availability and potential release to runoff, at levels which do not appear to affect crop yields. Further research is under way to explain these differences.

Fluidized-bed combustion ash was the only amendment that affected both P availability and EPC greatly, especially at the largest amounts used. The implication is that the $Ca(OH)_2$ component of this $CaSO_4$–$Ca(OH)_2$ mix effected a reduction in both soil-P availability and solubility. In contrast, CaS is most efficient in reducing P solubility but has little affect on availability. We would expect sulphide to oxidize to sulphate (SO_4), causing the soil solution to become more acid. The corresponding release of Al^{3+} to solution may cause solution P to decrease more rapidly than Bray-I P, which includes Al-P forms. We have not yet explored these interactions.

References
Bray, R.H. and Kurtz, L.T. (1947) Determination of total organic and available forms of phosphorus in soils. *Soil Science* 59, 39–45.
Nair, P.S., Logan, T.J., Sharpley, A.N., Sommers, L.E., Tabatabai, M.A. and Yuan, T.L. (1984) Interlaboratory comparison of a standardized phosphate adsorption procedure. *Journal of Environmental Quality* 13, 591–595.
Sharpley, A.N., Chapra, S.C., Wedepohl, R., Sims, J.T., Daniel, T.C. and Reddy, K.R. (1994) Managing agricultural phosphorus for protection of surface waters: issues and options. *Journal of Environmental Quality* 23, 437–451.

21.38 Occurrence and Effects of Phosphate-Saturated Soils
A. Breeuwsma, J.G.A. Reijerink and O.F. Schoumans
DLO – The Winand Staring Centre for Integrated Land, Soil and Water Research, Wageningen, the Netherlands

Introduction

Accumulation of phosphorus (P) in agricultural soils may cause increasing losses to surface waters, not only by surface runoff but also by leaching in areas with high water-tables. Although P is strongly adsorbed in most soils, the phosphate sorption capacity (PSC) of soil is not unlimited. Large application rates lead to partial saturation of the sorption complex and eventually to increased leaching rates. This study in the Netherlands determined both the occurrence of so-called P-saturated soils in areas with intensive livestock farming and P concentrations in groundwater and surface waters of a catchment area with a high percentage of P-saturated soils.

Methods

Assessment of phosphorus saturation at a regional scale. The degree of P saturation is defined as

$$\text{DPS} = 100 \left(\frac{\text{P-ox}}{\text{PSC}} \right)$$

where DPS is the degree of P saturation (%), P-ox is the total amount of P sorbed (extractable by oxalate) (mmol kg^{-1}) and PSC is the phosphate sorption capacity (mmol kg^{-1}). Van der Zee *et al.* (1990) calculated a critical value for DPS of 25% if P concentration in groundwater is not to exceed a value of 100 µg orthophosphate l^{-1}. The critical depth of soil was defined as the depth to the mean highest water-table (MHW).

The DPS and the area of P-saturated soils were assessed for areas with a manure surplus caused by intensive livestock farming. The PSC was calculated for each mapping unit of the national soil map, scale 1:50,000, using a soil database with aluminium (Al)-ox and iron (Fe)-ox data and a relation between PSC and Al-ox + Fe-ox (pedotransfer function). Data for P-ox were calculated from the natural P content and data on animal production, fertilizer use and crops. From 1970 onwards, data on production of animal manures were based on data from individual farms and corrected for manure transport to deficit areas. For reasons of confidentiality, data are presented for grids of 2.5 km × 2.5 km.

Phosphorus concentration in groundwater and surface waters. The impact of intensive livestock farming on water quality was studied in the catchment area of the Schuitenbeek (6800 ha). The sandy soils have a high degree of phosphate saturation. Groundwater samples were collected in 1988 in winter when water levels were usually within 50 cm of the surface, and in summer, when water levels were lower. Since 1976, surface-water quality has been monitored at the outlet of a brook, using a flux-proportional sampling device, by the Institute for Inland Water Management and Waste-Water Treatment (RIZA), Lelystad. Samples were analysed for dissolved orthophosphate and total P, according to Dutch NEN standards.

Results

The results show that over 70% of the soils in manure-surplus areas of the Netherlands are P-saturated (DPS > 25%). Some 20% of the area is strongly saturated (DPS > 50%) and 6% is very strongly saturated (DPS > 75%). These P-saturated soils lose P to surface waters of the order of 2.5 kg P ha^{-1} year^{-1} (Fig. 21.13), resulting in a mean annual concentration of total dissolved P in surface water of 1 mg P l^{-1}.

References

Reijerink, J.G.A. and Breeuwsma, A. (1992) *Spatial Distribution of Phosphate Saturated Soils in Areas with Manure Surpluses.* (In Dutch.) Rapport 222, DLO Winand Staring Centre for Integrated Land, Soil and Water Research, Wageningen, the Netherlands.

Van der Zee, S.E.A.T.M., van Riemsdijk, W.H. and de Haan, F.A.M. (1990) *Protocol*

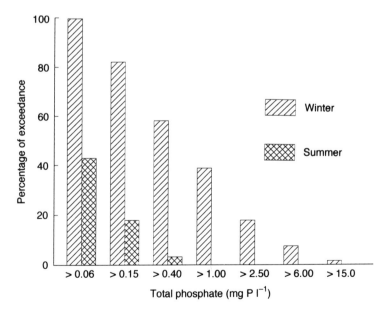

Fig. 21.13. Percentage of observations exceeding a given concentration of P: total P mg l^{-1} in groundwater (upper 20 cm) in the catchment area of the Schuitenbeek in 1988.

for Phosphate-Saturated Soils. (In Dutch.) Department of Soil Science and Plant Nutrition, Agricultural University, Wageningen, the Netherlands.

PHOSPHORUS RECOMMENDATIONS

21.39 Soil Phosphorus Levels in Dairy Farming

N. Culleton, J. Murphy and W.E. Murphy
Teagasc, Johnstown Castle Research Centre, Wexford, Ireland

Introduction

It is now widely accepted that high soil phosphorus (P) has the potential to cause P enrichment of water in rivers and lakes. Less than 50% of the fertilizer P applied is recovered in product (Tunney and Culleton, 1995) and the residue builds up in soil over time. There is also some evidence that optimum yields of crops can be achieved at lower soil P levels than are currently recommended (Tunney, 1990). However, there is also evidence that sugar beet responds to high P in soil (Herlihy and Hegarty, 1994), and that grass grows best in cold weather at high soil-P levels (Murphy, 1977).

The trial outlined in this study will examine the effects of soil P on grass growth, seasonality and chemical composition and milk yields and composition during a 6-year period to define more accurately the minimum soil-P levels for intensive dairy production without adverse effects on the environment.

Materials and methods

The grass-based dairy farm at Johnstown Castle occupies some 75 ha. Areas within the farm totalling 26.7 ha have been selected and divided into 36 paddocks of equal size (0.742 ha). The soil within each paddock was sampled to a depth of 10 cm and analysed for pH, P, potassium (K), calcium (Ca) and magnesium (Mg) (using Morgan's reagent). On the basis of the results, the paddocks were divided into three groups, each of 12 paddocks. The soil Morgan P levels in the low, medium and high paddocks averaged 4.2, 5.0 and 7.0 mg P kg^{-1} in November 1994. A herd of 22 cows is assigned to each treatment at a stocking rate of 2.5 livestock units (LU) ha^{-1}. Within each group of paddocks, there will be both grazing and silage-cutting and all slurries will be returned to the land from where the silage was cut. Slurry will be applied in March, June and late July. In March 1995, 50, 25 and 0 kg P ha^{-1} were applied to the high, medium and low P treatments, respectively. Similar treatments will be imposed in 1996, thus exaggerating the differences in soil-P contents between treatments. Grass growth on all treatments is being assessed by sampling pre- and postgrazing and presilage cuts and total yield, seasonal distribution and P content are measured. Soil-P levels, distribution of P in the profile and P forms in soil are being assessed. Cow performance in terms of milk yield, composition, animal health and reproductive performance will be determined. A range of small plot trials will also be conducted at times and rates of P application on soils with a range of P values.

Results

Table 21.6 outlines the various parameters that are being monitored. Initial results in the first year show no significant differences between treatments. This agrees with the results of Tunney (1995), who found that it took several years to deplete soil P to levels that caused a reduction in yield on a dairy farm.

Table 21.6. Effects of fertilizer P on soil P levels, milk yields and animal parameters.

	High P	Medium P	Low P	LSD ($P < 0.05$)
Cow wt at turnout (kg)	534	517	525	43
Cow wt on 1 August (kg)	550	523	530	47
Milk yield (to 1 August) (l cow^{-1})	3454	3595	3523	160
P (mg kg^{-1}) in soil Nov. 1994	4	5	7	2.3
P (mg kg^{-1}) in soil July 1995	4	9	15	3.9

LSD, least significant difference.

In the course of the trial, the effect of applied P on P levels in the soil, as measured by the Morgan's test in the immediate aftermath of spreading, was monitored. After 3 months, levels of P were still artificially high. This is similar to results published by Gallagher and Herlihy (1963).

References
Gallagher, P.A. and Herlihy, M. (1963) An evaluation of errors associated with soil testing. *Irish Journal of Agricultural Research* 2(2), 149–167.
Herlihy, P. and Hegarty, T. (1994) Effects of restrictions and prices of N, P and K fertiliser inputs and yield deficits of sugar beet. *Fertiliser Research* 39, 167–178.
Murphy, W.E. (1977) Management factors affecting growth in grassland production. In: Gilsen, B. (ed.) *Proceedings of the International Meeting on Animal Production from Temperate Grassland.* An Foras Taluntais, Dublin, pp. 116–125.
Thomson, N.A., Roberts, A.H.C., McCallum, J. and Johnson, R.J. (1993) How much fertiliser is enough? In: *Proceedings of the Ruakura Farmers Conference*, Ruakura, New Zealand.
Tunney, H. and Culleton, N. (1995) Phosphorus balance on a dairy farm in Ireland (abstract). *Irish Journal of Agricultural and Food Research* 35, 76–77.
Tunney, H. (1990) A note on the balance sheet approach to estimating the phosphorus fertiliser needs of agriculture. *Irish Journal of Agricultural Research* 29, 149–154.

21.40 Defining Critical Levels of Available Soil Phosphorus for Agricultural Crops

A.E. Johnston and P.R. Poulton
Soil Science Department, IACR Rothamsted, Harpenden, Hertfordshire AL5 2JQ, UK

Introduction
On the silty clay loam, pH above 6.5, at Rothamsted a comparison of appropriate soil extractants showed that 0.5 M sodium bicarbonate (NaHCO$_3$) at pH 8.5 (Olsen's method, Olsen phosphorus (P)) was best for assessing plant-available P in soil. This extractant is also suitable for many soils. Yields can be related to Olsen P to determine a critical value of Olsen P below which yield decreases appreciably as Olsen P decreases, causing a financial loss to the farmer, and above which yield increases little or not at all. Adding P to the latter soils is an unnecessary cost to the farmer and there is a risk of increased P losses from excessively enriched soils.

Methods

A number of long-term experiments on the silty clay loam at Rothamsted, a sandy clay loam at Saxmundham and a sandy loam at Woburn had large plots with different levels of Olsen P after many years of different P treatments. In recent years, these treatments have been modified to extend the range of Olsen P values. This recently added P, usually as superphosphate, was given some years to equilibrate within the soil and then a range of arable crops and grass were grown to assess the value of P residues in the soil and to relate yields to Olsen P.

Results

Arable crops and grass consistently yielded more when grown on soils with adequate reserves of P than on P-deficient soils (Table 21.7) (Johnston et al., 1970; Mattingly et al., 1970; Johnston and Poulton, 1992).

Yields of spring barley, winter wheat, potatoes and sugar (from sugar beet) all approached an asymptote at an Olsen P which rarely exceeded 25 mg P kg^{-1} (Fig. 21.14). Averaged over a number of years, this value varied little between crops on the same soil. In other experiments, there was surprisingly little difference between critical values for the same crop on different soils, provided there was no other overriding factor affecting yield. Examples of factors that can influence the critical value are given by Johnston and Poulton (see Section 21.41, this chapter).

In other experiments the critical level of Olsen P for potatoes and sugar beet did not change appreciably when yields increased two-fold in years of adequate rainfall (Johnston et al., 1986). For winter wheat and spring barley, the critical Olsen P value was not affected by the amount of fertilizer nitrogen (N) applied, which appreciably affected yield. Also, for winter wheat, there was no difference between first, second and third wheats after a break crop. In this cropping sequence, a serious decrease in the yield of the second wheat crop, probably due to take-all, caused by the soil-borne fungus *Gaeumannomyces graminis*, affecting

Table 21.7. Yields (t ha^{-1}) of spring barley, potatoes and sugar grown in three field experiments, each with two levels of Olsen P.

		Experiment and Olsen P (mg P kg^{-1}) in the 1960s					
		Agdell		Exhaustion Land		Woburn	
Crop	P applied (kg P ha^{-1})	4*	13	4	12	18	42
Spring barley grain	0	1.54	3.41	2.03	3.11	2.62	3.34
	56	2.89	3.88	3.49	3.48	2.86	3.61
Potatoes tubers	0	12.1	29.9	12.8	21.1	35.1	40.7
	56	25.4	38.2	32.6	32.6	38.2	43.4
Sugar (from beet)	0	3.38	5.77	3.84	5.67	5.17	5.90
	56	4.79	6.00	5.72	6.00	5.15	6.15

* Olsen P (mg P kg^{-1}); soils with least Olsen P had received no P since: Agdell, 1848; Exhaustion Land, 1852; Woburn, 1876.

rooting, might have increased the need for readily available P in soil but it did not do so.

Conclusions
Where P residues accumulate in soil in plant-available forms, Olsen P is a good indicator of that availability. There is a critical value of Olsen P below which there is a serious loss of yield. Above the critical value, yield increases little and it would be justified to recommend that farmers supply P, either as fertilizers or in organic manures, to maintain their soils just above the critical value.

References
Johnston, A.E. and Poulton, P.R. (1992) The role of phosphorus in crop production and soil fertility: 150 years of field experiments at Rothamsted, United Kingdom. In: Schultz, J.J. (ed.) *Phosphate Fertilizers and the Environment.* International Fertilizer Development Center, Muscle Shoals, USA, pp. 45–64.

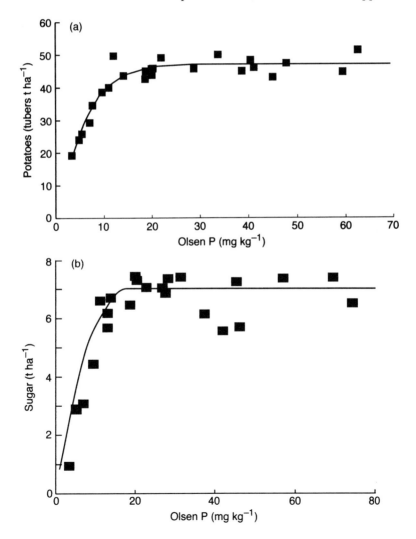

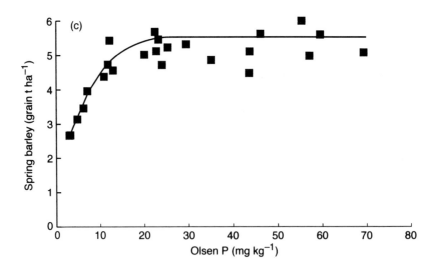
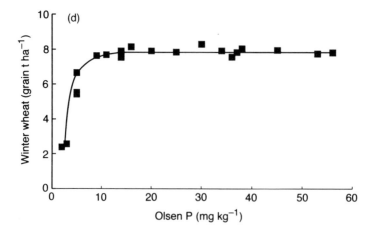

Fig. 21.14. Relationship between Olsen P and yields of (a) potatoes, (b) sugar (from sugar beet) and (c) spring barley grown on a silty clay loam with 2.4% organic matter and (d) winter wheat grown on a silty clay loam with 1.6% organic matter.

Johnston, A.E., Warren, R.G. and Penny, A. (1970) The value of residues from long-period manuring at Rothamsted and Woburn. IV. The value to arable crops of residues accumulated from superphosphate. In: *Rothamsted Experimental Station Report for 1969*, Rothamsted Experimental Station, Harpenden, Part 2, pp. 39–68.

Johnston, A.E., Lane, P.W., Mattingly, G.E.G., Poulton, P.R. and Hewitt, M.V. (1986) Effects of soil and fertilizer P on yields of potatoes, sugar beet, barley and winter wheat on a sandy clay loam soil at Saxmundham, Suffolk. *Journal of Agricultural Science, Cambridge* 106, 155–167.

Mattingly, G.E.G., Johnston, A.E. and Chater, M. (1970) The residual value of farmyard manure and superphosphate in the Saxmundham Rotation II Experiment, 1899–1968. In: *Rothamsted Experimental Station Report for 1969*, Rothamsted Experimental Station, Harpenden, Part 2, pp. 91–112.

21.41 Factors Affecting Critical Soil Phosphorus Values

A.E. Johnston and P.R. Poulton
Soil Science Department, IACR-Rothamsted, Harpenden, Hertfordshire AL5 2JQ, UK

Introduction

Dyer (1894, 1902) showed that 1% citric acid was a suitable extractant to distinguish between different phosphorus (P)-treated soils at Rothamsted. Subsequently, researchers worldwide have tested an ever wider range of extractants, seeking ever greater precision in estimating plant-available P in soil. Many converted the P concentration, determined by analysis, to an amount per unit area and made a fertilizer recommendation based on increasing the amount of P already in soil to that thought to be required by the crop. This was rarely successful and, by the late 1930s, soil analysis had largely fallen into disrepute.

It is more realistic to calibrate a soil-P test with crop yield and estimates of the likely response to a fresh application of P (Johnston and Poulton, Section 21.40, this chapter). This approach indicates a critical level of soil P below which soils are so impoverished that yields suffer or above which soils are so enriched that there is no economic justification to apply more P. For individual fields, changes in nutrient status can be followed over time and related to the fertilization policies adopted by the farmer.

Because crop roots do not explore the soil volume as efficiently as chemical extractants, calibrating yields with soil analysis data may not always give consistent results. Many factors, such as soil structure, affected by soil organic matter (SOM), depth of soil, stoniness and root diseases, affect root exploitation of soil for nutrients and these will affect the relationship between yield and Olsen P, because P has limited mobility in soil. Yields of arable crops have been related to Olsen P on three different soil types on sites managed by Rothamsted and are used to illustrate these points.

Methods

Experimental sites with plots with a range of Olsen P and, in some cases, different levels of SOM have been established on a silty clay loam (Rothamsted), sandy clay loam (Saxmundham) and a sandy loam (Woburn). Both clay loam soils have about 25% clay but it is more difficult to produce good seed-beds, especially in spring, on the sandy clay loam and this affects crop establishment. Larger amounts of SOM in both the silty clay loam and the sandy loam benefit yield (Johnston, 1991).

Results

The critical level of Olsen P in a poorly structured silty clay loam at Rothamsted was very dependent on the SOM content (Fig. 21.15). The structure of soils, and thus the ability of roots to grow through them, can depend on factors other than organic matter. The critical Olsen P for spring barley was less at Rothamsted than at Saxmundham because it was easier to produce a good seed-bed, which aided establishment, on the silty clay loam at Rothamsted (Fig. 21.16).

Combining results from two different experiments, both on a sandy loam at Woburn, suggested that spring barley yields responded to Olsen P up to 180 mg

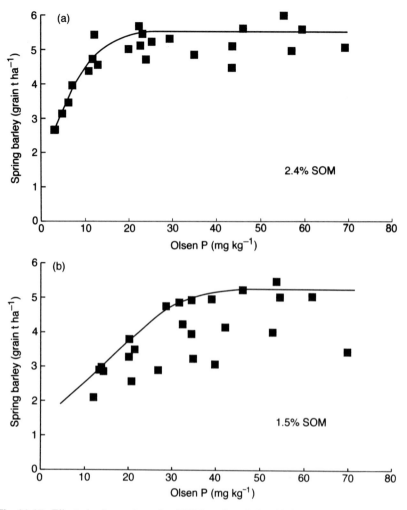

Fig. 21.15. Effect of soil organic matter (SOM) on the relationship between yields of spring barley and Olsen P in a silty clay loam soil, Agdell, Rothamsted.

P kg^{-1}. All other experience suggested this was a nonsense. Although no certain explanation can be offered, the most likely reason may relate to the different levels of organic matter in the two soils, which were 1.12% and 1.86–3.14%, respectively, in the two experiments. The sandy loam at Woburn is very liable to capping or crusting, especially in spring if heavy rain follows soon after cultivation. This can affect germination and early growth and the adverse effects are partially mitigated by extra SOM. On the soil with 1.12% organic matter, barley yields increased between 25 and 36 mg Olsen P kg^{-1}, in good agreement with other data on spring barley on poorly structured soils. On the other group of soils, which had much higher levels of both Olsen P and SOM, yield, Olsen P and SOM all increased together (Table 21.8). It is probable that barley yields were responding more to increasing SOM than to Olsen P.

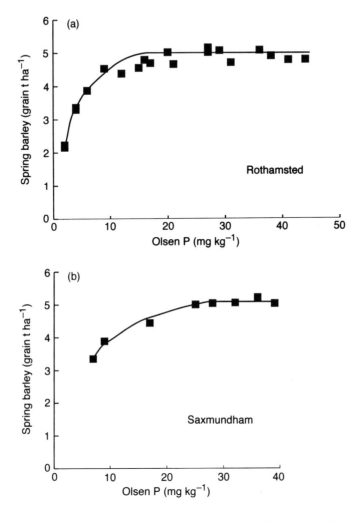

Fig. 21.16. Relationship between the yield of spring barley and Olsen P on a silty clay loam (Rothamsted) and a sandy clay loam (Saxmundham).

Table 21.8. Yield of spring barley, Olsen P and soil organic matter on a sandy loam soil.

Grain (t ha^{-1})	4.93	5.15	5.23
Olsen P (mg kg^{-1})	128	147	165
Soil organic matter (%)	1.86	2.48	3.14

Conclusion

Critical values for Olsen P for agricultural crops may vary between soils if the volume of soil explored by roots is greatly affected by soil structure, stoniness or depth of soil. On such soils, it may be necessary to set critical values somewhat higher than for soils with a larger soil volume available for root exploitation.

References

Dyer, B. (1894) On the analytical determination of probably available 'mineral' plant food in soils. *Journal of the Chemical Society Transactions* 65, 115–167.

Dyer, B. (1902) *Results of Investigations on the Rothamsted Soils.* Bulletin of Official Experimental Stations, US Department of Agriculture No. 106. Washington DC,

Johnston, A.E. (1991) Soil fertility and soil organic matter. In: Wilson, W.S. (ed.) *Advances in Soil Organic Matter Research: the Impact on Agriculture and the Environment.* Royal Society of Chemistry, Cambridge, pp. 299–313.

21.42 Distribution of Available Phosphorus in Soil under Long-term Grassland

W.E. Murphy and N. Culleton
Teagasc, Johnstown Castle Research Centre, Wexford, Ireland

Introduction

The phosphorus (P) in water-soluble phosphatic fertilizers added to soil forms compounds in most soils that have extremely low solubility in water and therefore the P is held in the soil against leaching by drainage water. It is readily mobile only in some peat soils and sands. When P is applied to heavier-textured soils, its distribution in the rooting zone of the soil profile depends on tillage operations and the action of soil organisms, including plant roots. In the case of grassland, tillage operations are absent and soil P distribution is dependent on processes within the soil. This can lead to stratification of available P, with the larger concentrations near the surface.

Method

In a long-term grazing trial at Johnstown Castle, a mainly grass sward was divided into three areas receiving 0, 15 and 30 kg P ha^{-1}. The areas were divided into paddocks and a rotational grazing system was practised. The cattle returned to each paddock every 3 weeks. After 20 years, soil samples were taken from the 0–20, 20–40, 40–80, 80–120, 120–160 and 160–200 mm soil horizons. The samples were analysed for available P by using both Morgan and Olsen reagents.

Results and conclusions

The amounts of extractable P in the soil from different depths and the three P treatments are shown in Table 21.9.

Both Morgan P and Olsen P show that residues from the applied P have accumulated in the top 160 mm and that there has been enrichment with P throughout this depth either by leaching or by translocation in roots followed by their death. The differences between treatments when measured in terms of available P were not significant at depths below 160 mm. The concentration of P in soil exposed to water flowing over the soil surface is therefore higher than would be indicated by tests on soil samples taken from 0 to 100 mm, as is the practice for grassland soils in this country.

Table 21.9. Morgan P and Olsen P at different depths within soil as a result of different P application rates for 20 years.

	P application rate (kg P ha^{-1} year^{-1})					
	0	15	30	0	15	30
Depth (mm)		Morgan P			Olsen P	
0–20	4.6	13.1	27.8	33.6	44.8	63.6
20–40	3.5	8.2	18.4	29.6	35.3	56.0
40–80	3.1	6.8	15.1	24.1	29.9	44.4
80–120	2.8	5.6	12.6	23.3	26.7	47.9
120–160	2.2	3.7	7.6	20.0	21.8	29.6
160–200	1.4	1.4	4.0	16.3	17.6	20.4
LSD for phosphorus rate		1.96			3.87	
LSD for depth		0.86			2.67	
LSD for rate by depth		1.49			4.63	

LSD, least significant difference.

21.43 Comparison of Fertilizer Phosphorus Recommendations in Ireland and England and Wales

P.R. Poulton,[1] H. Tunney[2] and A.E. Johnston[1]
[1]*IACR-Rothamsted, Harpenden AL5 2JQ, UK;* [2]*Teagasc, Johnstown Castle, Wexford, Ireland*

Introduction

On a national scale, the concentration of readily soluble phosphorus (P) in Irish soils appears to be increasing (Tunney, 1990), which is in contrast to a lack of appreciable change in the P status of soils in England and Wales during the last 21 years (Salter *et al.*, Chapter 21, this volume). This may, in part, be related to the disparity between fertilizer P recommendations given by the Agricultural Development Authority (Teagasc) in the Republic of Ireland and the Ministry of Agriculture, Fisheries and Food (MAFF) in England and Wales. The relationship between yield and soil P is important for the financial viability of farming (Johnston and Poulton, Chapter 21, this volume), but there is also concern that agricultural soils may be a diffuse source of P to rivers and lakes by leaching, surface runoff and soil erosion. Thus soil P levels should not be unnecessarily high.

Methods

Recommendations for P applications to crops given by Teagasc (1994) and MAFF (1995) are based on soil analysis, and both advisory systems divide the range of analytical values into indices. Unfortunately, however, the two organizations use different extractants. Teagasc use Morgan's method, extracting soil with 10% sodium acetate buffered at pH 4.8 (Morgan P), while MAFF use Olsen's method, extracting with 0.5 M sodium bicarbonate (NaHCO$_3$) at pH 8.5 (Olsen P). The relationship between the two methods was estimated by analysing 100 grassland soils from Ireland, which had a range of Morgan P from 0 to 15 mg P l^{-1}. Morgan

P was determined at Johnstown Castle and Olsen P at Rothamsted. Data to illustrate the relationship between yield and Olsen P are taken from Rothamsted experiments.

Results

Morgan's method extracts much less P than Olsen's method, but there was a linear relationship with the following equation:

$$\text{Olsen P} = 5.8 + 2.91 \text{ Morgan P} \quad (r^2 = 0.67)$$

Using this relationship the upper and lower values for Morgan P in each Teagasc P index were converted to Olsen P. The indices, comparable Olsen P values and the P recommendations for grazed grassland and sugar beet in both systems are given in Table 21.10.

Only two indices have analytical values which are directly comparable, index 2 in both systems and Teagasc index 5 and MAFF index 4. Teagasc puts all soils below index 2 into one index, while MAFF has two indexes, 0 and 1, in what many field experiments have shown to be a critical range of soil-P values. Above index 2 Teagasc has two indexes, 3 and 4, but MAFF has only one for the range 25–45 mg Olsen P l^{-1}. Again, field experiments in England and Wales have shown that few crops respond to soil P above 25 mg Olsen P kg^{-1} (Johnston *et al.*, 1986). For example, yields of sugar, from sugar beet, and grass were related to Olsen P as shown in Fig. 21.17. For mineral soils which are air dried and ground to pass a 2 mm sieve, the bulk density, when measured by volume for soil analysis in many laboratories, is about 1; hence mg P kg^{-1} is approximately equivalent to mg P l^{-1}.

Much more fertilizer P is recommended by Teagasc than by MAFF. For example, for both grazed grassland and sugar beet (Table 21.10), there is a

Table 21.10. Morgan P and equivalent Olsen P, and P index used by Teagasc and MAFF for P fertilizer recommendations, with examples for grazed grass and sugar beet in both systems.

Soil analysis						
Teagasc index		1	2	3	4	5
Morgan's P (mg P l^{-1} soil)		0–3	3–6	6–10	10–15	15+
Equiv. Olsen P (mg P l^{-1} soil)		0–14.5	14.5–23.5	23.5–35	35–49.5	49.5+
MAFF index	0	1	2	3		4
Olsen P (mg l^{-1})	0–9	9–15	15–25	25–45		45–70
P recommendations (kg P ha^{-1})						
Grazed grass						
Teagasc (10 cm)*		40	30	10	0	0
MAFF (7.5 cm)	26	17	9	0		0
Sugar beet						
Teagasc (10 cm)		75	55	40	20	0
MAFF (15 cm)	45	34	22	22		0

* Recommended soil sampling depth.

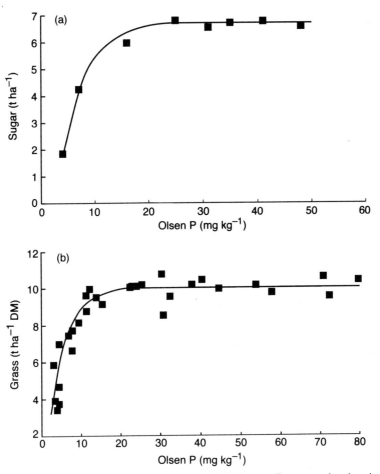

Fig. 21.17. Relationship between Olsen P and the yield of sugar (from sugar beet) and grass on a silty clay loam soil, Rothamsted.

twofold difference below index 2 and at index 2 a threefold difference for grazed grass and a 2.5-fold difference for sugar beet.

When soils are below critical Olsen P values, there is good reason to apply sufficient P to build up soil fertility. However, there is little justification for appreciably exceeding the critical value. The aim should be to maintain soils just above the critical value, using either fertilizer or organic manures. The appropriateness of P fertilization policies can be checked by periodic soil analysis.

Conclusions

There is a concentration of readily soluble P in soil above which yield does not increase as soluble P increases. Below this value, there is a serious risk of loss of yield. Keeping soils much above this critical value is an unnecessary financial burden for the farmer and there may be a risk of a greater loss of P by leaching and soil erosion. Phosphate recommendations for both fertilizers and manures should be based on maintaining readily soluble soil P just above the appropriate critical value. Timing the application should aim to minimize the risk of surface runoff. The success of such a policy can be checked by periodically analysing the soil in each field.

References
Johnston, A.E., Lane, P.W., Mattingly, G.E.G., Poulton, P.R. and Hewitt, M.V. (1986) Effects of soil and fertilizer P on yields of potatoes, sugar beet, barley and winter wheat on a sandy clay loam soil at Saxmundham, Suffolk. *Journal of Agricultural Science, Cambridge* 106, 155–167.
MAFF (1995) *Fertiliser Recommendations for Agriculture and Horticulture Crops*. RB 209, Ministry of Agriculture, Fisheries and Food, HMSO, London, 112 pp.
Teagasc (1994) *Soil Analysis and Fertilizer, Lime, Animal Manure and Trace Element Recommendations*. Johnstown Castle, Wexford, 36 pp.
Tunney, H. (1990) A note on a balance sheet approach to estimating the phosphorus fertilizer needs of agriculture. *Irish Journal of Agricultural Research* 29, 149–154.

21.44 The Impact of Fertilizer Strategies on the Phosphorus Status of Arable Soils in England and Wales

J.L. Salter,[1] B. Higgs[1] and C.J. Dawson[2]
[1]*The Fertilizer Manufacturers Association, Thorpe Road, Peterborough PE3 6GF;* [2]*Chris Dawson and Associates, Westover, Ox Carr Lane, Strensall, York, UK*

Introduction
Recommendations for fertilizer use have to meet two criteria: to maximize economic benefit to the farmer and to minimize adverse environmental impact. The second criterion has increased in importance recently. Good advice to meet both criteria must be based on knowledge of the behaviour of plant nutrients in the soil. However, assessing how well farmers adopt this advice is difficult. An attempt has been made here in respect of their use of phosphate fertilizers on arable crops in England and Wales.

Methods
Data for phosphate fertilizer application rates, national average yields and the concentration of phosphorus (P) in harvested crops have been used to estimate a national P balance for the main arable crops in England and Wales over a 21-year period. The likely effect of this P balance on Olsen-extractable P (Olsen P) levels in soil was then related to changes in Olsen P, measured independently by the representative soil-sampling scheme (RSSS), over the same 21-year period.

Phosphate fertilizer use. The survey of fertilizer practice for England and Wales has provided annual information on fertilizer use in mainland Britain since 1969 (Church and Webber, 1971). For this study, applications of P to cereals, oilseed rape, potatoes and sugar beet were used for 1974 to 1994.

Annual yields, phosphorus content of harvested crops and phosphorus balance. From the Ministry of Agriculture, Fisheries and Food (MAFF) data on areas and yields of all crops grown in England and Wales, total annual yields of each crop were calculated. These data, together with the amount of P removed by each crop, were used to determine the P balance, weighted according to the area of crop grown.

Representative soil sampling scheme. The survey of fertilizer practice also offered an opportunity to collect information about the nutrient status of soils in England and Wales. The aims of the RSSS were to obtain unbiased estimates of the nutrient status of soils and to monitor changes. Here, changes in Olsen P between 1974 and 1994 were related to the P balance as determined above. For details of the scheme and various changes, see Church and Skinner (1986) and Skinner *et al.* (1992).

Results

The average annual overall fertilizer P balance in England and Wales for cereals, oilseed rape, potatoes and sugar beet is shown in Fig. 21.18. Poor yields, due to drought in 1976, produced a large positive P balance; otherwise it remained roughly constant between 1977 and 1988 before declining to less than 5 kg P ha^{-1} from 1990.

To relate a positive P balance to changes in Olsen P (P extracted by 0.5 M sodium bicarbonate (NaHCO$_3$) at pH 8.5) in soil, it has been assumed that 15% of the P balance remains soluble in this extractant after a period of 12 months in the soil. (This is based on data from long-term experiments at Rothamsted.) A positive balance of 5 kg P ha^{-1} would increase soluble P in soil by 0.16 mg P l^{-1} assuming 2.0 million l of cultivated soil per hectare. Thus, the annual P balances shown above would be expected to raise Olsen P only marginally. This expectation is borne out by data from the RSSS, shown in Fig. 21.19.

There has been no appreciable increase in the proportion of soils in each P index in the last 25 years. There is, however, slight evidence that the percentage of soils in the higher indices has declined. Perhaps more importantly, the data suggest that there is scope for an increase in P use on soils currently in index 0 and 1.

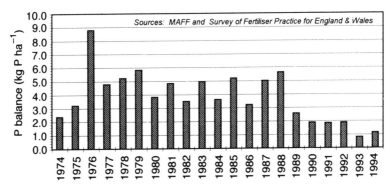

Fig. 21.18. The average annual positive fertilizer P balance in England and Wales, based on the average amounts of fertilizer P applied to, and estimated removals in, the average national yields of cereals, oilseed rape, potatoes and sugar beet.

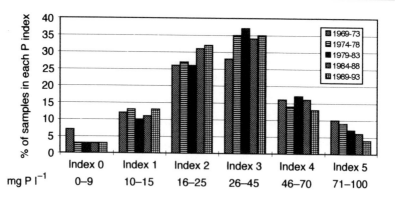

Fig. 21.19. The proportion of soils in each P index in five 5-year periods from 1969 to 1993, based on Olsen P values (mg P l^{-1}).

While for England and Wales the national P balance suggests that overall there is no excessive use of P fertilizers, there may be exceptions for particular crops or locations. For example, over the last 15 years, the average annual positive P balance for the potato crop has been 165 kg P ha^{-1} from fertilizers. Where potatoes are grown, there could be a build up of Olsen P in the soil, especially if organic manures are also applied. Thus, there is a need for individual farmers to consider carefully their P fertilization policies throughout the rotation and to use soil analysis as a guide to the changing P status of their soils.

Conclusion
Collecting data for phosphate fertilizer use, crop yields and P off-takes has allowed P balances to be estimated and compared with an independent assessment of soil P status. Such comparison suggests that on a national scale in England and Wales farmers have, by and large, followed advice on P fertilization of arable crops. While scope remains for increased P fertilizer use on some soils, there has been a slight decrease in the proportion of soils with unnecessarily high levels of Olsen P, but a small proportion of soils with a very high P status may pose problems locally.

Acknowledgements
Thanks are due to A.E. Johnston (Lawes Trust Senior Fellow, Rothamsted) for comment and encouragement, MAFF for permission to use RSSS data and Alan Todd (Rothamsted) for his assistance in making the data available. Data on annual yields were taken from MAFF sources.

References
Church, B.M. and Skinner, R.J. (1986) The pH and nutrient status of agricultural soils in England and Wales. *Journal of Agricultural Science, Cambridge* 107, 21–28.
Church, B.M. and Webber, J.C. (1971) Fertilizer practice in England and Wales, a new series of surveys. *Journal of the Science of Food and Agriculture* 22, 1–7.
Skinner, R.J., Church, B.M. and Kershaw, C.D. (1992) Recent trends in soil pH and nutrient status in England and Wales. *Soil Use and Management* 8, 16–20.

22 Phosphorus Loss from Agriculture to Water: Synthesis and Summary

A.E. Johnston,[1] H. Tunney[2] and R.H. Foy[3]

[1]IACR-Rothamsted, Harpenden, Hertfordshire AL5 2JQ, UK; [2]Teagasc, Johnstown Castle Research and Development Centre, Wexford, Ireland; [3]Agricultural and Environmental Science Division, Department of Agriculture for Northern Ireland, Newforge Lane, Belfast BT9 5PX, UK

There are adverse effects on the flora and fauna of freshwater lakes, and consequently on the overall biological balance, as a result of excessive nutrient enrichment. In most cases, the availability of phosphorus (P) usually controls the level of primary productivity, and excessive concentrations of P can lead to imbalances. During the last few decades, water-quality monitoring programmes have shown steadily increasing P enrichment of surface waters, despite considerable financial investment to decrease point-source pollution in many regions of the world. As further point-source control becomes less cost-effective, attention must be directed towards P losses from agriculture, not least because increasing P enrichment is being reported more frequently in what were pristine lakes draining rural or agriculturally dominated catchments.

Agronomically, P is essential for crop growth and farmers are understandably reluctant to risk the financial viability of their farms by using too little P. In part, this is because, until recently, advisory and educational programmes encouraged farmers to increase plant-available soil P from the inherently low levels in most soils. Such advice was agronomically acceptable, but little thought was given as to whether there were upper limits to which soil P should be raised. This was because it was thought that there was more risk for the farmer in underproviding P than in overproviding it; crops were not adversely affected by large amounts of P in soil; soils were considered to lock up P in unavailable forms and measured losses of P from agricultural soils were so small as to be economically insignificant. However, off-site, these small quantities of P can increase P concentrations in water bodies to levels that have considerable biological impact and consequent economic implications. In some cases, this impact can occur many kilometres from the source of the P, often from regions not controlled politically or administratively by those responsible for the rivers and lakes where pollution is acute. It may take many years for any adverse impact to be fully manifest and many more years before any improvement in water quality occurs following the introduction of remedial measures.

Action must be taken, therefore, to minimize P losses at source. To achieve this requires a better understanding of how P losses are related to land-management systems, soil characteristics, topography and climate and an appreciation of how P inputs to the farm can be managed to maintain economic viability with minimum environmental risk. The identification of threshold levels of P in water must also address the biosensitivity of a water body to P inputs and its desired trophic state. This will require the development of environmental indicators – both to identify high-risk areas for P losses and to assess whether agriculture is an important source of P to surface water – and of a cost–benefit analysis of possible remedial measures. Having decided on remedial measures, it may be necessary to require farmers to implement them, especially if the proposed strategies could be ineffective unless all farmers within a catchment cooperate. Achieving compliance may require a mix of voluntary and mandatory measures, but at present there is some uncertainty about where one or the other would be appropriate.

Currently, discussions about the effect of P enrichment of water bodies on their flora and fauna are impeded by the diverse nomenclature and methodology used to estimate the various fractions of P lost from soil and thus identify their biological significance in both the short and the long term. Urgent action is required internationally to standardize nomenclature and methodology and thus resolve this situation. A list of definitions used at the University of Pennsylvania (Table 22.1) is given as an example which could be the basis of further discussions to arrive at an agreed list.

In the larger environmental context, agriculture is considered as a diffuse source of P to water. In reality, however, there are both point and diffuse sources. Point sources, such as seepage from slurry and silage storage facilities and washings from farmyards, can be recognized and controlled by improved management. Within an overall catchment control strategy, it would be most appropriate to first ensure cost-effective remedies for such point sources.

Of general concern is the widely held belief within agricultural circles that soils strongly retain P applied to them. This has led to the view that P polluting surface water was unlikely to be coming from agricultural soils. However, in the light of the small quantities of P that are limnologically significant, this view cannot be upheld unless there is absolutely no risk of any P being lost from soil.

Much evidence was presented at the workshop to show that P can be transported from agricultural land to water bodies by various pathways, the importance of which will differ from region to region. These pathways

Table 22.1. Suggested terminology for P lost in runoff and drainage water.*

Phosphorus form	Abbreviation	Methodology†

Total phosphorus (total amount in dissolved and particulate phases)	TP	Digestion of unfiltered water sample
		  Kjeldahl procedure
		  Acid ammonium persulphate
		  Perchloric acid

Total dissolved phosphorus (dissolved inorganic (orthophosphate) and organic P)	TDP	Acid persulphate digestion of filtered sample
Dissolved orthophosphate (immediately algal-available)	DP	Murphy and Riley on filtered sample‡

Bioavailable phosphorus (dissolved orthophosphate and a portion of the particulate P that is algal-available)	BAP	Extraction of unfiltered sample with
		  NaOH
		  NaCl
		  Anion exchange resin
		  Ammonium fluoride
		  Iron-oxide paper strips

Molybdate-reactive phosphorus (dissolved orthophosphate and acid extractable particulate P (possibly algal-available))	MRP	Murphy and Riley on unfiltered samples
Particulate phosphorus (inorganic and organic P associated with or bound to eroded sediment)	PP	By difference TP − TDP
Dissolved organic phosphorus§ (includes polyphosphates and hydrolysable phosphates)	DOP	By difference TDP − DP

* Based on a list prepared by A.N. Sharpley and H. Pionke, Pennsylvania State University, USA, and presented at the workshop. The list does not imply that all analyses must be done on all samples.
† Not an inclusive list of appropriate methods that can be used.
‡ Filtered sample represents that passing through a 0.45 μm filter.
§ If dissolved organic P constitutes more than 25% of TDP, it might be necessary to consider measuring polyphosphates and hydrolysable phosphates.
NaOH, sodium hydroxide; NaCl, sodium chloride.

include soil eroded by water and wind, surface runoff and leachate from soil, especially where soils are freely draining or are artificially drained.

Soil erosion and surface runoff are important in hilly areas, especially where soil characteristics, farming systems and management singly or in combination cause a decrease in water-infiltration rate. Such losses are strongly related to storm events, and P lost in surface runoff can be greatly increased where animal wastes, especially slurry, have been applied just prior to a storm or prolonged rainfall. Although this pathway has been studied at the experimental-plot and field scale, much more needs to be known about how catchment hydrology controls P export. Both hydrology and chemical processes controlling P release are dynamic and highly variable both temporally and spatially. Thus, all fields do not necessarily contribute to P losses from a catchment. Often, much of the total annual export of P from a catchment may come from only a small area within it and then only during a small number, perhaps as few as three or four, storm events. Thus, for this pathway of loss, runoff and erosion potential are likely to be more important, though less amenable to control, than soil-P content in determining total P losses. Initially, therefore, decreasing P losses by erosion and runoff within a catchment must target hydrologically active source areas. Similarly, the use of buffer or riparian zones to decrease losses of P from soil to streams may vary in effectiveness depending on how the riparian zones affect local hydrology. Much research is still needed on both the short- and long-term bioavailability of P strongly adsorbed on soil particles transported to lakes by erosion.

Perhaps the most contentious issue was whether P can be lost by leaching. Historically, the belief that such losses did not occur developed for a number of possible reasons. In a few long-term experiments, a large proportion of the applied P could be accounted for and the available analytical techniques were unable to estimate with accuracy P lost in drainage. Later, many studies showed that, when water-soluble P in fertilizers was applied to soil, complex insoluble compounds were formed. The belief that soils cannot lose P by leaching also ignores the evidence that, on many soils, P reserves accumulate in plant-available forms. Plants take up P from the soil solution, so that P in the soil solution may be at risk to loss by leaching and runoff.

Sufficient P to cause eutrophication in surface water may be lost by leaching from two diverse groups of soils. One group consists of sandy-textured and organic soils, typically very productive when intensively managed, and frequently found in valley bottoms or where soils have been reclaimed from the sea. In many cases, there is little depth of soil to the water-table and then it is frequently observed that the water is excessively enriched with P. The second group tend to be heavy-textured, clayey soils on flat or gently undulating sites. They are often artificially drained to improve removal of surface water and aid workability. The drains discharge to surface ditches. Although the subsurface horizons of such soils often have a large capacity to adsorb P, increased levels of P in drainage have been reported where surface soils were excessively P-enriched. This could be explained by preferential or bypass flow down cracks and fissures leading to rapid movement of P to the drains. In contrast to surface runoff, the soluble P fraction forms a significant proportion of the total P transported in the leachate.

Whether P is lost by soil erosion, in surface runoff or in drainage, the potential risk of P pollution must increase as soil P accumulates. For those soils where some part of the residues of P applied in fertilizers and manures accumulates in plant-available forms, the extent to which soil-P levels should be increased has not been widely addressed. In part, this is because soils enriched with P reserves invariably yield more than impoverished soils and agronomically this was perceived to be a benefit, especially as plants were not adversely affected at elevated levels of soil P. In part, it was because of the widely held view that P is held firmly within the soil matrix. All other factors being optimum, it would be expected that, as plant-available soil P increases, yield will approach an asymptote, plant-available P being determined by a suitable analytical method. Thus there will be a critical value of soil P, which may vary with crop and soil, below which yield will decrease appreciably, resulting in a financial loss to the farmer. There is no financial incentive for the farmer to appreciably increase soil P above the critical value, but, provided there is no unnecessary environmental risk, there are good reasons to maintain soils just above this value. Adopting such a policy requires an appropriate method of soil analysis and the use of nutrient-management planning.

Nutrient-management planning is applicable to all nutrients, not just P. For P, it aims to balance off-farm inputs of P in feed and fertilizers with outputs in produce, while maintaining adequate soil-nutrient reserves. For example, acute P pollution from agricultural sources is often in areas with intensive livestock production. The problem is exacerbated by applying manures under less than ideal conditions and/or when there is too little land available for their proper disposal, i.e. to make efficient use of the nutrients in the manures. Although the problem may arise from the quantity of P involved, there are also many questions as to whether P is transported in organic forms or whether transport is related to other attributes of the added organic matter, which also accumulates in such soils. The plant and environmental availability of P in manure and inorganic fertilizer needs further investigation.

Complete nutrient-management planning on intensive livestock farms is especially important. The use and management of nitrogen (N) to drive manure management has allowed P to accumulate in soil and increased P losses in surface runoff and drainage. In many cases, there could be severe financial penalties from decreasing intensive livestock production. Such penalties would apply not only to farmers but to all those servicing the food chain. Thus decisions to implement major changes could involve large-scale economic restructuring. This has led to various lines of research aimed at decreasing the amount of P in animal farming systems and these need to continue. They include manipulating the dietary P intake of livestock, enzyme additives for animal feed to increase the efficiency of P uptake during digestion and questioning the need to use inorganic fertilizer P if soil-P levels are satisfactory. The suggestion to dispose of surplus manure off the farm may appear to be a viable option, but the full environmental costs would need to be evaluated.

Maintaining soil-nutrient reserves at adequate levels is an essential part of nutrient-management planning. Most current methods of soil analysis should be satisfactory for predicting the need to apply P for a crop, and some, with appropriate data from field experiments, could be used to estimate critical soil-P values, as defined above. A very disturbing feature which became apparent at the workshop was the discrepancy between countries in their view of the amount of P that needed to be applied to crops grown on soils with the same concentration of soil P. While it might prove impossible to get all countries to use the same method of soil analysis, correlations between methods could be sought. Data on critical levels of soil P for different crops could then be compared across countries and P recommendations at comparable soil-P levels unified or differences satisfactorily explained.

Not all the P added to soils is extracted by those reagents currently used for agronomic tests, and some part of these residues may be bioavailable if the soil is transferred to water. Would an environmental soil-P test be useful? The advantages would need to be carefully considered before research was undertaken to find an appropriate method. It would have to estimate the bioavailability of soil, sediment and runoff P to aquatic organisms and the long-term capacity of a soil to retain P against leaching. Reliable estimates of the P loading capacity of a soil alone could perhaps aid the development of a P management strategy. The concept of P saturation of soils has been developed for a very limited range of soil types, using a not too complicated analytical methodology. The usefulness and applicability of this or other methods and a more rigorous definition of P saturation to encompass a wider range of soils require further research.

The diverse pathways by which P can be lost from soil suggest that there will be no simple, one-off remedy applicable to all situations in all regions. However, to test P losses for a range of management options, even for a few major farming systems, soils and climates, is impractical at present. Thus there is an essential need for multidisciplinary teams to collect reliable data from a few catchments and develop models to describe P losses. Such models can be validated as part of an ongoing process to minimize P losses in an ever-increasing number of catchments.

The workshop discussed many of the factors that are essential inputs to the development of agricultural systems that are both sustainable and environmentally benign. For some systems good information already exists, but for all systems further research is essential. The fertilizer industry responded positively to the suggestion that there was a need to supply appropriate products to make nutrient-management planning easier. However, the workshop also recognized that success in diminishing P losses from agriculture to water will require the active participation of farmers. They will need reassurance that the concept of decreased or zero P inputs where appropriate can be substantiated by field-based experimental data. Currently, they are often confused by apparently conflicting information coming from scientists. Equally importantly, educational programmes will be needed to convince them of the importance of the wider environmental benefits occurring to humankind as a result of modifying the nutrient inputs on their farms. Some element of compulsion to implement remedies seeking to minimize P losses may well be required. Such measures could range from the obligatory adoption of codes of good agricultural practice to fiscal measures. The suggestion that some proportion of existing subsidy payments could be used to purchase professional advice on nutrient-management planning and to pay for soil analysis appeared to offer a possible way forward. The farmers in the European Union countries apply about 2 million tonnes of P fertilizer annually at a cost of 2 billion European currency units (ECU). There is scope, therefore, for considerable financial savings if much of this P is being applied to soil already so enriched with plant-available P from past applications that crops will not respond to further increases in the P content of the soil. The time is now opportune to maximize the efficiency of P use from both fertilizers and manures by the use of nutrient-management planning and thus to minimize the risk of excessive losses of P from soil to water.

ACKNOWLEDGEMENTS

The authors wish to acknowledge the contribution of the following people, who helped summarize the workshop discussions on which this chapter is, in part, based: A. Breeuwsma, R. Grant, L. Heathwaite, W. Magette, M. McGarrigle and A. Sharpley.

Index

adsorption 6, 55, 140, 209, 212
agriculture 55, 64, 77, 81–82, 98, 111, 225, 311, 329–330, 351
    code of good practice 216, 219
    sustainable 180, 193–194, 196, 273
    watershed 248–249
aids, financial 332
algae 90, 98, 340, 354
    blue-green 1
animals 187, 213, 344
    confined operations 5
    feeds 285–286
    nutrition 283
    P deficiency 288–289, 290
    stocking rates 58
applications 23, 189, 313
Australia 182, 184
Austria 314, 379–381, 406
    soil analysis 179
availability 5, 7, 11, 143, 315, 431, 436

balance 5, 19, 187, 249, 275, 313, 345
    animal production 291
Baltic Sea 77
Barton Broad 97
Ballinderry River 57–58, 60, 64, 70
Belgium 342, 398, 400, 403, 417

fertilizer consumption 192
    soil analysis 179, 183
bioavailable phosphorus (BAP) 11, 25, 29, 86–88, 152, 163–164, 206, 213, 217, 225, 231, 243
biomass 120
    marine 121
Bray-1 15, 160, 168, 169, 227
Bray and Kurtz No.1 102, 182
budget 218, 299

calcium : phosphorus ratio in animal diet 284
calcium chloride 263
Canada, field experiments 155–156
catchments 225, 356, 383
    Irish 57, 59, 358
    management 2, 391
cattle 187, 216, 299
cereals, winter 213, 244
chlorophyll a 35, 120
commission
    Oslo 339
    Paris 339
Community Research Programmes 333
cows, milk 287

crops 280
    cover 28
    requirement 144, 145
    yield 177
cyanobacteria 1, 122
cycling, microbial 216

deforestation 98
Denmark 77, 81, 104, 142, 179, 189, 192, 383, 392
deoxygenation, lake 122
desorption 55, 209
detergents 330
diet 22
dietary phosphorus, bioavailability 286
dissolution 140
dissolved
    organic phosphorus (DOP) 57, 59, 196, 207, 259
    phosphorus (DP) 11, 80, 89, 152, 160–162, 166, 217
    reactive phosphorus (DRP) 57, 61, 356
drainage 206
    artificial pipes 210, 211
    subsurface loss 145, 221

ecology, freshwater 120
ecosystems 312, 329
England 367, 370, 452
    soil analysis 179
environment 194, 197
erosion, soil 15, 85, 152, 167, 180, 196, 209, 226, 228, 231, 313, 323, 330, 344, 374–381
Esthwaite Water 103
Europe 103, 317
    fertilizer consumption 192
    field experiments 155–156
    soil phosphorus tests 179, 182, 183, 184
European Fertilizer Manufacturers Association (EFMA) 311, 317
eutrophication 2, 55, 77, 90, 96, 184, 197, 205, 243, 301, 305, 329, 340, 363
excretion 292

export 59, 211, 218, 246, 247, 299

farming 299, 440
    intensive 193, 425, 430
    profits 20
fertilizer 8, 139, 153, 177, 184, 186, 189–190, 192, 213, 232, 253, 277
    grass yield, effect on 178
    mineral 59, 212
    reactions 141
    recommendations 193, 449
    residual behaviour 143
    slurry 59
    strategies 178, 188, 452
    taxes 278
Finland 77, 81, 354–356, 389
    soil analysis 179
fish, population 126
fixation 139, 195–196, 279
Florida 36
flow 369
    base 3
    stream 3
    subsurface 12, 213
    tile 12, 19
France 342
    fertilizer consumption 192
    soil analysis 179, 183

geochemical (P) 97–98
Geological Information System 406, 412
Germany 342, 425, 430
    fertilizer consumption 192
    soil analysis 179, 183
grassland 191, 214, 216, 351, 372, 448
    low intensity 55
grazing 22, 66
Greece 313

hydrograph, storm 226
hydrology 412
    catchment 13

import 299

index 218
infiltration, capacity 209
inorganic phosphorus 6, 207, 210, 215
input 59, 111, 122
inventories 417
    lake 125
Ireland, Republic of 314, 357, 448
    fertilizer consumption 192
    soil analysis 179, 183
Italy
    fertilizer consumption 192
    soil analysis 179, 183, 189

lake 319
    enrichment 119
    input–output budget 122
    loading model 57
Lake Champlain 29, 31
Lake Geneva 243
LaPlatte River basin 31
land
    arable 213
    riparian use 219
leaching *see* runoff, subsurface
    (leaching)
legislation 333, 407, 410
liming 5
limitation, light 121
livestock 212
loading 77, 80
loss
    control strategies 19
    evaluation 403
    phosphorus 18, 55, 77, 146, 196,
        253, 357
    summer, 64
    winter 64
Loughgrove 127, 128
Lough Augher 107
Lough Ennel 127
Lough Erne 108, 121, 128
Lough Neagh 57, 59, 60, 108, 124, 128, 357
Lough Patrick 103, 105, 108

macrospores 13, 210–211, 221
management 324

lake 95
land 213
manure 20, 26
nutrient 297, 298
programmes 39–42
source 21, 237
strategies 217, 221
tillage 20
manure 177, 185, 192–193, 232, 253, 258, 273, 298, 313, 318, 322, 342, 344
    beef 8, 18, 31
    poultry 9, 18, 304, 431
    storage 17, 80, 344
    swine 9, 18, 304
manurial phosphorus 7, 212, 425
Mehlich-1-P 182
Mehlich-3-P 161, 165, 166, 168, 182
microbial 6
mineralization 180, 207, 212, 312
mobility 97
models 96, 125, 194, 196, 219, 361, 412
    diatom 97, 99
    runoff, soil moisture 226, 227
molybdate reactive fraction (MRP) 57, 187, 207, 210, 255, 256, 260, 265, 386, 387
Morgan 181–182
movement 257, 369, 422
    indicators 263, 267

Netherlands 314, 342, 361, 363, 407, 410, 438
    fertilizer consumption 192
    soil analysis 179, 183
New Zealand 70, 182, 184–185, 358
nitrogen 59, 305
North Sea 341
Northern Ireland 104, 107, 356
    soil analysis 179
Norway 77, 314, 342

ocean, nutrient cycles 129
Olsen 162, 182, 189, 256, 260, 263
    change point 263
organic phosphorus 6, 138, 185, 207
OSPAR convention 339

overfertilization 194
oxygen, demand 126

PARCOM
    recommendations 341, 342
particulate phosphorus (PP) 57, 59, 87–88, 152, 207, 211, 385
pasture 70, 195
pathway, hydrological 206, 209, 210, 221, 349
persulphate, digestion 207
phytase 22, 285, 292
phytate P 285, 290
phytic acid 285
phytoplankton 57, 120, 130
pigs 289
    intensive systems 283
plant
    aquatic growth 15
    uptake 209
ploughing 213
policy, European Water 331
pollution
    freshwater 215, 249
    marine 339
    non-point 302
    point source 98
    water 180, 320, 331
population
    rural 58, 124
    urban 58, 124, 273, 277, 281
Portugal 313
poultry 290
    intensive systems 283
precipitation 140

rainfall 207
retention 139, 142, 420, 422
recycling 276
    internal 35
replacement 279, 280, 281
residual phosphorus 7
resin 162
rivers 129, 319, 386
Rothamsted 318, 322, 367, 369
    Broadbalk 259, 260, 261, 262
    Woburn 374

ruminants 287
runoff 3, 152, 226, 254, 298, 313
    agricultural 82, 84, 206, 247
    base-flow 56
    ground water 138
    overland flow 137
    remedial measures 29
    storm 56, 210, 216, 218, 244, 249
    surface 18, 56, 90, 137, 145, 159, 161, 164, 166–167, 180, 208, 213, 227, 229
    subsurface (leaching) 180, 195, 258, 313, 321, 323, 349–360, 361–374

saturation degree (PSD) 264, 398, 400, 417
Scotland 427
Seale Hayne
    fertilizer studies 207
Secchi depth 35
sediment 12, 96, 122, 126, 214, 219, 246
sewage 206, 273–274, 392
    treatment works 61
sheep 289
Slapton land use study 207
sludge 274, 318
slurry 66, 213, 216, 257, 263
soil 111, 321, 343
    analysis 178
    available phosphorus 6
    chemical forms 427
    clay 189, 267, 367
    critical level 151, 152, 155, 232, 319, 441, 445
    environmental programmes 167
    extracted solutions 145
    fissures 210
    management guidelines 168
    organic 258
    peaty 210
    residual phosphorus 185, 212
    sandy 210, 267
    status 151–153, 415
    structure 210, 214
    test 177, 298, 305
    test methods 23, 38, 151–152, 154, 157–158, 160, 161, 180–187, 215

threshold levels 38
types 83, 182
soluble reactive phosphorus *see*
    dissolved reactive phosphorus
    (DRP)
sorption
    capacity 195, 263, 361, 438
    Langmuir maximum 166
    saturation 165–166
sources 120, 205, 225
    anthropogenic 123
    atmospheric 123
    critical area 228, 236
    diffuse 120, 243
    natural 123
    non-point 205, 218, 243
    single 79
Spain 313
strips 15, 163–164, 228
    vegetation 343
    vegetative filter 15, 219
St Albans Bay 34
Sweden 77, 81, 273, 342
Switzerland 193, 314, 342, 372
    soil analysis 179, 183

tanks, septic 58
tillage 31, 343, 379
    contour 28
topsoil, mineral 138
total dissolved phosphorus (TDP) 59, 245, 247–249, 258–259
total particulate phosphorus (TPP) 259
total phosphorus (TP) 59, 97, 110, 210, 245–246, 248–249, 254, 260
transport 377, 420
    phosphorus 11, 85, 147, 206, 254
    preferential 15

treatment, waste-water 79

United Kingdom 314, 349, 351, 396, 422, 449
    fertilizer consumption 192
    soil fertility 183, 190
urine 276–277
USA 228, 391, 436
    field experiments 155, 156, 181, 184

vegetation, macrophyte 126
vitamin D 284

Wales 108, 452
    soil analysis 179
waste 273, 329–330
water
    clarity 122
    drainage 66, 253, 255, 367
    loading rates 56
    quality 95, 180, 184, 197, 300
    quality criteria 36
    quality deterioration 221
    quality standards, EC 70
    sampling 56, 394
    surface 55, 193, 389, 398
watershed 300, 301, 406
wetland 244
White Lough 103, 128
weathering, chemical 212

zones 3, 28, 90, 331
    critical 226
    riparian 3, 28, 219, 221, 233
zooplankton 126